Danai Theodoraki

High-Resolution Climatic Changes at Early Neolithic Franchthi

LEIZA Publications · Band 13

Danai Theodoraki

High-Resolution Climatic Changes at Early Neolithic Franchthi

LEIBNIZ-ZENTRUM
FÜR ARCHÄOLOGIE

LEIZA Verlag · Mainz 2026

Die vorliegende Arbeit wurde vom Fachbereich 07 - Geschichts- und Kulturwissenschaften der Johannes Gutenberg-Universität Mainz im Jahr 2025 als Dissertation zur Erlangung des Grades eines Doktors der Philosophie (Dr. phil.) angenommen (D77)

The present work was accepted in 2025 by the department 07 - Historical and Cultural Sciences of the Johannes Gutenberg University Mainz as dissertation to attain the academic degree of a Doctor of Philosophy (Dr. phil.) (D77)

Redaktion und Satz: Claudia Nickel, Marie Reiter (LEIZA)
Umschlaggestaltung: Nicole Viehöver (LEIZA);
Umschlagfoto: Irina Shishkina on unsplash. - The cover image presents specimens of *Patella vulgata*, serving as a general representation of the genus, rather than the specific species *Patella caerulea* investigated in this work.

Bibliografische Information der Deutschen Nationalbibliothek

Die Deutsche Nationalbibliothek verzeichnet diese Publikation in der Deutschen Nationalbibliografie; detaillierte bibliografische Daten sind im Internet über **https://dnb.dnb.de** abrufbar.

ISSN 2942-5700 (Print)
ISSN 2942-5719 (Online)
ISBN 978-3-88467-387-4 (Hardcover)

Propylaeum
FACHINFORMATIONSDIENST
ALTERTUMSWISSENSCHAFTEN

Diese Publikation ist auf http://www.propylaeum.de dauerhaft frei verfügbar (Open Access).
URN: urn:nbn:de:bsz:16-propylaeum-ebook-1770
DOI: https://doi.org/10.11588/propylaeum.1770
ISBN: 978-3-96929-512-0 (PDF)

Elektronische Ausgabe publiziert bei
Universität Heidelberg / Universitätsbibliothek, 2026
Propylaeum - Fachinformationsdienst Altertumswissenschaften
Grabengasse 1, 69117 Heidelberg
https://www.uni-heidelberg.de/de/impressum
E-Mail: ub@ub.uni-heidelberg.de

Ludwig-Lindenschmit-Forum 1, 55116 Mainz
E-Mail: verlag@leiza.de

Druck: Memminger MedienCentrum Druckerei und Verlags-AG,
Fraunhoferstraße 19, 87700 Memmingen
E-Mail: info@mm-mediencentrum.de

Table of Contents

Appendices (DOI: 10.11588/propylaeum.1770c26319)

Acknowledgements

This PhD thesis was completed as part of the Emmy Noether SEAFRONT project under the direction of Dr. Niklas Hausmann and was funded by the Deutsche Forschungsgemeinschaft (German Research Foundation, project number: 439799406). The project was based at the Leibniz-Zentrum für Archäologie (LEIZA). Research activities were also carried out at the Institute of Electronic Structure and Laser of the Foundation for Research and Technology-Hellas (IESL-FORTH) in Herakleion and at the North East Amino Acid Racemization Laboratory (NEaar) at the University of York.

I acknowledge support from the FINCO project of IPERION HS, funded by the European Commission H2020-INFRAIA-2019-1 and GA no. 871034.

I would like to acknowledge the invaluable help of the following individuals and institutions in various capacities:

I would like to express my gratitude to my supervisor, Dr. Niklas Hausmann, for his unwavering support, guidance, and encouragement throughout the course of my doctoral work. I am deeply grateful for the opportunities and knowledge he has provided me over the years.

Regarding permits, I appreciate the sampling and export permit (ΥΠΠΟΑ/Φ77/367840) granted by the Department of Applied Research, the Directorate of Conservation of Ancient and Modern Monuments and the General Directorate of Antiquities and Cultural Heritage of the Greek Ministry of Culture. I also thank the Argolid Ephorate of Antiquities.

I am also thankful to Prof. Stefanos Paraskevopoulos, Spyros Iossifidis, and the boat crew of the Management Body of the National Marine Park of Alonissos Northern Sporades for enabling fieldwork in the area and facilitating the collection of modern samples. I thank the Greek Ministry of the Environment and Energy for granting the survey permit (protocol number: ΥΠΕΝ/ΔΔΔ/29195/996).

I extend my gratitude to Dr. Stella Katsarou from the Ephorate for Palaeoanthropology and Speleology and Yannis Karliambas from the Ephorate of Antiquities of Thessaloniki City.

For fieldwork and modern limpet sample collection, I am grateful to Prof. Marcello Mannino for his invaluable contribution to the SEAFRONT project's trip to Sicily. I am indebted to Angeliki Kita who accompanied me to the Northern Sporades fieldwork and Areti Polymenidi who accompanied me to Franchthi. I also thank Tatiana Andreovits and Manos Papageorgiou for the collection of limpets from Crete and Panagiotis Keramidas for limpets from Thessaly.

For their help regarding the Franchthi excavations, I acknowledge Em. Prof. Catherine Perlès and the late Em. Prof. Karen Vitelli. For providing detailed zooarchaeological data, I thank Prof. Natalie Munro and Prof. Mary Stiner.

From the Institute of Electronic Structure and Laser of the Foundation for Research and Technology-Hellas (IESL-FORTH) in Herakleion, I appreciate the contributions of the members of the group of Prof. Demetrios Anglos, Dr. Victor Piñón, Dr. Panagiotis Siozos, Andreas Lemonis, Giannis Psyllakis and Mihalis Andrianakis regarding the LIBS system. I am also grateful to Dr. Aggelos Philippidis for conducting Raman analysis and to Aleka Manousaki for conducting SEM analysis.

From the Institute of Geosciences at Johannes Gutenberg University in Mainz, I am grateful to Prof. Bernd R. Schöne, and his group members Christoph Johannes Gey, Dr. Katharina Elena Schmitt, and Michael Maus.

From the Max Planck Institute for Chemistry in Mainz, I am grateful to Dr. Hubert Vonhof for the analysis of water samples.

I would like to acknowledge the invaluable assistance and support provided by the North East Amino Acid Racemization Laboratory (NEaar) at the University of York headed by Prof. Kirsty Penkman. I am particularly grateful to the laboratory members Dr. Samantha Presslee, Dr. Lucy Wheeler, and Dr. Martina Conti for their expertise and assistance with the laboratory work during my research stay there.

My thanks also go to my colleagues at LEIZA, especially to Prof. Roland Schwab.

For fruitful discussions regarding the outcomes of my work, I am thankful to Dr. Asier García-Escárzaga and Dr. Rosa Maria Arniz-Mateos. For help in the identification of limpet species, I am grateful to Dr. Rena Veropoulidou, while any errors on this matter are solely my responsibility.

Finally, I extend my gratitude to my family and my friends Marianthi Tzortzi, Marilena Konstantinou, Angeliki Kita, Tatiana Andreovits, Elena Dimou, Manthos Gkogkos, Eftychia Bikou, Vicky Itsiou, Miranda & Zoi Papadopoulou, Georgia Ilia and Iasonas Kouloumpis for their invaluable support during these years.

Chapter 1: Introduction

This thesis examines human-environment interactions in coastal sites in the eastern and central Mediterranean in the context of the Neolithic dispersal across the region. Pioneer farmers advance westwards from the Near East and Anatolia across the Mediterranean and into Europe. The expansion of agriculture took place relatively quickly and was neither regular nor uniform but rather arrhythmic. The period when staple crops and herd animals of the Neolithic spread to Europe constitutes a significant moment in European prehistory. In the context of this movement, the first farming communities appeared in the Aegean and Greece around 7,000 BC. Various hypotheses exist for the driving forces of this movement among which that of demographic pressure, social stress and environmental stress due to climatic deterioration (e. g. Demoule/Perlès 1993; Cauvin 2000, 1–3; Price 2000, 2–3; Guilaine 2000–2001, 272; Perlès 2001, 38; Guilaine 2015, 85).

In the present work, the latter hypothesis is examined in order to address the phenomenon of the Neolithic dispersal. The climatic conditions and their potential impact on the newly-founded domestic economies are studied in order to identify climatically-induced changes on past human communities. However, to study the interplay of the climate and the environment on the people of the past, there are certain methodological problems. These have to do with the quality of the environmental record and its comparability with the archaeological record. Often the climatic and environmental archives are highly variable and refer to variable temporal and spatial scales, and thus do not allow for chronologically reliable comparisons with the archaeological record. This inconsistency leads to a disconnect between the archaeological and the climatic data, hindering the ability to obtain reliable results. In the present work, I address this disconnect, and suggest a way to link the different lines of evidence in order to test how climate and human agency interlink. The proposed approach allows the synchronisation of the archaeological record with the palaeoclimatic record, as proposed by Berger (2021, 1374).

This thesis is thus ascribed in the wider context of the Neolithic dispersal and the question of the interlink of climate and human agency. To explore this question I analysed archaeological shell samples from the site of Franchthi which was chosen as a case-study site (see **fig. 1**). By producing seasonally-resolved climatic data through high-resolution analyses of mollusc shells, this research aims to shed light on the relationship between climate and human activity during the Neolithic.

1.1. Research Questions and Objectives of the Thesis

This thesis investigates archaeological questions by using archaeometry techniques, such as sclerochronology, spectroscopy and amino acid dating, contributing to the field in multiple ways. The Neolithic transition, an important and multi-faceted concept in the study of the prehistoric past, has been examined using various approaches, with new data continually enriching the archaeological archive (e. g. Childe 1925; 1958; Theocharis 1967; 1973; Ammerman/Cavalli-Sforza 1984; Kotsakis 2001; Weninger et al. 2014; Reingruber 2018a). Previous research often focused on long-term narratives (Weiberg et al. 2016, 41) rather than shorter, human-scale histories. Large-scale climate proxies (e. g. pollen records, marine/ice cores, tree rings, corals, and marine/lake sediments) are typically used for studying human-environment interactions. However, they are not effective for analyzing short-term, site-specific climate variability (Leng/Lewis 2016, 298; West et al. 2018, 191). In contrast, the sub-annual resolution of molluscan records enables climate reconstructions at time-scales perceptible to prehistoric populations. Subsequently, enhanced resolution across larger shell assemblages can significantly improve the palaeoclimatic results produced. At the same time, mollusc shells offer the advantage of site-specific climatic data further enabling the investigation of the local climate.

Sclerochronology, the study of the hard parts of organisms as they evolve through time (including shells), requires extensive isotope sampling. This process is costly, which limits how many samples can be analysed. The Laser-Induced Breakdown Spectroscopy (LIBS) technique facilitates fast, efficient season-of-collection estimates on large assemblages, enabling the study of seasonal exploitation of limpets across different excavation units and chronological periods (Hausmann et al. 2019a, 7). This method allows the assessment of climate impacts on large samples, enhancing understanding of past coastal exploitation. Incorporating the principles of the theory of *time perspectivism* (Bailey 1981; 2007; 2008), I analysed archaeological shell material covering longer time spans, and in finer time-resolutions. Another goal of this work is refining chronological sequences by performing dating analyses on archaeological shell samples, especially from the Early Neolithic contexts of Franchthi Paralia (Perlès et al. 2013, 16; Vitelli 1993, 44), as the resulting palaeoclimate record requires a reliable chronology.

Furthermore, this thesis aims to integrate different scales of analysis both temporal and spatial, which is to say from different chronological periods, and different areas of the site under study. Also, the approach employed in this work aspires to reconstruct local and short-term expressions of the climate, which would have been perceived by the prehistoric occupants of the Franchthi.

Thus, this thesis explores the interplay between climate and human agency during the Neolithic in the eastern and central Mediterranean. It examines human-environment interactions at coastal sites, focusing on the impact of climatic conditions on the newly-founded domestic economies. Key questions addressed include: Can we detect short-term climate fluctuations locally? Have these fluctuations affected human behaviour or caused agricultural risk? By producing high-resolution shell-derived palaeoclimatic data, this research aims to identify climatically-induced changes in the human communities of the past.

To specifically answer the above research questions, the following objectives are carried out:

1. Conduct a pilot study on modern reference specimens by analysing them using LIBS and stable isotope analysis.
2. Apply the same methodology on the archaeological samples from Franchthi to produce site-specific seasonally-resolved climatic data.
3. Conduct direct climate-subsistence comparisons to understand the potential impact of the climate on the Neolithic people.

To reconstruct the seasonal and palaeoclimatic record at Franchthi, I first conducted a modern pilot study on reference specimens (1st objective) to verify the devised methodology. A large number of modern limpet specimens from different coastal locations in the Mediterranean were collected and analysed using LIBS and stable isotopes, which allowed me to verify the methodology and facilitate the interpretation of subsequent archaeological climatic proxies.

In this manner, I analysed the modern shell assemblage in order to corroborate the efficiency of the method employed before proceeding with the archaeological samples. Specimens that were collected nearby the archaeological sites under study and with a known season of collection, functioned as control data regarding the employed methodology. The strong correlation between the results of the two techniques applied to modern specimens from known environmental conditions establishes a solid foundation for their combined use in the subsequent analyses of the archaeological finds.

Following the analysis of the modern samples, the archaeological samples from Franchthi are analysed (2nd objective). The results of the analyses are twofold and aim to reconstruct the palaeoclimate and the seasonality of collection patterns. Primarily, an emphasis is on the reconstruction of the long- and short-term palaeoclimatic record of the site and the investigation of climatic changes related to the advent of the Neolithic period. In addition to that, seasonality of collection is associated with season-

Fig. 1 Present-day view of the site of Franchthi seen from the north. The cave entrance is visible, the present shoreline, while the Paralia trenches situated at the base of the slope, are covered by vegetation. – (All photos, graphics and tables by D. Theodoraki, unless otherwise stated).

ality of occupation at the site and new evidence is presented on the matter regarding the Palaeolithic and the Mesolithic period.

Finally, to evaluate the impact of the climate on the Neolithic people and their practices, direct layer-by-layer comparisons between the shell-derived palaeoclimatic parameters and the broader contextual subsistence information are considered (3rd objective). Emphasis is placed on the Neolithic period and the potential changes in subsistence patterns to determine the impact of climatic changes on the site's human populations. The problem of comparability due to different scales of analysis is acknowledged, necessitating the synchronisation of archaeological and climatic data. Directly addressing the problems identified in Berger (2021), this thesis directly compares stratigraphically correlated shell-derived palaeoclimatic data with subsistence archaeological data to explore the interplay between palaeoclimate and human practices.

1.2. Structure of the Thesis

Followed by the introductory chapter, Chapter 2 (Archaeological Background) of the thesis discusses the different approaches to the Neolithisation phenomenon. The theoretical approaches to the study of bioarchaeological remains and shells are presented briefly as well as the subject of the different temporal scales of analysis. Subsequently, a concise description of the Final Palaeolithic, the Mesolithic and the Neolithic period in Greece is presented alongside important sites of the three periods as well as the question of the Neolithic expansion, which constitutes a central theme of this monograph, are addressed here. This chapter concludes with the detailed presentation of the previous studies on Franchthi, which include the history of research and overview of the site, the presentation of the modern and the submerged landscape, the pottery and the lithics assemblage, the zooarchaeological and the archaeobotanical evidence, the shell assemblage and the recent excavation work.

Chapter 3 (Methods and Materials) describes the different methods of analysis employed in this thesis (section 3.1) as well as the modern (section 3.2) and archaeological samples (section 3.3).

Here I make use of: sclerochronology, stable isotope analysis, LIBS, Amino Acid Racemisation (AAR) technique and their purposeful combination. Each of these techniques is presented alongside its uses and applications in relation to the specific material under study.

The modern samples analysed for the calibration study are presented in section 3.2 (Modern Samples) together with information regarding their provenance, sample size, and the season they were collected. The collection process is also described in this section.

Lastly, section 3.3 (Archaeological Samples) presents in detail the archaeological specimens from Franchthi that were selected for analysis: The provenance of the species in terms of the stratigraphy and the excavation units, the number of specimens per unit, and the units' spatial and temporal distribution. Furthermore, the morphological characteristics of the assemblage together with post-depositional observations that were recorded are presented in this section. The detailed recording of the molluscan assemblage prior to its analysis, facilitates the interpretation of the material in the specific temporal and spatial context.

Chapter 4 (Results) consists of the results of the analyses that were performed on the samples. First, the results of the elemental and isotope analyses on the modern reference samples are presented by location. Second, the elemental and isotope analyses on the archaeological samples are described in detail and presented by trench and excavation unit in chronological order. At the end of the chapter, the results of the AAR analysis are reported.

Chapter 5 (Discussion) consists of the discussion section, where the outcomes of the results are considered in comparison to the previously known data from Franchthi and other sites. A synthesis of the overall findings of this work is carried out in relation to the general archaeological background and prior results from publications on material from the site. The findings of this work are further integrated in the wider research questions on the matter of human-environment interactions in the context of the Neolithic dispersal.

Finally, Chapter 6 (Conclusions) synthesises the key findings of the research, highlighting how the aims and objectives have been met. It offers a reflection on the implications of the results and discusses their relevance and contribution to the broader scientific field, and ties together the research questions, methodologies, and outcomes, presenting a cohesive narrative of the study's significance as well as directions for future research.

Chapter 2: Archaeological Background

2.1. Introduction

In this section, I discuss the topic of the Neolithisation process. To begin with, around 7,000 BC, the first farmers in the modern Greek territory began to create permanent settlements, and introduce domesticated plants and animals (Demoule/Perlès 1993; Perlès 2001, 38). Thus, when we speak about the Neolithic way of life we refer to the cultivation of domesticated plant species, the herding of domesticated animal species and the establishment of sedentary life, elements which gradually lead human populations into new and complex farming communities (Cauvin 2000, 1–3; Price 2000, 1). This moment in time represents a period in the circum-Aegean area when a major shift in human lifeways occurs and the traditional Mesolithic hunter-gatherer ways of life are abandoned. The process of this transition, is referred to as the »Neolithisation« process or the »Neolithic transition«, and produced a large body of scientific work for many decades (e. g. Childe 1936; Renfrew 1973; Ammerman/Cavalli-Sforza 1984; Cauvin 2000; Perlès 2001). The Mesolithic-Neolithic transition is not only a change of lifestyle but also a boundary between »two models of man« where human beings purposefully behave towards socially-defined goals (Thomas 1988, 59).

The »Neolithic« is a term that is often contested and new meanings and interpretations are attributed to it (Çilingiroğlu 2005, 1; Reingruber 2018b, 4). Although this differs throughout European countries, agriculture constitutes one of the key components while the existence of polished stone or pottery is also interpreted as proof of Neolithic lifeways (Fowler et al. 2015, 4). However, this period should only be considered through an economic or subsistence strategies lens. The Neolithic has also been defined as a way of thinking and reorganising the world and the social and symbolic connotations of the dynamic process are put centre stage (Hodder 1990) or as the process of transformation of agricultural social relations and the adoption of innovations by communities (Thomas 1988, 64).

In the following sections, important aspects of the Neolithic will be discussed that are necessary to understand the scope of this study and the meaning of its results for the wider archaeological context. The different approaches to the Neolithic dispersal from the Near East into the Aegean are presented alongside certain climate-related theories. Subsequently, the archaeological background that is related to the site is concisely presented. Additionally, the archaeological evidence of the Final Palaeolithic, the Mesolithic and the Neolithic period of the study area are presented alongside the climatic and environmental context. Finally, the priorly published archaeological record from the site of Franchthi is presented in order to set the scene for the presentation of the analyses that follow in the next chapters.

2.1.1. Different Approaches to the Neolithic Transition

Discussion on the matter of the Neolithic transition is characterised by the dichotomy on the origins of the Neolithic between »autochthonous/indigenist« and »diffusionist/migration« approaches. More specifically, the autochthonous approach advocates for an exclusively local process with no external influences; the cultural diffusion approach supports a local process ignited by exchanges of foreign artefacts and techniques; the demic diffusion approach the influx into Greece of foreign groups of farmers and herders; while a mixed approach contends that interactions among local populations following the Mesolithic lifeways and incoming external farmers led to the Neolithic way of living (Kotsakis 2001, 64; Perlès 2001, 38).

Theocharis (1967; 1973, 34–35) was the first to question the paradigm set by Childe of the predominance of the »Fertile Crescent« (Childe 1925) and consider the possibility of an indigenous course of the Neolithic transition. The indigenist stand was supported also by Gimbutas who excavated Achilleio together with Theocharis (1973, 35). The demic diffusion model was first supported by Childe (1958) and later on following this line of reasoning Ammerman and Cavalli Sforza (1984) proposed their »wave of advance« model. The latter however assumes a sweeping unidirectional east-to-west trend that neglects the heterogeneity of the archaeological record.

The »Neolithic package« is a term that gained the favour of archaeologists since the 1970s, refers to economic products or objects that played a predominant role in the domestication processes. It is a term that was often criticised, but however showed the need of prehistorians to group and contextualise the Neolithic assemblages (Çilingiroğlu 2005, 2). In this model, changes in the archaeological archive were often interpreted as colonising events. In this colonisation process the package appeared simultaneously. On this matter Reingruber proposes that attention needs to be given to evidence of continuity in the archaeological record apart from the evidence of change. The Neolithisation process in the Aegean can also be perceived as the result of connectivities in time (with the Mesolithic) and in space (circum-Aegean) (Reingruber 2018a, 8).

More conceptual approaches stress the more social aspects of the mechanisms and conditions of the Neolithic transition while emphasising the social complexities and strategies of the Neolithic communities that have to be taken into account (e. g. Hodder 1990). Kotsakis argues that the indigenous/diffusion dichotomy is old-fashioned and overly simplistic, and that the Neolithic transition has to be viewed in a different manner, by adopting an approach that relies on archaeological facts and cultural affinities, and emphasises the interplay of agency and structure in the process of neolithisation. He also stresses that socially embedded practices such as competition, conflict, and group identity and their traces in the archaeological record play an important role in the process, whereas diffusionist models neglect the social context of the processes, which constitutes a basic aspect of contemporary archaeology. The diffusion of the Neolithic although it is a physical phenomenon which takes place in space and time, it is also a social one. A social process that takes place in a social place and time and this social element of the process is often not examined when the subject is examined (Kotsakis 2001, 64–65).

2.2. Neolithic Dispersal into the Aegean: Mechanisms and Models

Having presented the different approaches to the Neolithisation process, this section presents the mechanisms and proposed routes of the Neolithic dispersal from Anatolia and the Near East into the Aegean. The prevailing evidence suggests that this dispersal was not a monolithic event but rather a complex, stepwise process, likely involving various trajectories (Weninger et al. 2014, 24). The routes this movement followed have been hypothesised to include a large variety of potential trajectories running east to west and may have been terrestrial and/or maritime. The latter is advocated by Horejs et al. (2015) who argue for a maritime colonisation from the eastern Mediterranean to the Aegean via previously established sea networks.

Recent studies have explored the dispersal of Neolithic farmers through various methods, including agent-based modelling and genetic analyses, providing insights into the dynamics of population movement. To better understand the complex dynamics of this expansion, researchers increasingly employ agent-based modelling (ABM) (e. g. Bernabeu Aubán et al. 2015; Zanotti et al. 2015; Romanowska et al. 2021). These models are a type of computer simulation that track the behaviour of individual, heterogeneous »agents« (e. g. individuals, households, settlements, etc.) and their interactions with their environment and each other over time (Romanowska et al. 2021). Applied to the Neolithic dispersal, agent-based models can simulate different scenarios, such as the speed of movement, the role of population growth and the impact of resource availability. For instance, studies like those by Bernabeu Aubán et al. (2015) for the Western Mediterranean or by Wirtz and Lemmen (2003) for a global scale model, demonstrate how varying parameters – such as subsistence ways, adoption of technologies, or the preference for specific environmental features – help evaluate the plausibility of various dispersal hypotheses against the archaeological record. Ultimately, these models help explore the underlying processes, often revealing complex outcomes influenced by factors like trajectory, pace, direction, ecological conditions, demography and social structure (Zanotti et al. 2015).

Genetic studies, particularly those using ancient DNA (aDNA), analyse the genomes of ancient human remains from the Near East and across Europe, including the Aegean and this way researchers can trace ancestral lineages and patterns of gene flow. Early genetic analyses indicated a significant demographic contribution of Near Eastern farmers to the European gene pool, largely supporting a demic expansion model for Anatolian farmers (Hofmanová et al. 2016). More recent high-resolution genomic studies have provided finer-grained insights into this complex process. These analyses confirm that farming was first introduced to Europe in the mid-7th millennium BC by migrants from Anatolia, who initially settled in the south-east before spreading throughout the continent. Ancient DNA datasets revealed complex population dynamics, beyond simple migration, including an indigenous hunter-gatherer ancestry. Research has also demonstrated that south-eastern Europe remained a crucial nexus between east and west even after the arrival of farmers, with intermittent genetic contact with steppe populations occurring up to 2,000 years earlier than the major migrations from the steppe that ultimately replaced much of the population of northern Europe (Mathieson et al. 2018; Olalde et al. 2018; Skourtanioti et al. 2023). These findings show that beyond a strong genetic continuity with European Neolithic farmers, genetic diversity was also shaped by local processes, leading to varied sources and proportions of hunter-gatherer ancestry across different regions and through time (Lipson et al. 2017; Mathieson et al. 2018). While these studies confirm a dominant role for migration, they also highlight regional variations in admixture with local Mesolithic hunter-gatherer populations, suggesting that the Neolithisation process was far from uniform, involving diverse degrees of cultural exchange and genetic integration (Skourtanioti et al. 2023; Marchi et al. 2022). Genetic insights often complement archaeological interpretations, yet aDNA, like any scientific proxy, carries inherent limitations and interpretive ambiguities (Sykes et al. 2019; Brami 2023).

It is important to note that certain phases of this Neolithic dispersal have been associated with, or coincided with, significant changes in the climate, a topic that will be further detailed in section 5.3.2. For instance, the abrupt movement of early Neolithic farmers from the Aegean into the Adriatic and south-eastern Europe coincided with the end of the Rapid Climate Change event at 8,000 BP (Rohling et al. 2019, 41). Weninger et al. (2014) further argue that this climatic event played a major role in the neolithisation process and they identify two major chronological steps in the Neolithic dispersal. The first step, from Central Anatolia into the Aegean and the southern Marmara region, coincides with the onset of the Rapid Climate Change conditions (approximately 8,600 cal BP). This appears to have happened on a short time-scale. The second step, involving the dispersal of the Neolithic further north-west, starts at 8,050 cal BP, towards the end of the Rapid Climate Change conditions, and lasted approximately 200 years (Weninger et al. 2014, 24). The synchronicity of these steps with the beginning and end of the specific climatic conditions provides an argument for the association between the climatic event and the north-westward movement of the Neolithic farmers (Krauß et al. 2017, 37).

2.3. Theoretical Framework

In this section, I discuss how shells, though previously undervalued, have gained recognition in archaeological research for their ability to reveal detailed environmental and cultural insights. The evolution of archaeological theory from *New Archaeology* to *Post-processual Archaeology* has highlighted the importance of integrating broader social theories and focusing on the active role of material culture in interpreting past human-environment interactions.

Shells can contribute to archaeological research in many ways even though they have been understudied or ignored during the second half of the 20th century (Claassen 1998, 4). Increased interest in the study of shells coincided with the appearance of increased interest for the palaeoenvironment and palaeoeconomy (Gassiot 2005, 40). The ecological approach of Clark in the 1950s brought forth the importance of ecology and the environment for archaeological research and many specialists, such as zooarchaeologists or archaeobotanists, and contributed to the reconstruction of the prehistoric environment and diet (Clark 1952). In the 1960s and 1970s, *New* or *Processual Archaeology* contended that culture is the means of adaptation to the natural environment. Binford gave emphasis on the interpretation of archaeological remains and cultural behaviour (Binford 1968). They also turned to scientific rather than historical approaches, to the study of the natural environment and the investigation of the social and economic aspects of the societies of the past. For example, before the 1960s, the natural environment was viewed as a passive background to human activity, neglecting the relation and the interaction of people with the environment (Evans 2003). *New Archaeology* sought to *interpret* and not solely describe by analysing past civilisation as systems that consist of subsystems (Willey/Phillips 1958). This line of thought employed positivist paradigms, using hypotheses and results as theoretical tools, as well as sampling and statistics, creating models in order to interpret the archaeological data (Trigger 2006, 386–444; Renfrew/Bahn 2011, 38–39). Furthermore, *New Archaeology* encouraged the use of ethnoarchaeological research (Trigger 2006, 405). The important contribution of *New Archaeology*, became nonetheless a subject of criticism as a normative and functionalist approach, which brought excessive emphasis on matters of the environment, ecology, demography and survival rather than on social or cognitive factors (Renfrew/Bahn 2011, 489). This tendency against positivism and behaviourism was widespread in the humanities and social sciences during that period (Trigger 2006, 446). Other critiques considered the system approach of *New Archaeology* too rigid and strict (Trigger 1978) and stressed the need to focus on the ideological and symbolic aspects of past societies (Flannery 1967).

In the mid-1980s, several critics of *New Archaeology*, such as I. Hodder, M. Shanks and C. Tilley, influenced by French marxist anthropology, attempted to surpass its limitations. *Post-processual* or *Contextual Archaeology*, intends to incorporate archaeological theory into a wider range of theoretical matters of the social sciences (Preucel/Hodder 1996, 7) emphasising on the symbolic and ideological dimensions of ancient societies, while highlighting the active dimensions of material culture. According to this, people and societies shape their own social reality and material culture is an integral part of this structure (Hodder 1982; Trigger 2006, 450; Hodder/Hutson 2010, 43; Renfrew/Bahn 2011, 505). In this manner, through lines of archaeological and anthropological thought, notions such as agency and material culture (Dornan 2002), symbolism, or the ways of perceiving the environment (Ingold 2000), are incorporated into the archaeological discussion.

Agency in archaeology means that deterministic models of human behaviour are challenged by recognizing that individuals intentionally act and shape the external world through their actions (Shanks/Tilley 1987, 116; Dornan 2002, 304). For example, the matter of agency can be applied in the sense of perceiving material culture not as passive isolated objects, but rather as artefacts created by a person or a group of people. In this sense, archaeology seeks the relation of this artefact with society. People are not mere players in a predetermined system, but actively play a role in shaping material culture embedded in social meanings and symbolisms (Hodder/Hutson 2010, 48–49). Another example of agency and symbolism in archaeology is drawn from ethnoarchaeology and the Lozi people in Zambia. There, similarities in pottery decoration did not reflect passive networks, but were used to create social differentiations and alliances within the village forming part of the shaping of social strategies and »internally generated symbolic schemes« (Hodder 1982, 186).

Just as human agency shapes material culture and social structures, it also profoundly influences how people relate to their environment and other species. One way to approach these human-environment and human-animal relations is by examining the ways in which people engage with their surroundings through subsistence procurement (Ingold 2000, 9). Among the numerous examples on this matter that can be drawn from the influential work of Ingold is that of the role of the environment in the shaping of the individual and social identity, where the natural environment is not a mere background to human activities (Ingold 2000, 54). For example, groups of indigenous people such as the cultivators called Bette Kurumba and Mullu Kurumba or the South Indian hunter-gatherers called Nayaka, perceive the forest as their parent that offers food to its children, while if they misbehave the spirits of the forest as well as the spirits of their ancestors will respond accordingly by inflicting pain and sorrows (Bird-David 1990, 190).

These intricate relationships between humans and their environment can also be explored through the palimpsests found at archaeological sites. A palimpsest, in this archaeological sense, refers to a record or site where evidence of past activities is layered over time, much like a parchment that has been written on, erased, and written on again, with traces of the original still visible. In this context, shells are an excellent example of such a palimpsest. Each shell specimen incorporates multiple temporal layers of sub-decadal or sub-annual scales as I will exhibit through the methods extensively in Chapter 3. The shell record contains cumulatively many more temporal as well as spatial layers of information. The seasonal and climatic information contained within shell specimens offers significant insights into human agency. By examining this information, we can discern the actions and decisions of individuals and communities over short time periods, revealing how they interacted with their environment and adapted to changing conditions. People were not merely passive recipients of environmental influences but actively engaged with their surroundings, making deliberate choices that shaped their world. The study of their subsistence practices opens a window into these human-environment interactions.

The multi-temporality of the shell specimens brings us to the theory of *time perspectivism*, which consists of »the belief that different time-scales bring into focus different features of behaviour, requiring different sorts of explanatory principles« (Bailey 1981, 103). This means that by re-adjusting our scale(s) of observation, we can make phenomena amenable to our set of questions, and go further than the large-scale data. Bailey advocates for the investigation of the multiple scales of phenomena and different types of processes (Bailey 2008, 13).

From the shorter temporal scales (at the individual shell specimen level) it is possible to expand to the larger scales (at the excavation unit or the site level) via the analysis of a large number of samples. This can lead to the understanding of the interplay between climate change and human agency on a larger scale by integrating diverse lines of evidence and by investigating short-term, site-specific climatic variations that directly affect human activities. To this end, it is essential to synchronise archaeological findings with high-resolution climatic records (Berger et al. 2016), such as those derived from shells at specific sites. By doing so, we can more accurately reconstruct how past communities, like those at Franchthi, responded to environmental changes. This holistic approach not only illuminates localised human-environment interactions but also offers a framework for understanding larger-scale human reactions, or lack thereof, to broader climatic shifts, thus providing a comprehensive view of human resilience and adaptation throughout history (e. g. Weiss et al. 1993; deMenocal 2001; Weninger et al. 2006; Clare et al. 2008; Flohr et al. 2016; Krauß et al. 2017; Ivanova et al. 2018; Roffet-Salque et al. 2018; Kaniewski et al. 2020; Glais et al. 2023).

More specifically, in the larger discussion of »collapse« and »resilience« in archaeology, there are three different perspectives in current literature on how prehistoric communities responded to climatic changes during the Holocene in the Eastern Mediterranean. First, there are studies that focus on larger geographical contexts and shorter time-scales, the Early Neolithic, and conclude that cultural adaptation followed regional and climatic gradients. This is the case of the papers by Krauß et al. (2017) and Ivanova et al. (2018), discussed in detail in section 5.3.2. Krauß et al. by comparing coeval sites across space, they argue that the Early Neolithic settlements in south-eastern Europe are located in specific biogeographic zones that share similar environmental and climatological characteristics (Krauß et al. 2017).

Second, there are numerous authors and studies advocating for climate-induced unrest, societal breakdown, widespread migrations and major changes due to intense climatic events (e. g. Weiss et al. 1993; deMenocal 2001; Weninger et al. 2006; Roffet-Salque et al. 2018; Kaniewski et al. 2020). Such an example from south-western Turkey is the study by Clare et al. (2008) on climate-induced warfare based on a case-study of the site of Pisidia. The paper focuses on the 8,200 cal BP climate event using data from core LC21 (see section 5.3.1) and on other contextual evidence indicating breaks in the cultural sequences, population collapse and warfare. Based on this evidence, they make the hypothesis of linking the climatic and archaeological evidence and conclude that the changes in the archaeological record were triggered by the altered climatic conditions (Clare et al. 2008).

Third, there is literature that contends that the communities of the past adapted in the face of climatic change, such as the study by Glais et al. (2023) or the work by Flohr et al. (2016). The authors of the latter study do not detect any absolutely dated evidence for widespread collapse, large-scale site abandonment or migration events in the farming communities of south-west Asia during the severe and abrupt 9,200 and 8,200 climate events. A number of interpretations are given for this conclusion: First, absolute effects may not have been severe and thus may not have altered vegetation and harvest to a degree that would have impacted significantly food procurement. Second, another explanation is that plants, people and animals had already adapted to adverse conditions and coping mechanisms were in place to face changing climatic conditions. Third, the early farming communities were resilient and were not impacted by the extreme climatic events. Resilience may have relied on socio-cultural, economic and political factors as well as on adaptation strategies that would have entailed resource diversification and storage (Flohr et al. 2016, 35). This subject of »collapse« and »resilience« is further discussed in section 5.4.

2.4. Climate

This section focuses on the description of the modern climate and the climate during the Early Holocene to better understand human-environment interactions during the Neolithic dispersal. Understanding the climatic and environmental conditions of this period is crucial for examining the hypothesis that climate played a significant role in the spread of agriculture and the establishment of early farming communities around 7,000 BC. By describing the modern and prehistoric climate, I set the scene for what is currently known before delving into the analysis of the archaeological samples.

2.4.1. Modern Climate in Greece

The Mediterranean region is highly susceptible to local and global climate changes. It has a long history of environmental and cultural biodiversity and is an archive of a large amount of climate information of the past (Roberts et al. 2011; Mercuri/Sadori 2014). Due to the fragmentation and variation of the landscape, the region exhibits a high degree of regionality where many micro-regions exist (Horden/Purcell 2000, 12–15).

While global-to-hemispheric atmospheric circulation patterns affect the characteristics of the modern climate and weather, these are also controlled by smaller-scale regional-to-local circulation systems (Maheras et al. 2006; Maheras/Anagnostopoulou 2003; Lionello et al. 2006).

Greece has a typical Mediterranean climate, characterised by warm to hot, dry summers and mild to wet winters, according to the Hellenic National Meteorological Service (EMY). Its climate features mild and moist winters, relatively hot and dry summers, and generally long periods of solar radiation through most of the year. Within Greece's different regions, a large variety of climatic types can be observed due to country's diverse topography and the alternating bodies of land and sea. Thus, the dry climate of Attica and eastern Greece differs significantly from the humid climate of northern and western Greece. Striking climatic differences can also be observed in very close proximity (https://emy.gr/en).

The modern climate of Greece also varies in rainfall and, to some extent, in temperature. Regional differences occur from west to east and from north to south, depending on topography and elevation. On the Mediterranean scale, there is also an environmental meridional gradient in Greece. More specifically, the climate is more humid, with cold, rainy winters in the north and west. Frost is less common in central Greece, where mild, rainy winters and hot, dry summers differ in stark contrast. Southern Greece has a semi-arid environment where frost is uncommon and mean yearly temperatures are higher (Demoule/Perlès 1993, 359).

From a climatic perspective, there are two main seasons: The cold and rainy winter season (mid-October to late March) and the hot and rainless summer season (April to October). January and February are the coldest months of the winter period, with mean temperatures fluctuates between 5 and 10 °C in coastal areas, 0 and 5 °C on the mainland, and even lower temperature in northern areas.

Even during winter, rains in Greece do not last for several days, and the sky is not continuously clouded. Bad winter weather is often interrupted during January and the first half of February by fair, sunny weather, and these days are called since ancient times *the halcyon days* (Carapiperis 1957). An etiologic myth existed in antiquity for this weather phenomenon (as for so many other phenomena) that occurred yearly. According to mythology, Alcyone was the daughter of Aeolus and wife of Ceyx. They were so happy about themselves that they dared to compare themselves to Zeus and Hera, an act of *hybris*. Annoyed at their excessive pride, the gods turned them into birds, a halcyon and diver respectively. Alcyone laid her eggs on the edge of the sea, but the waves kept destroying them. Zeus took pity on her and commanded that the waves should be calm for seven days when the halcyon hatches her eggs. Sailors called these days the halcyon days because they knew there were no storms (Grimal 1986, 34).

Returning to the modern climate, winter is milder in the Aegean and the Ionian islands than in the northern and eastern areas of the country. During the hot and rainless season the weather is stable with fair skies and abundant sunshine, with the rare exception of quick rains or short-lived storms. The hottest period takes place during the last ten days of July and the first ten days of August when the mean maximum temperature fluctuates between 29 °C and 35 °C. During this season, the high temperatures are eased by the cool sea breeze in the coastal areas and the northern winds that blow mostly in the Aegean. Spring is relatively short, as winter is long and summer begins early. Autumn is long and warm, sometimes lasting until mid-December in southern Greece (https://emy.gr/en).

The annual weather patterns of the Peloponnese are also typical of the Mediterranean, with hot, dry summers and moderate, wet winters. More than 80 % of the annual precipitation falls between October and April, while precipitation is orographically influenced by the mountain masses that transverse the central part of the Peloponnese from north to south (Weiberg 2016, 43).

The climate of the southern Argolid is one of the driest in the country and is characterised by hot and dry summers and cool and moist winters. The climate is influenced by orographic and marine factors and local climatic variations are large as well as large annual temperature variations. Snow is rare. During winter, winds from the west or the north-west are common, keeping the east side of the peninsula protected. The summer brings strong periodic or »etesian« winds from the north and the north-east which can last for many consecutive days, often obstructing marine traffic (van Andel 1987).

2.4.2. Early Holocene in the Eastern Mediterranean

The end of the Pleistocene is characterised by the Younger Dryas climatic event. The Younger Dryas (12,800–11,600 BP), a cold and arid event, is identified in pollen diagrams in the Eastern Mediterranean (Bottema 1995). The Younger Dryas prevented humid conditions from developing and was more pronounced in the northern hemisphere. It is attested in Western Europe and Greenland but absent from Antarctica and New Zealand. In the Eastern Mediterranean, this cold event is associated with a drop in sea surface temperature and the end of the African Humid Period. Pollen records from Northern Greece and Lake Kopais are associated with this phenomenon (Ehrmann et al. 2007, 54; Geraga et al. 2010; Gogou et al. 2007; Kotthoff et al. 2008; Kuhnt et al. 2007; Rossignol-Strick 1995; Triantaphyllou et al. 2009).

Recent climate reconstructions show that the Eastern Mediterranean region did not appear to be substantially affected by the Younger Dryas (Cruz-Diaz et al. 2023, 8). Over a large portion of this area, vegetation changes were not clearly noticeable (Bottema 1995, 890).

2.4.2.1. Early Holocene Rapid Climate Change Events

The end of the Younger Dryas marks the beginning of the Holocene. Complex patterns of climatic change were seen across the Mediterranean at the onset of the Holocene, 11,800 cal BP. The beginning of the Holocene is characterised by rapid warming with rising temperatures in a few decades, while rainfall was also higher (Arz et al. 2003; Roberts 1998, 117–120; Robinson et al. 2006, 1530).

This rapid warming is also characterised by periods of 150–500 years of Holocene Rapid Climate Change (RCC) events (Mayewski et al. 2004). These events are identified in core LC21 of the Aegean record and are dated to approximately 8,600–8,000 cal BP, 6,300–5,500 cal BP, and 3,300–2,600 cal BP. The cause of these multi-decadal cooling events is thought to be the increasing intensity and frequency increase of cold polar and continental air outbreaks during the winter that spread across the eastern Mediterranean and the Aegean, passing through mountain ranges to the basin's north. Wintertime sea surface temperature dropped by 2–4 °C as a result of these episodes (Rohling et al. 2002, 590; 2019, 39). These Rapid Climate Change events would manifest as periods of extreme cold and precipitation, with particularly cold winters and highly variable climatic conditions (Rohling et al. 2019, 50). In addition, certain authors contend that such climatic anomalies were more frequent and intense than they are today (Clare et al. 2008, 70).

The 8,600–8,000 climatic event that coincides with the period of interest for the present work, lasted five or six centuries (Mercone et al. 2000; Casford et al. 2007). During this event, there was Aegean-Levantine cooling, together with deep-water ventilation, similarly to other Rapid Climate Change events. The specific 8,600–8,000 event represents a time of important changes in the region (Rohling et al. 2019, 41).

Understanding how past societies navigated these climatic events is crucial for archaeologists. Numerous authors have examined the varied societal responses in the face of these climatic events, as already discussed in section 2.3. For instance, Weiss et al. (1993) and deMenocal (2001) highlight instances of societal disruption and even collapse in response to abrupt climate change. Conversely, Flohr et al. (2016) and Glais et al. (2023) emphasize the remarkable resilience and adaptive strategies, such as resource diversification, that allowed communities to confront environmental challenges without widespread collapse. This study will discuss this topic in Chapter 5 in relation to the evidence from the Early and Middle Neolithic period of Franchthi which approximately coincides with the timing of the climatic events, contributing to these ongoing debates about human responses to climate change.

2.4.2.2. The 8,200 Event

The climatic 8,200 event was first detected in the Greenland cores and is described as a cold, arid, and windy event in and around the North Atlantic basin (Alley et al. 1997, 485). This event is associated with a final massive glacial meltwater outflow into the North Atlantic from Hudson Bay (Barber et al. 1999). Aridity increased south of the 42nd latitude, while precipitation decreased by approximately 17 % (Berger/Guilaine 2009; Pross et al. 2009; Abrantes et al. 2012). This climatic deterioration is part of the long-term anomalies that take place during the Holocene and seem to be associated with solar output fluctuations (Rohling/Pälike 2005, 978).

The 8,200 event is a sharp, short-lived event (approximately 150 years) that happened within the Rapid Climate Change event of 8,600–8,000 BP. Thus, the 8,200 event is ascribed to the major multi-century Rapid Climate Change event and the one should not be confused with the other (Rohling et al. 2019, 41).

2.4.3. Prehistoric Environment

The environment during the Late Pleistocene and the Early Holocene was very different from today. The different regions are characterised by different vegetation and climatic traits. Several zones are present: The Lowlands (< 700 m a. s. l.) can be distinguished in accordance with their location in the Lowlands Northern belt (moist and cool), the Lowlands Ionian belt (moist and warm) and the Lowlands Aegean belt (dry and warm). Thessaly, central Greece, northern and eastern Peloponnese and the Aegean islands belong to the Aegean belt. During the Late Glacial Maximum, vegetation in Northern and Central Greece consisted of sagebrush steppe of *Artemia* and Chenopodiaceae indicating a cold and dry climate. There are also small amounts of oak pollen suggesting the presence of deciduous wood patches in certain locations. Around 10,000 BP, gradual climatic improvement led to the rapid spread of open forests dominated by junipers and deciduous oaks, characterised by the presence of *Pistacia*. This indicates warm conditions with wet summers and winters (Wijmstra 1990; Jameson et al. 1994, 166). A forest emerged, featuring species like *Olea*, hornbeam (*Carpinus* and *Ostrya*), elm (*Ulmus*), and other members of the Mediterranean deciduous forest. This type of forest also covered the Peloponnese (van Zeist/Bottema 1982, fig. 14.11). Although evidence is sparse, pollen data from a core from Lake Lerna in the Argive Plain confirms the existence of oak woodland prior to Neolithic deforestation (Jahns 1993).

During the Neolithic, according to palynological and anthracological analyses, in Northern Greece there was dense forest coverage from deciduous or semi-deciduous oak trees (Bottema 1982; Ntinou 2014, 413). Forest vegetation included oak, which would have been the prevailing species. Cores from Epirus and Macedonia indicated a progressive reforestation after the Pleistocene and a climatic woodland of deciduous oak, elm, ash, lime and hazel, with pine on the slopes. It is thus considered that early farmers confronted a densely forested landscape (Bottema 1979; Creig/Turner 1974; Halstead 1989, 28; Demoule/Perlès 1993, 360; Perlès 2004, 14). Around 5,000 BP, the vegetation across different regions of Greece began to diverge due to the increasingly dominant human influence on the landscape (Jameson et al. 1994, 166).

2.5. History of Research

2.5.1. The Late Upper Palaeolithic

The Palaeolithic record of Greece is under-investigated and characterised by a high degree of fragmentation and discontinuity in time and space. Nonetheless, in the last decades Palaeolithic research has intensified with new excavations, surveys and the reassessment of old findings under the light of recent discoveries (Tourloukis/Harvati 2018; Tourloukis 2021, 72). Here, I will briefly present some of the sites of the later phases of the Upper Palaeolithic in Greece that are coeval to Franchthi (**fig. 2**).

The Aurignacian period is best known from the carbon-dated sites of Franchthi, Klissoura Cave 1 and Kolominitsa Cave (Douka et al. 2011; Koumouzelis et al. 2001; Darlas/Psathi 2016). Klissoura Cave 1, in the eastern Peloponnese, presents one of the more complete Palaeolithic sequences in Greece, that spans from the Middle Palaeolithic to the Mesolithic period. The Aurignacian material from Klissoura is dated to ~33–31,000 BP and has many similarities with the European Early Aurignacian. A large number of ecological niches would have been exploited, while the faunal evidence includes increased numbers of fallow deer remains and to a lesser extent of European ass, red deer, boar and ibex as well as small game (Koumouzelis et al. 2001; Kaczanowska et al. 2010; Kuhn et al. 2010). In addition, the analysis of tooth eruption indicated that the site would have taken place during all the seasons (Starkovich 2017, 64–78). A similar observation has also been made for Kephalari Cave (Starkovich pers. comm. as reported in Elefanti/Marshall 2018, 12). Similarly to Franchthi, the sites of Klissoura and Kephalari, represent long-lived sites, visible in the landscape, that functioned as »home bases« exploiting a large variety of the site territory, environments and resources (Vita-Finzi et al. 1970, 6; Elefanti/Marshall 2018, 14).

Furthermore, the Gravettian, Epigravettian and Final Upper Palaeolithic sites are caves and rockshelters, while none are open settlements. Among them are Asprochaliko, Kastritsa, Klithi, Megalakkos and Boila in Epirus (Bailey et al. 1983; Adam 1989; Bailey 1997; Kotjabopoulou et al. 1999), Grava in Corfu (Adam 2007; 2016), Theopetra in Thessaly (Adam 1999b), Klissoura, Kephalari and Franchthi (Kaczanowska et al. 2010; Reisch 1976; Perlès 1999), as well as the caves in the Mani peninsula Melitzia, Kastanis, Skini 3 and 4, and Tripsana (Darlas/Psathi 2016). The main primary materials used for lithics during the Upper Palaeolithic in Greece are of local origin, mostly flint and radiolarite (Adam 2016, 36). Evidence for longer distance procurement of obsidian from the island of Melos is attested at Franchthi during the Late Upper Palaeolithic (Perlès 1987, 176–177), while at lower densities Melian obsidian is also found at Kephalari and Klissoura during the end of the period as well as the Mesolithic (Kaczanowska et al. 2010).

The function of these sites has been interpreted as hunting stations that are occupied seasonally and are characterised by discontinuity in their occupation. In addition, subsistence practices would have consisted of diverse resources that were exploited by hunter-gatherer groups. This diversity would have provided alternatives that would minimise risk during periods of shortages or in competitive habitats (Tourloukis/Harvati 2018, 57–61). Other researchers have proposed a »tethered forager« model for the area of Argolid in which people moved between prominent locations, within daily foraging distances based on the evidence of Franchthi, Klissoura and Kephalari (Elefanti/Marshall 2018).

2.5.2. The Mesolithic

In this section, I showcase a few key Mesolithic sites in Greece – Franchthi Cave, Sidari, Theopetra Cave, Cyclops's Cave, and Kerame I – each chosen to illustrate specific aspects such as cultural continuity, maritime activity, subsistence strategies, and the diverse environmental adaptations of Mesolithic communities. The term »Mesolithic« was almost unknown in Greek bibliography until the 1950s as was also relative evidence from sites. The first to

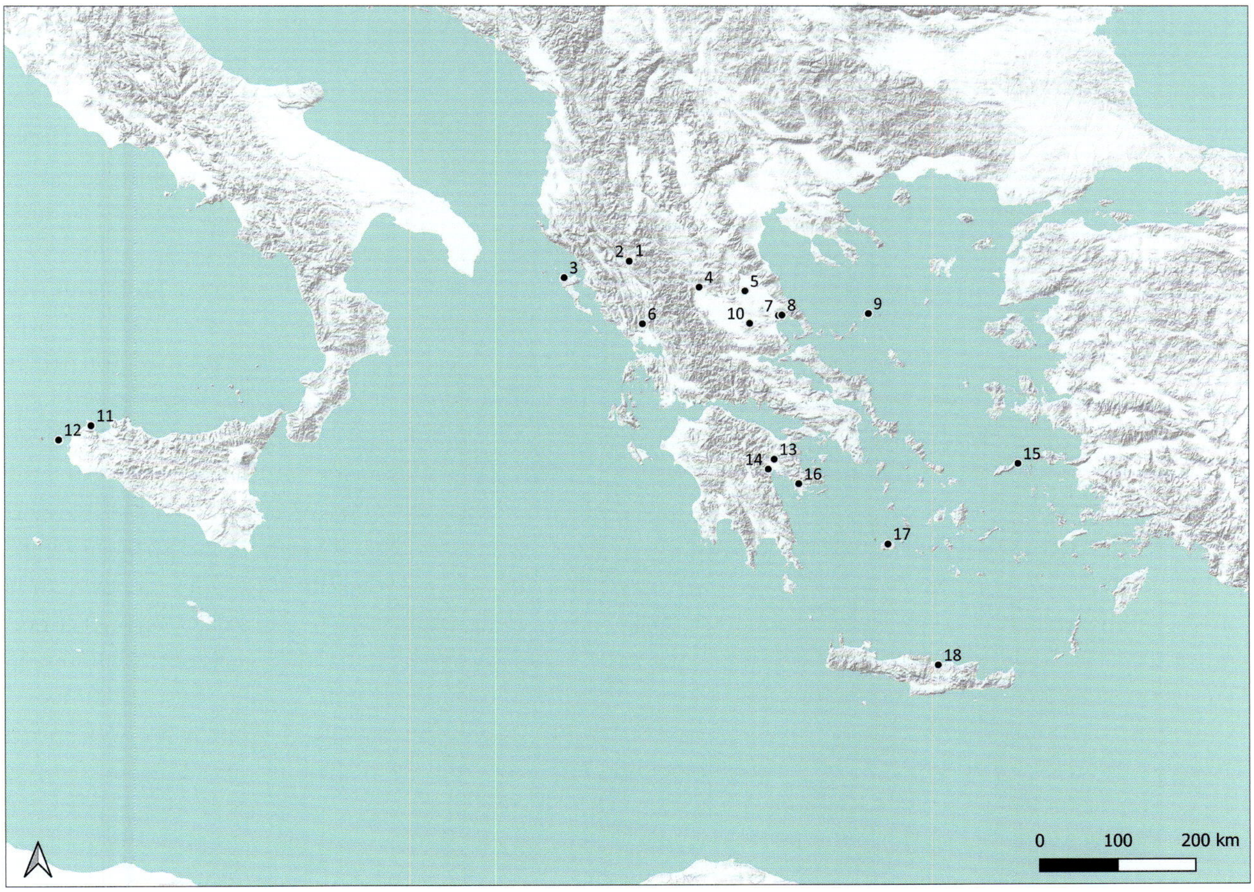

Fig. 2 Map of the central and eastern Mediterranean showing the location of key archaeological sites discussed in the text: **1** Kleidi. – **2** Boila. – **3** Sidari. – **4** Theopetra. – **5** Argissa. – **6** Asprochaliko. – **7** Sesklo. – **8** Dimini. – **9** Cyclops's Cave. – **10** Achilleio. – **11** Grotta dell'Uzzo. – **12** Grotta d'Oriente. – **13** Klissoura. – **14** Kephalari. – **15** Kerame. – **16** Franchthi. – **17** Melos. – **18** Knossos.

use it was Theocharis (1959; 1981) in his attempt to describe finds that didn't belong either to the Palaeolithic or the Neolithic. Theocharis insisted in his work on the significance of this chronological phase, as well as on the cultural continuity and the origins of the Neolithic.

The earliest investigations in Mesolithic layers took place in the 1920s and 1930s by Markovits, who was not aware at the time of this period and compared the finds from Zaimi and Ulbrich Cave with those from the Mesolithic cultures of Northern and Western Europe. Similar attempts were carried out by Sordinas with material from Sidari in Corfu. Sidari was also the first systematically excavated Mesolithic site (Sordinas 1969; 1970).

The Mesolithic record of Greece has been considered to be a sparse, geographically discontinuous and variable record (Galanidou/Perlès 2003, 30–31). The large variability in the characteristics of each site does not allow for the definition of a »cultural definition« of the Greek Mesolithic. Moreover, opinions diverge on the exact starting point of Mesolithic, often set between 10,500 and 9,000 cal BP, or at the Pleistocene-Holocene boundary, while new AMS dates would offer a better understanding of the chronology (Galanidou 2011, 222. 233). The sites with well-stratified deposits and radio-carbon dates are Franchthi, Klissoura, Theopetra and the cave of Cyclops (Facorellis 2003). These sites are situated on various different locations and altitudes, both close to the present coast or on the mainland. In the north-west, the occupation of upland caves (Grava, Megalakkos and possibly Boila) shows continuity with the late Upper Palaeolithic, while to the east (Cyclops's and Maroulas) and the south (Franchthi), sites are characterised by a maritime character (Galanidou 2011, 222. 236).

Regarding the climate of the Mesolithic, the extreme conditions of the Late Glacial had already changed prior to the advent of the Mesolithic. Average temperature had increased (by up to 10 °C) alongside the ocean level (by 100–130 m), while precipitation increased as a result of higher ocean evaporation (Bottema 2003, 33).

Regarding the reconstruction of the vegetation of this period, it presents a highly diverse picture which differs depending on the area. Vegetation diagrams show that from 13,000–10,000 BP steppe vegetation prevailed at lower elevations. At increasing heights of about 500 m, open oak woodland was present, while above that level a variety of deciduous and coniferous trees was present (Bottema 2003, 46).

The excavation at Franchthi in the 1960s and the discovery of Mesolithic layers brought forth in the archaeological discussion the subject of the Mesolithic period. The presence of Melian obsidian in the Mesolithic layers enabled the discussion on long-distance sea-travels (Perlès 1987, 176–177). Systematic surveys took place by American archaeologists in the Berbati-Limnes region and the larger area of the cave (Wells et al. 1993; Jameson et al. 1994; Runnels 1996). The Mesolithic layers at Franchthi span from ~10,300 to 9,600 BP. The Mesolithic deposit is very thick (3 m) and accumulated quickly over a short period (Farrand 2003, 74). Runnels (1995, 726) contended that evidence on Mesolithic activities were scarce and discontinuous without signs of a long-standing tradition and suggested a migration via the sea from the Middle East. Perlès (1990, 128–129) on the other hand, interpreted the stone assemblage of Franchthi as a result of economical adaptation based on the temporal habitation of the cave and focused on transformation activities rather than hunting activities.

The coastal Mesolithic side of Sidari on the island of Corfu was excavated in 1965–1966 and has been characterised as a shell midden by the excavator A. Sordinas. It was one of the first systematic Mesolithic excavations carried out in Greece after those of Markovits in the 1920s and 1930s. Six strata were revealed, among which stratum D was deemed to be the most significant dating to the late phase of Mesolithic (7,515–6,027 cal BC/9,464–7,976 cal BP), which Sordinas named »Sidarian«. The layer contained a large number of shells, fish bones and microlithic tools (Sordinas 1969; 2003).

The cave of Theopetra, a limestone massif of the Cretaceous era, is located on the western part of the Thessalian plain, 100 km from the Aegean. Even though some of the Mesolithic and Neolithic layers have been disturbed, a layer dating to the Lower Mesolithic (10,601–10,122 cal BP/8,652–8,173 cal BC and 10,224–9,559 cal BP/8,275–7,610 cal BC) has been found. Archaeobotanical remains of *Hordeum vulgare* ssp. *exastichum*, *Triticum boeoticum* and wild legumes (*Lens* sp., *Vicia ervilia*) have been found in the Mesolithic layers of Theopetra. Ovicaprids have also been identified in the same layers rendering the distinction from the archaeozoological remains of the Early Neolithic layer problematic. A Mesolithic burial of a young woman has also been found and possibly dates to the Upper Mesolithic (Kyparissi-Apostolika 2003).

Cyclops's Cave lies on the now uninhabited island of Youra in the Northern Sporades. The shores of Youra are largely steep and there aren't many sheltered bays. The cave is located on the SW part of the island. The cave yielded deposits dating from the Early Holocene to the Late Roman period, however, its significance mostly lies in its Mesolithic occupation. Radiocarbon dating on marine shells indicated that the site was used from the Mesolithic to the Final Neolithic period (10,550–5,450 cal BP/8,600–3,500 BC) as well as during the first centuries AD (Facorellis 2011). The Mesolithic chipped stone industry included local flint and Melian obsidian, suggesting an obsidian trade network already since the 8th millennium BC. The cave's deposits yielded a highly specialised bone tool industry, among which were many fish hooks (Moundrea-Agrafioti 2003; 2011). In addition, it appears that there were intensive processing and preservation activities of inshore and migratory fish (Mylona 2003; 2011).

The extensive number of fish bones, sea shells, land snails, mammal bones and bird remains has led the excavators to believe that the cave was occupied on a seasonal basis by hunter-gatherers, specialised in fishing and bird hunting who also consumed game, seafood and land snails. These populations seem to have followed the migratory itineraries of birds and fish and had a developed seafaring activity and long-distance voyages.

The Mesolithic site of Kerame I on the island of Ikaria is located on a high coastal cliff at an elevation of 20 m, north of Ayios Kirikos. It constitutes an unusually extended site, which would have been even larger if the erosion of the peninsula is taken into account. The only reliable absolute dates based on obsidian flakes with the use of the SIMS-SS obsidian hydration dating method date the site to 17,698–8,179 cal BP/15,749–6,230 cal BC and 43,998–6,399 cal BP/42,049–4,450 cal BC (Laskaris et al. 2011, tab. 1). Apart from a few thousand stone tools made from obsidian, flint and quartz, no organic remains were found nor visible constructions, probably as a result of former intensive cultivation. The lithic industry of Kerame is large and homogeneous. The style of debitage and tool morphology is very similar to the Mesolithic of the Western Cyclades. Apart from Maroulas, Ikaria is the second island in the Aegean with a site/outdoors occupation (Sampson et al. 2012; Sampson 2014).

2.5.3. The Neolithic

The advent of the Neolithic in the Greek territory dates to the early 7th millennium BC. Relative chronological sequences of the Early Neolithic in Greece were developed based on limited contextual data in the 1950s. Following the identification of the Pre-Pottery Neolithic, terms like »Aceramic« or »Preceramic« Neolithic were adopted from the Near Eastern examples in the 1950s and 1970s. The first stage of the Neolithic phenomenon was characterised early on as »Preceramic Neolithic« or »Aceramic Neolithic« period, due to the lack of ceramics in Neolithic layers. This was the case in the sites of Argissa, Sesklo, Soufli Magoula, Gentiki and Achilleio in Thessaly, in Knossos in Crete and Franchthi (Theocharis 1981). However, in the 1980s and more recently, the existence of this Aceramic period in Greece was contested by many researchers, as it then became evident that ceramics did exist in the layers of some of these sites and the term »Initial Neolithic« was preferred and gradually gained more ground (Papathanasopoulos 1996).

In this section, I will focus on the Aegean and the territory of the modern Greek state. The history of research of the Aegean Neolithic dates back to the 19th century as Neolithic settlements attracted the interest of travellers such as W. Leake, who described several prehistoric sites. The first systematic catalogue of Thessalian sites however was authored by Tsountas in 1908, who worked in Thessaly between 1899 and 1906 and conducted pioneering work at Sesklo and Dimini (Tsountas 1908). Wace and Thompson continued the work of Tsountas in the area, documented 23 sites in »Prehistoric Thessaly« (1912), and carried out excavation work at Rachmani and Tsangli among other sites. Heurtley's work on Macedonia (1939) is also part of the earliest investigations in the Aegean in the 1920s and 1930s. Research resumes in the 1960s in Greece, however on a limited scale, while large-scale investigations are carried out in the 1970s at Sesklo by D. Theocharis and at Dimini by G. Hourmouziadis (Demoule/Perlès 1993).

For long, archaeological research on the Neolithic focused on the mounds that were visible landmarks in flat landscapes, such as the Thessalian plain. Since the early 1990s, large-scale salvage excavations throughout the country contributed to the advance of the (generally under subsidised) Neolithic research. Salvage excavations continue to be a driving force in Greek archaeology today, bearing however the limitations that such work sites have for scientific research, such as the partial recovery of archaeological material due to time constraints and commercially-driven decisions. Due to the general lack of funding in archaeology, rescue excavations are often some of the few instances that large-scale excavations take place. Also, large-scale excavations enabled the exposure of extended areas that provided valuable information for the advance of the analytical framework; new evidence of habitation, building forms, construction techniques, residential elements among others changed the established preconceptions on the Neolithic by revealing a growing diversity of the archaeological archive regarding settlement organisation (Kalogiropoulou/Sarris 2018, 2).

While earlier investigations focused on establishing grand narratives in search of cultural and national identities (Tsountas 1908; Mylonas 1929), 20th century research gradually shifted towards the examination of social structures (Theocharis 1967; Chourmouziadis 1979) to find itself today into the quest of human-centred approaches. An emphasis is placed on the regional scale and the unique characteristics of each region (e. g. Pappa et al. 2004; Nanoglou 2008; Souvatzi 2008; Kotsakis 2014; Kalogiropoulou/Sarris 2018, 1–2).

2.5.3.1. Settlement Patterns

During the Neolithic period, habitation also took place in caves (for example Theopetra or Franchthi) where in some cases it continued from earlier periods. Evidence for occupation also exists on islands such as Crete, Euboea, and Youra during the Early and Middle Neolithic. As for the Cyclades, Ftelia on Mykonos is dated to the Late Neolithic period (Papathanasopoulos 1996, 39–40).

As for the Neolithic settlements, their form varies in flat or extended settlements and mounds or tells depending on the way they are developed. The formation of the flat settlement is extensive in space while the formation of the tell is extensive in height. Tell formation is a cumulative process and is due to the continuous use of mud for construction and the continuous and stable use of habitational space for long periods of time (Sherratt 1983). Extended settlements are characterised by their large size which can occupy many hectares, while their average height is very small and often are not distinct in the landscape (Kotsakis 1999).

Interestingly, research has attributed to tells (or *magoules* in Thessaly and *toumbes* in Macedonia), symbolic connotations as well as a specific importance in the Neolithic landscape. In Thessaly for example, where hundreds of tells are located, the settlements would be considerable landmarks in the landscape (Halstead 1999, 87). The recurring rebuilding in the same place is deemed to attribute a symbolic meaning to the temporal continuity of a household, a settlement and its inhabitants. On the household level, the Neolithic household becomes more autonomous from the community and may compete with other households, whereas on the societal level the monumentality of the tell becomes an ideological mechanism and there is an emphasis on the temporal dimension and the ancestry of settlement and its inhabitants. These are among the elements that represent a changing social reality. These elements are also absent from the extended settlements (Kotsakis 1999, 73–74).

2.5.3.2. Geographic Context

Until recently, 917 Neolithic sites have been recorded in Greece, including both settlements and caves. 276 of the sites are situated in Thessaly, where some of the larger excavation campaigns of the 1960s and 1970s took place. In the geographical department of Macedonia, so far 143 sites are known to date, while in Thrace 17 sites, in Epirus 11 sites. In Central Greece and Euboea 186 sites and only 86 sites in the Peloponnese. In Crete, there are 135 Neolithic sites while fewer sites are located on the rest of the islands: 7 in the north-eastern Aegean, 45 in the Dodecanese, 22 in the Cyclades, 10 in the Ionian islands and 4 in the Sporades (Papathanasopoulos 1996). This uneven distribution of Neolithic sites reflects partly the uneven distribution of surface investigations of these regions where the quantitative data are drawn from. These numbers are certainly only indicative since on-going research brings to light more new sites.

The Greek territory is characterised by its mountainous relief while more than two-thirds of the country are above 300 m a. s. l. This relief creates constraints on inland communications. A characteristic element is the Pindus mountain range, which trends north-northwest to the south-southeast, from Albania to Central Greece and the north of the Peloponnese. Pindus essentially separates the western coast of the Ionian sea from western Macedonia and Thessaly. Throughout prehistory, it constituted both a climatic and topographic barrier and according to researchers it contributed to varying occupational settlement patterns and traditions. As rugged as the mountainous relief of Greece may be, it was not a barrier that could not be passed and interregional interaction was a widespread phenomenon (Perlès 2004, 10).

In general, the landscape is characterised by steep slopes, often eroded and bodies of water. The lowlands comprise about 10 % of the country, where agriculture takes place (Demoule/Perlès 1993). Neolithic site distribution is subjected to natural factors, such as the topography and the hydrography of a region (Bintliff 1977), as the setting of the settlement is closely related to the productive activities of its inhabitants. One could argue that Neolithic farmers would prefer regions where farming and herding could take place. However, topography or the economy are not the sole factors that shape site distribution as non-productive factors are also at play such as the control of the region by neighbouring communities, competition or conflict among groups, and symbolic dimensions of the landscape (Kotsakis 2004, 59).

2.6. Franchthi

2.6.1. Introduction

In this section of the thesis, I will present the published work on the site as it is known so far from the published »Excavations at Franchthi Cave, Greece« volumes and various other scientific articles. My own contribution on the site material is presented in the following chapters.

The site was excavated extensively from 1967 to 1979 by Indiana University and the American School of Classical Studies at Athens, and a large number of studies on the different materials recovered have been published. Between 2012 and 2019 significant underwater investigations have taken place in the vicinity of the site, while more recently in 2011 and 2012, small-scale rescue excavations have been carried out nearby the site (see section 2.6.2).

Franchthi has long attracted interest of researchers as it has revealed an impressive continuum of human occupation, which extends from the early Upper Palaeolithic to the Final Neolithic. Another reason why it became an important site in the Eastern Mediterranean was the discovery of Melian obsidian already since the Upper Palaeolithic, placing it in extensive exchange networks (Perlès 1987, 141). In addition, the Franchthi excavations carried out pioneering work for their time. One of the novelties of their research was the application of the water sieving technique in some of the trenches which had been applied before in a few sites. At Franchthi, flotation and water-sieving were applied in trenches FAN, FAS, H1A and H1B. Selected samples from the Paralia trenches were also water-sieved, e. g. most of L5. Trenches were dug in shallow passes or »units« following variations in the sediment and the material from all the trenches was dry sieved (Diamant 1979, 207; Hansen 1991, 24).

2.6.2. History of Research

Excavations in Franchthi were conducted by the team of Indiana University under the aegis of the American School of Classical Studies at Athens. A total of eight field campaigns took place in Franchthi in 1967, 1968, 1969, 1971, 1973, 1974, 1976, and 1979. Initially, excavations began in the summer of 1967 as part of the larger survey program the »Argolid Exploration Project« conducted by the University of Pennsylvania and Indiana University under the direction of M. H. Jameson and T. W. Jacobsen. The project initially at the time focused on the classical site of Halieis in nearby Porto Kheli, however the research interest quickly turned to Franchthi (Jacobsen/Farrand 1987, 2).

Between 2012 and 2019, further investigations took place in the submerged vicinity of Franchthi. New areas were investigated using different and improved techniques which entailed detailed swath bathymetry, side-scan sonar imaging and sub-bottom profiling building on the previous work that had been conducted in the earlier decades. This work revealed new submerged structures and sites and reconsidered previous work on the site and its submerged environment (Beck/Koutsoumba 2015; 2016; 2017; Beck et al. 2021).

In addition, the Ephorate for Palaeoanthropology and Speleology (EPS) in 2011 and 2012 conducted rescue excavations in Franchthi Cave to develop access and tourist facilities on site, to improve the access to the site and to install informative signposts. Thus, the project did not have research goals but aimed at rendering the site accessible to visitors (Katsarou, pers. comm.).

The rescue operations involved topsoil digging, surface arrangements and the cleaning and conservation of the old trenches and wall sections. An information kiosk and various sign posts were installed in the interior of the cave and its surroundings and some of the old trenches (FA, G and HA) in the cave were covered and filled. Small retaining lines of stones were also placed in order to prevent further material from rolling down the hill.

More specifically, part of the operations focused on the path from Lambayana beach to the cave entrance, which runs alongside the coast. The 185 m long path was widened and arranged, shallowly dug and divided into ten sectors for the sake of operations. These sectors cover the path from the small pier to the informative sign posts outside the cave's entrance.

Rescue work in the area close to the site revealed that activities and traces of occupation expanded in a wider area around the cave, while the excavators consider many of the finds of these rescue excavations to be *in situ*. Among the finds from sectors 1–9 that had not been covered by the 1970s excavations and were dug both along the path that leads to the cave as well as near the Paralia trenches, were pottery, shells (*Cerastoderma*), bones, flake tools and a stone structure (01/2012 diaries). The pottery studied by S. Katsarou is dated to the Middle, the Late and the Final Neolithic and belonged to primary depositions (Katsarou, pers. comm.). Two undisturbed burials were also found, one near the entrance of the cave and one near trench O5. Lastly, various human remains have been found close to the surface and to the Paralia trenches, usually without grave goods (Papathanasiou, pers. comm.).

2.6.3. Surrounding Landscape

Nowadays, the limestone headland of Franchthi is located in a region with numerous low mountain ranges (up to 700 m) such as the Dhidhima Range in the west or the Adheres Range in the east. Among these ranges, two plateaus are created; the Dhidhima plateau (north of Franchthi) and the Iliokastro plateau (north-east of Franchthi). The area is also characterised by lower limestone ridges. There are many streams in the area, but they are generally dry unless heavy rains occur. The coastal peninsula of the southern Argolid is surrounded by bays, while there are sheltered embayments in Kilada and Porto Kheli. The coasts are generally steep with cliffs (van Andel 1987, 3). Natural vegetation in the area is characterised by the presence of pine woodland (*Pinus halepensis*), *phrygana* and herbaceous vegetation, open maquis vegetation and grassland (Gavrielides 1976, 145; van Andel 1987, 8). *Juniperus communis* is the dominant species, while there are also other species such as *Olea europea*, *Quercus coccifera*, *Pistacia lentiscus*, *Coridothymus* sp. among others (van Andel 1987, tab. 1). Modern cultivated vegetation includes olive groves, cereal fields, citrus orchards and vineyards (Gavrielides 1976, 145–148). In terms of geology, the area of Franchthi belongs to formations of the Upper Palaeozoic and consists of cretaceous limestone (Gaitanakis/Fotiadis 1991, 325–326).

Contrary to its present surroundings the site was not always coastal in prehistory (see **fig. 1**). The cave of Franchthi was located 6 km inland during the Late Pleistocene as the sea gradually flooded the coastal plain between 8,500 and 6,300 cal BP (**fig. 3**; Surdez et al. 2018, 14). Thus, the prehistoric landscape was very different from the present-day environment.

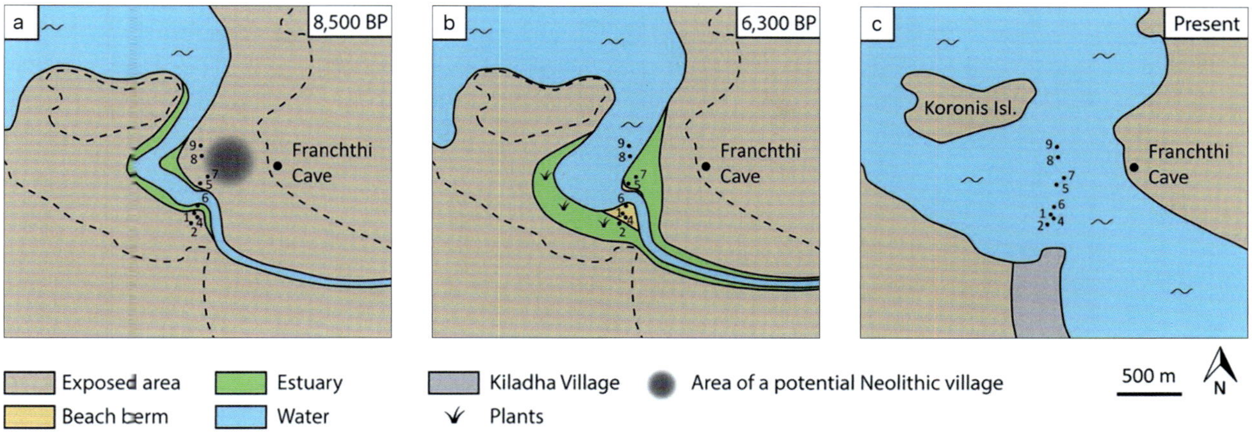

Fig. 3 Reconstruction of the landscape evolution in the bay of Kilada: **a** 8,500 cal BP. – **b** 6,300 cal BP. – **c** Present state. – (After Surdez et al. 2018, fig. 8, under CC BY 4.0).

The palaeogeography of the submerged landscape of the Kilada Bay and the wider gulf was investigated already in the 1980s by van Andel and Lianos (1983; 1984) who carried out marine geophysical surveys and coring activities in order to reconstruct the coastal geography and evolution of the Late Pleistocene and the Early Holocene landscape. This innovative work for that time period also included high-resolution seismic profiling survey in the eastern Argolic Gulf and indicated the presence of a palaeo-riverbed and a river terrace in the coastal plain. The lowest glacial shores were also identified at −115 and −118 m. Pottery was also found *in situ* in cores in sediments offshore the cave which was radiocarbon-dated to the Neolithic period and indicated that human occupation took place on the banks of the river (van Andel et al. 1980; van Andel 1987; Gifford 1990).

Between 2012 and 2019, a collaboration between a team of scientists from the University of Geneva, under the auspices of the Swiss School of Archaeology in Greece, the Ephorate of Underwater Antiquities and the Hellenic Centre for Marine Research conducted further work on the submerged area of Franchthi. The profiling of the sea bottom identified palaeo-shorelines, palaeo-sea level indicators, an active faulting network and submerged structures confirming and mapping more precisely the submerged environment's main features.

Among the finds of the recent surveys geomorphological features, a palaeo-shoreline was identified that dates to the Last Glacial Maximum (18,000 BP) at depths between 116 and 127 m. A second prominent feature between 159 and 173 m could represent the sea level during the Oxygen Isotope Stage 6 (140,000 BP). Furthermore, a ~5 m deep marine terrace was mapped in front of the cave which may indicate the sea level during the Neolithic period. Another terrace at 11 m depth below sea level is incised by a meandering riverbed. Between Franchthi Cave and the port of Kilada a feature was identified that could possibly represent a cultural deposit indicating prehistoric human presence. At least two submerged Neolithic sites, 150 and 430 m away from the cave, were identified in the bay, while their function and precise date remains unknown and another site dating to the Final Neolithic/Early Bronze Age I off the nearby beach of Lambayanna north of Franchthi (**fig. 4**). Thus, even though the contemporaneity of these structures is not clear, it may be possible that they are a succession of sites and a reflection of the interactions of the prehistoric people with the migrating shoreline (Sakellariou et al. 2015; Surdez et al. 2018; Beck/Sotiriou 2020; Beck et al. 2021, 135–138).

Pollen data was analysed from core OK85/11 retrieved from the bay of Kilada from a submerged site offshore from Franchthi (Bottema 1990). The chronology of the specific core only provides evidence

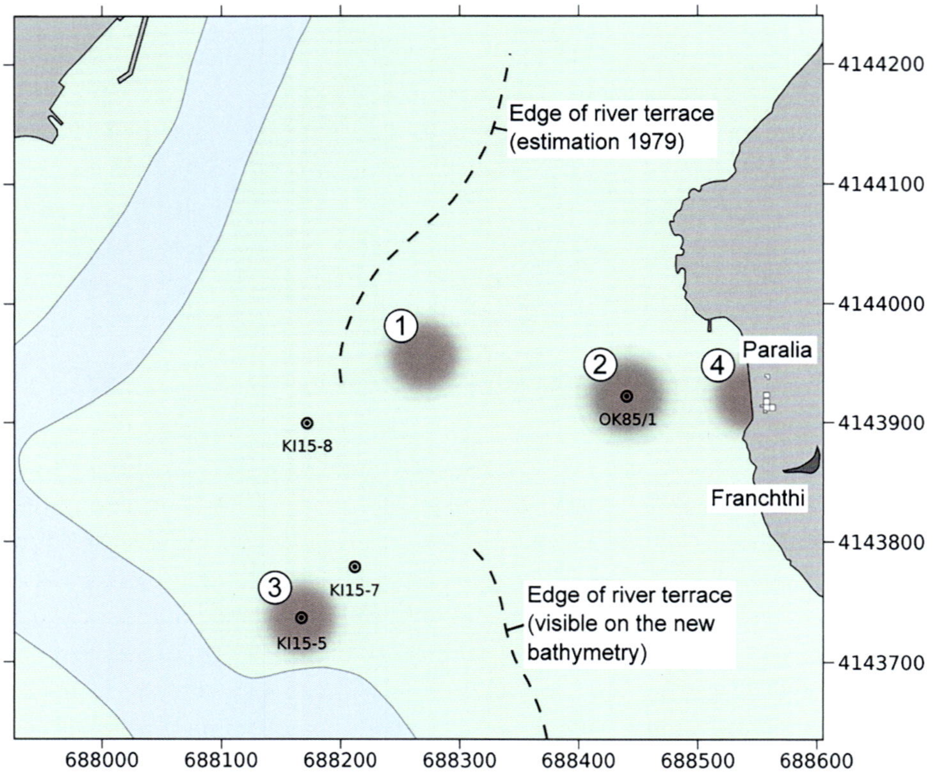

Fig. 4 Depiction of the recent offshore survey results with the locations of the palaeo-riverbed, the river terrace, the location of the submerged Neolithic sites (**1–4**) and the cores (OK85/1, KI15-8, KI15-7, KI15-5). – (Reproduced from Beck et al. 2021, fig. 2, under CC BY 4.0).

on the Late Neolithic period as dates acquired range between 7,940–7,506 cal BP/5,991–5,557 cal BC and 7,956–6,959 cal BP/6,007–5,010 cal BC. According to the analysis, an open vegetation was present during the Late Neolithic as the low arboreal pollen present is dominated by *Quercus cerris* type (Bottema 1990, 120). Generally, the pollen data informs on vegetation of a broader region, but the marine sediments may contain unrecognised gaps and the chronology lacks precision due to limited radiocarbon dates (Jameson et al. 1994, 167).

2.6.4. General Overview of the Site

Starting with its geographical location, Franchthi is located in the southern Argolid in the area of Ermionis. The rounded limestone headland of Franchthi prevails in the landscape. On the opposite and very close to the shore there is the island of Koronis and across the bay to the east from the cave lies the small village of Kilada (Jacobsen/Farrand 1987).

Two main sectors were designated during the Franchthi excavations, the cave and the Paralia (Greek for »shore«) because of its proximity to the shore. According to Jacobsen and Farrand, the cave site and the Paralia constitute two interconnected sites, »clearly related but distinct« (1987, 2). Franchthi Cave constitutes a karstic formation measuring 150 m in length and a mouth that opens to the north-west about 15 m above the present shoreline. The roof of the cave has collapsed and boulders cover about two-thirds of the interior rear part of the cave. This collapse consequently restrained the extent of archaeological investigations in the rear part of the cave. Thus, trenches inside the cave lie in the posterior part of the cave where a sequence of more than 20,000 years was revealed. The site is dated to the Upper Palaeolithic, the Mesolithic and the Neolithic period (c. 35,000 BP or 40,000–39,000 cal BP to 3,000 cal BP) while evidence of later occupation doesn't belong to secure stratigraphic contexts (Jacobsen 1973a; Jacobsen/Farrand 1987, 2; Farrand 2000, 75–78; Douka et al. 2011, 1140–1143).

100 m from the cave's mouth and along the shoreline lies the second site, the Paralia. The Paralia excavations revealed a number of terraces and free-standing structures dating to the Neolithic period. These investigations covered a relatively small area of approximately 113 m^2 on a short strip of land. The site, however, covered a larger area since investigations in the adjacent bay and the wider area brought to light further evidence (Jacobsen/Farrand 1987, 2; Jacobsen 1976; van Andel et al. 1980; van Andel 1987; Gifford 1990; Beck/Koutsoumba 2014; 2015; 2016; 2017; Beck et al. 2021). Vitelli further suggests that the naturally protected cave and the exposed slopes of the Paralia may have been areas that were exploited for different activities (Vitelli 1993, 31).

The stratigraphic sequence occupation was not continuous as certain hiatuses occurred. The earliest stratified material belongs to the Upper Palaeolithic period. According to Jacobsen, the sparse remains of that period indicated that the site was sporadically used by small hunter-gatherer groups on a seasonal basis. Sea level was 100 m less than today and the shore several kilometres afar. Tools consist of bladelets and flint scrapers and there is evidence of small-scale game hunting. There is no evidence of wild plants during this period. After this period, a depositional hiatus occurs, suggesting a temporary abandonment of the site around 16,950 cal BP which is associated with the Last Glacial Maximum, while it is not clear if the cave was occupied during this time (Jacobsen 1981, 305–306; Perlès 1999, 312).

Subsequently, more intense activity takes place at the cave during the Final Palaeolithic period. Hunting of large game, predominantly of red deer is attested in the faunal remains, while there is evidence small-scale fishing, shell-collecting (Shackleton 1988) and the exploitation of wild plants, such as pulses, nuts and wild cereals (Hansen 1991). Microlithic tools of Melian origin also appear (Perlès 1987, 142). According to Jacobsen, during this period Franchthi is used during certain times of the year (spring, summer and autumn) (Hansen 1991), though he also considers the possibility of year-round occupation (Jacobsen 1981, 306).

After a hiatus in the occupation which coincides with the Last Glacial Maximum, evidence for human occupation at the site re-emerges in the Mesolithic (Perlès 1999, 314). During the Lower Mesolithic, the prehistoric occupants of the cave continued to rely on the hunting of red deer and the gathering of wild fruits, cereals and legumes (Jacobsen 1981, 307). Furthermore, the cave of Franchthi possibly becomes a site of symbolic importance as collective rituals, burials and cremations take place on the site. The Mesolithic remains represent between 15 and 34 different individuals that belong to all age groups and sexes (Cullen 1995, 274).

The Upper Mesolithic period, present in trenches FAS, FAN, H, H1B, H1A and G1, is dated between 8,400 and 7,400 cal BC by radiocarbon dating (Perlès et al. 2013, 18). During this period intensive tuna fishing takes place and it appears that fishing was an important activity of the occupants of the cave (Rose 1995), while the lithic assemblage is characterised by microliths (Perlès 1990, 46). The archaeobotanical remains also reflect an intensified exploitation of available resources (Hansen 1980).

During the Final Mesolithic period, excavated in trenches FAS and FAN, the tuna remains and non-geometric microlithics are replaced by transverse arrowheads, while subsistence lies in wild fruit, legumes, cereal, land snails and wild game (Munro/Stiner 2015, 597). Discontinuity between the Mesolithic and the Neolithic is attested by wild cereals, oats and barley, completely disappearing from the botanical record before the appearance of domesticated cereals (Hansen 1991, 76). During the Final Mesolithic period, the lithic evidence is quite crude and resembles that of Early Mesolithic, while during the Initial Neolithic phase the lithic assemblage remains close to the Mesolithic traditions (Perlès 1990, 85. 104–105; 1999, 317).

During the Initial Neolithic, the first evidence of domestication takes place with the appearance of domestic sheep, goats, wheat (*Triticum dicoccum*) and barley (*Hordeum distichum*) in small quantities (Payne 1975; Hansen 1991), while there is a clear drop in the hunting of red deer and the fishing of tuna. Trapezes on pressure-flaked blades appear in the lithic assemblage (Perlès 1990). Two domestic seeds of *T. dicoccum* from units FAN 162 and 163, which belonged to the interface of the rocky layer X1 and the »grey clay« (X2), were dated successfully between ~6,800 and 6,400 cal BC, that is the first half of the 7th millennium BC. Although there is some intermixing in the Final Mesolithic and Initial Neolithic layers (Perlès 1990, 91–93), the lack of a wild ancestor of wheat in Greece indicates that wheat was brought in as a domesticated species during the Neolithic. Therefore, these domestic seeds can only be attributed to the Neolithic period, not the Mesolithic (Perlès 2013, 1007).

During the Neolithic period, activity also expanded outside of the cave at the base of the present-day slope. Excavations at the Paralia area brought to light a number of substantial stone walls. The interpretation of this complex of walls remains unanswered by the excavators. They either belong to small domestic structures, representing an expansion of the settlement in the cave, or they may represent agricultural terraces on the slopes in front of the cave (Jacobsen 1976, 83; 1981, 309; Vitelli 1993, 71). Furthermore, during the Neolithic, there is an increased abundance of Melian obsidian, andesites, as well as the first appearance of marble and other exotic raw materials on the site. Objects of personal adornment also become more common and variable during this time, sometimes found as grave goods. Mortuary evidence as well as the archaeological record of the Neolithic do not suggest the presence of social stratification based on wealth or status, but of a rather »egalitarian« model of socio-political organisation (Jacobsen 1981, 310).

2.6.5. Stratigraphy

As described in section 2.6.1, the site consists of two areas, the cave and the Paralia (**fig. 5**). The trenches excavated at the cave are named A to H (A, B, BE, C, D, E, F, F1, FF1, FA, FAN, FAS, G, G1, H, H1, H1A, H1B, H2, H2A H2B) and at the Paralia L to Q (L5, L5NE, O5, O5N, O5NE, P5, PQ5, Q4, Q5S, Q5N, Q6N, Q6NE, QR5) (Jacobsen 1987, tabs 1–2).

The lower deposits of Franchthi Cave, strata P through V in the FA trench which is the deepest trench in the cave, represent the Upper Palaeolithic sequence (**tab. 1**). At the bottom of the sequence lie some possibly Middle Palaeolithic artefacts. Volcanic ash (Campanian tephra) from stratum Q, has been dated to 39,280 ± 110 by $^{40}Ar/^{39}Ar$ (De Vivo et al. 2001). The accumulation rate during this period was very slow as well as of an intermittent nature and lasted more than 24,000 years (Farrand 2000, 83–86). The radiocarbon dates point towards major periods of little or no sedimentation during this chronological period, thus indicating major hiatuses during this time period (Farrand 2003, 74). The Upper Palaeolithic apart from trench FAS (units 227–220), is also present in trenches H1A (units 220–167) and H1B (units 215–150), G1 (units 66–67), H (units 57–60, 70–71) (Perlès 1987, 46 Document I.1).

Densely accumulated land snail middens (stratum T) are present in trench HH1 in stratum T near the end of the FA hiatus, ~13,000 BP. Upper Palaeolithic artefacts, mostly flint or chert bladelets, have been found in these sediments together with large mammal bones, mostly of equids and bovids (Perlès 1990; Jacobsen 1973a, 59). This sequence ends with a rocky deposit with sparse artefacts (stratum V) at 11,000 BP which is followed by a hiatus of ~650 years (Farrand 2003, 74).

The Mesolithic sequence is 3 m thick and spans from ~10,300–9,600 BP. The Mesolithic deposits were accumulated in a short time period and the deposition was rapid. The Mesolithic deposits contained numerous animal and fish bones and botanical remains. The sequence is followed by the rocky layer (stratum X1) and subsequently by a hiatus of approximately 600 years (Farrand 2003, 74). The Mesolithic is present in layers W and X1 of trenches FA and H1. More specifically, the Lower Mesolithic is further present in trenches A, H1A (units 165–122), H1B (units 148–122), FAS (units 195–172), FAN (230–198), H (units 62–67), G (units 65–45). The Upper Mesolithic is present in trenches H1A (units 119. 117–100), H1B (units 117–107), FAS (units 170–155), FAN (units 198–175), H (units 61. 50–53), F/F-1 (units 44. 43), G1 (units 45–27. 39–34), H2A (units 103–199. 163–153). The Final Mesolithic is present in trenches FAS (units 151–148), FAN (units 172–164), G1 (units 26–22) (Perlès 1987, Document I.1; 1990, tab. XV).

The Neolithic deposits (strata X2 through Y3) are 3–4 m thick. Stratum X2 represents the basal Neo-

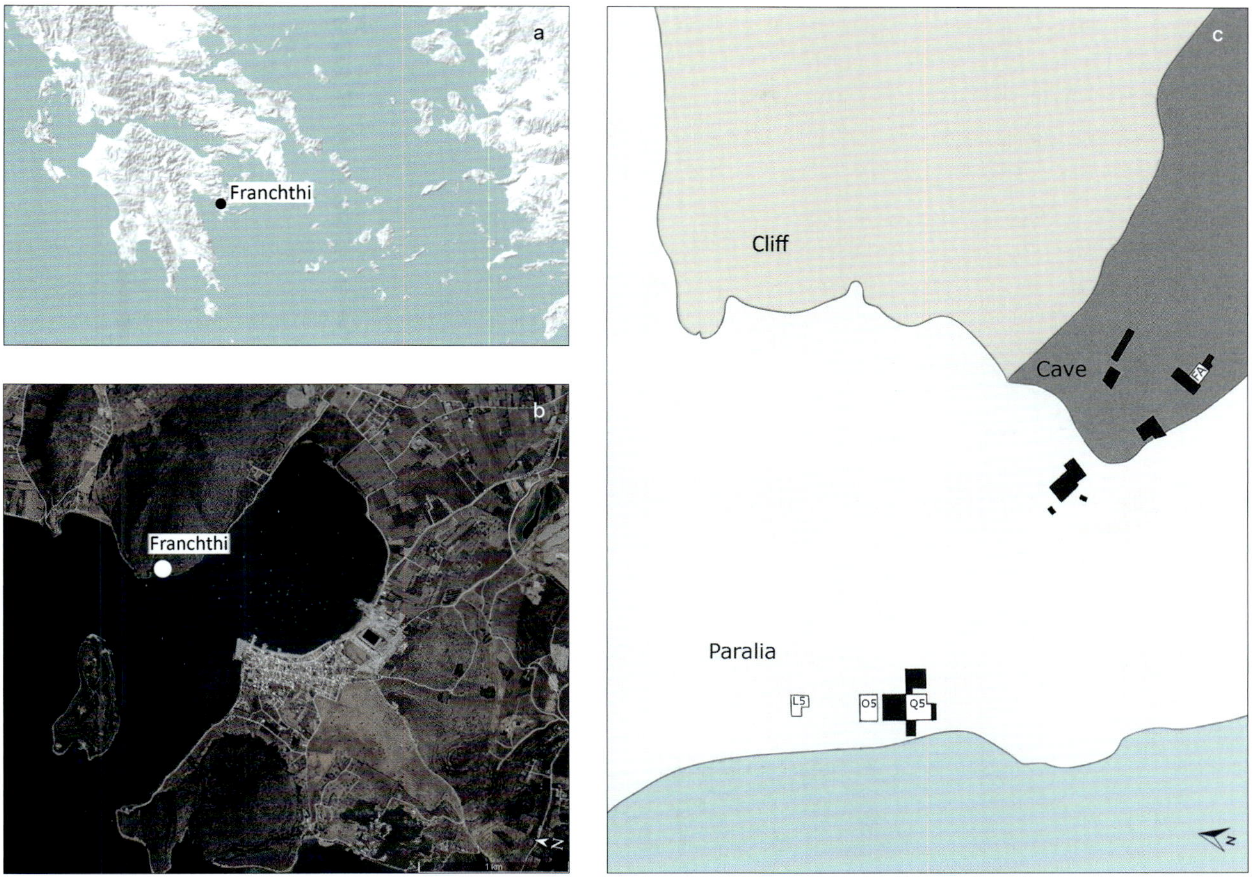

Fig. 5 Geographic setting and site layout of Franchthi: **a** Regional context map based on Esri basemap data. – **b** Satellite view of the bay of Kilada derived from Google Earth imagery. – **c** Schematic representation of site sectors and excavated trenches from which material was sampled.

lithic deposits, where sherds have been found (Vitelli 1993, 33). During this initial Neolithic phase human presence is reduced. Later Neolithic deposits (strata Y1, Y2 and Y3) follow. Larger rocks occur in these deposits, while it is possible that the rocks in stratum Y2 are associated with a major rockfall of the cave brow. The archaeological material from these deposits is rich in ceramics and ovicaprine bones while land snail shells (abundant in the Mesolithic and the Upper Palaeolithic) are reduced. The sedimentation rates are relatively high, but less than the Mesolithic (Farrand 2003, 74). The Neolithic is present in trenches H1A (units 99–80), H1B (units 99–73), FAS (units 147–59), FAN (units 163–59), F/F-1 (units 42–1), B/E (units 21–1), F/A (units 58–1), H2A (units 152–41). In the Paralia area it is present in trenches L5, L5NE, O5, O5NE, P5, P5/Q5, Q5N, Q5S, Q5W, Q6N, Q/R (Perlès 1987, Document I.2).

2.6.6. Pottery

In comparison to other ceramic assemblages, the Franchthi ceramic assemblage derives from a chronological sequence that covers most of the Neolithic. In addition, most of the excavated ceramic material has been retained and all sediments were sieved and material larger than 5 mm was also recovered, a practice that was not common up to that point in time in Greece (Vitelli 1993, 3).

The Early Neolithic assemblage is dominated by the lime monochrome burnished ware, togeth-

Lithostratigraphic layer	General date range (cal BP)	Cultural attribution and key characteristics	Trenches present
Y3 (Neolithic)	6,100-5,900	Later Neolithic deposits	A (units 21-26, lower 40), FA (units 46-49), FAS (units 59-74), FAN (units 59-63, 70-72, 75, 78, 79), FF1 (units 7-10)
Y2 (Neolithic)	7,700-6,900	Possibly major rockfall	A (units 28-39, 41-58), FAS (units 84-127), FAN (units 73, 76, 85-137), FF1 (units 11-42B1 and 40A1), FA (units 49A-57), H1 (units 40-71), H1B (units 72-74), H (units 13-28, 32-39), H37P, H2A (units 20-134)
Y1 (Neolithic)	8,000-7,700		A (units 59-60), FAS (units 129-133), FAN (units 142-146), FF1 (units 43B1, 2 and 40A2-41A1), G (units 25, 26), GG1 (units 19), H1B (units 75-79), H1A (units 72-73), H (units 40, 37P-EE), H2A (units 136-148), H2A Ped (units 180-188)
X2 (Initial Neolithic)	8,900-8,500	Sherds present; reduced human presence	A (units 61-64), FAS (units 136-145), FAN (units 150-160), FF1 (units 41A2-A4, 44B1-B4, 40D-42D), G (units 27-34), G1 (units 20-27), H1B (units 80-114), H1A (units 74-108), H (units 41-49), H2A (units 150-151), H2A Ped (units 189-197). Paralia: L5, L5NE, O5, O5NE, P5, P5/Q5, Q5N, Q5S, Q5W, Q6N, Q/R (Neolithic generally)
X1 (Final and Upper Mesolithic)	9,450-8,950	Mixed deposits	A (unit 65), FAS (units 148-151), FAN (units 163-172), FF1 (units 45B1-B4), G1 (units 26-22), H1A (units 110-115), H (units 50, 46-47), H1B (mixed), H2A (missing)
W3 (Upper Mesolithic and Mixed Mesolithic)	10,900-10,000		A (unit 66), FAS (units 152-159), FAN (units 174-184), G (units 35-42), G1 (units 28-37), H1B (units 117-127), H1A (units 117-131), H (units 49-51), H2A (units 152-163), H2A Ped (units 197-203)
W2 (Lower Mesolithic)	10,900-10,000		A (unit 67), FAS (units 163-192), FAN (units 187-230), G1 (units 38-59), H1B (units 128-141), H1A (units 132-155), H (units 53A1-54A7, 61B1-63B4)
W1 (Lower Mesolithic)	10,900-10,000		FAS (units 195-197), G1 (units 60-65), H1B (units 141-149), H1A (units 155-160), H (units 54A5-55A3, 63B4-65B)
V (Upper Palaeolithic)	Undated	Rocky deposit with sparse artefacts; followed by ~650-year hiatus	FAS (units 199-202), G1 (units 66-67), H1B (units 148-150), H1A (units 161-168), H (units 55A2-A4, 56A1-2, 65B)
U (Upper Palaeolithic)	11,700-12,500	Final Palaeolithic	FAS (units 202-206), H1B (units 151-154), H1A (units 169-174, 175?), H (units 56A3-57A, 67B)
T (Upper Palaeolithic)	~13,000 (for snail middens) / 13,300-12,700	Epigravettian	H1B (units 155-166), HH1 (stratum T), H1A (units 175-189), H (units 58A1-A6, 59A1-A3, 67B5-6, 69B-70B, 71B1-4?). Not present in trench FAS
S2 (Upper Palaeolithic)	14,000-15,000	Epigravettian	H1B (units 164-172), H1A (units 186-204, 190-204), H (units 60A2-10, 71B1-4?). Absent in FAS
S1 (Upper Palaeolithic)	Undated	»Gravettoid«	H1B (units 173-180), H1A (units 205-210), FAS (unit 208)
R (Upper Palaeolithic)	26,000-27,000	»Gravettoid«	H1B (units 181-207), FAS (units 209-217), H1A (units 212-220)
R (Upper Palaeolithic)	36,000	Aurignacian (carenated end-scrapers, twisted bladelets)	H1B (units 208-212)
Q (tephra)	39,280 ± 110 ($^{40}Ar/^{39}Ar$) / 39,000	Aurignacian, Campanian tephra	FAS (units 218-222), H1B (unit 213)
P	> Age of tephra	Sparse, non-diagnostic industry; possibly Middle Palaeolithic artefacts	FAS (units 224-227, 222, 223), H1B (units 214-215), FA (bottom of sequence)

Tab. 1 Summary of stratigraphy, dating, and occupation phases at Franchthi. – (Douka et al. 2015; Perlès et al. 2013; Stiner/Munro 2011; De Vivo et al. 2001; Farrand 2000; Perlès 1987; 1999; 2004; Vitelli 1993; 1999).

er with a few other varieties, which is represented in Vitelli's Franchthi Ceramic Phase (FCP) 1 (Vitelli 1993, 39–41). Hole-mouthed jars and deep hemispherical bowls are most frequent, as well as open bowls and jars with low collars. Decoration includes both painted and relief decoration (Jacobsen 1973b, 262–263). On the basis of the limited pottery of this phase inside the cave, Vitelli suggests that Early Neolithic activities may have been more concentrated on the Paralia, even though there is ceramic evidence of this phase also in the cave. Thus concludes that both the cave and the Paralia were utilised during that time but may not have been occupied in their entirety or continuously through that phase (Vitelli 1993, 47).

During the Middle Neolithic, the Neolithic Urfirnis ware appears in the ceramic sequence in five varieties marking the beginning of a new ceramic tradition on the site. This ceramic phase (FCP 2) is the most widely distributed and well-represented ceramic phase (Vitelli 1993, 49–50). This ware is characterised by finer paint and a lustrous character, while an increasing number of new shapes are used. The monochrome Urfirnis characterises a large percentage of the ceramic assemblage while the patterned and black burnished Urfirnis wares gradually increase. The latter is characteristic of the final phase of the Middle Neolithic (Jacobsen 1973b, 264–266). Evidence suggests that pottery is produced locally even though evidence is not conclusive. It appears that lime, sandy, andesite, serpentine and Urf wares are produced locally, while the gritted ware may have been produced in the Corinthia. In addition a common recipe, rich in calcium carbonate temper, is consistently used by potters in the Early and Middle Neolithic pottery (Vitelli 1993, 208). Vitelli highlights the symbolic significance of pottery in the Early Neolithic, suggesting that it was used in ceremonial and ritual contexts that fostered social cohesion. She also proposes that women potters/shamans held a specific social status and bearers of an important social status (Vitelli 1993, 217–218).

Ceramic Phases 3 to 5 comprise the Late Neolithic. During the Late Neolithic, the burnished ware declines and the black burnished ware gradually increases. Shapes include carinated bowls, cups and bowls with incurved and offset rims, broad open bowls with thickened or inverted rims and necked jars. More rare are fragments of four-legged vessels. During this period matt-painted pottery increases in quantity. FCP 3 mostly includes locally produced low lime burnished pottery (Vitelli 1999, 22–23). FCP 4 pottery represents a small amount in the ceramic assemblage but it is characterised by diversity in wares and forms (Vitelli 1999, 61). The coarse-ware vessels sometimes bare plastic decoration of relief bands, while others bare incised decoration. The last stage of the ceramic sequence (FCP 5) dated to the Final Neolithic is dominated by a coarse ware (Jacobsen 1973b, 266–275). Pottery in this phase is highly variable and is made of diverse techniques and styles, while most of it is locally produced (Vitelli 1999, 88–93. 97).

2.6.7. Lithics

Interpretation of the cultural affinities and continuities at Franchthi has been based to a large extent, apart from the pottery, on the lithics assemblage, studied by Perlès (1987; 1990; 2004). The lithic finds of the most ancient levels of the cave comprise lithic phase 0. This phase is poor in lithics finds and is characterised by twisted bladelets, calcareous bladelets and flakes, and butt-facetted flint flakes. These finds represent evidence of sporadic occupation of the cave during that period (Perlès 1987, 85–88).

During the Upper Palaeolithic (39,000 cal BP), lithic phase I is typologically close to an Aurignacian phase. The lithic assemblage is composed of scrapers, notched tools, denticulates, tools retouched on the laterals, backed tools, simple and geometrical truncated tools. The data are not abundant, while the diagnostic types present are the carinated endscrapers, and the twisted bladelets (Perlès 1987, 89–90; 1999, tab. 29.1).

Lithic phase II (26,000–27,000 cal BP) is differentiated from the previous one by a discontinuity. The Aurignacian elements disappear and the backed edges element prevails. The assemblage includes backed bladelets single and obtuse, double backed and pointed bladelets. The material is poor and each excavation unit contained few pieces (Perlès 1987, 98; 1999, 312).

Lithics of phase III are also scarce, but exhibit traits of cultural continuity with the previous phase. During lithic phase III, obtuse and pointed backed bladelets both double-backed and single-backed are the diagnostic types manufactured (Perlès 1987, 108; 1999, 312).

Lithic phase IV dates to the 14,000–15,000 cal BP, is rich in material. Lithic phase IV entails changes in the assemblage, while backed bladelets continue to be a predominant component. The diagnostic types include microburins, »La Mouillah points«, bladelets, obtuse, truncated bladelets. New techniques appear such as the microburin technique, which is now used for pointed backed bladelets. Among the changes observed in the assemblage is the appearance of numerous »La Mouillah points«. There is an increase in typological diversity and a more balanced structure which suggests a more intensive use of the cave and slightly more varied activities (Perlès 1987, 115; 1999, 314).

The diagnostic features of the lithics assemblages of the Final Palaeolithic cease to be present in the assemblages of the Early Mesolithic levels. One of these features is the microlithic features, that is the microburins, the backed bladelets and geometrics such as triangles, trapezes and segments (Perlès 1987).

Lithic phase V belongs to approximately 13,300–12,700 cal BP. In this phase, there is a tendency towards a diversification of the toolkit and tool types increase. There are single backed bladelets, mostly obtuse, and double-backed pointed bladelets as well as geometrics (triangles) and microburins. These changes in the lithic assemblage point, according to Perlès, towards a different cultural tradition, while the hiatus between phases IV and V could support this hypothesis (Perlès 1999, 314).

Despite a brief temporal hiatus between phases V and VI, there is a cultural continuity among these phases. The lithic assemblage of lithic phase VI, radiocarbon dates to 11,700–12,500 cal BP, is rich and more diverse. This phase is dominated by backed edge and geometric tools and a tendency towards hypermicrolithism. The changes are only typological and the introduced new types are few. In lithic phase VI, there are backed bladelets, obtuse bladelets, triangles, microburins and new geometric types such as segments and trapezes. Melian obsidian is also present for the first time, while there are also rare bone tools and grinding implements (Perlès 1987, 141; 1999, 314).

During the Lower Mesolithic (lithic phase VII), dated to the first half of the 10th millennium BP, the lithic assemblage undergoes drastic changes. The backed bladelets and geometrics decrease, microlithics and microburins disappear. The assemblages consist of »transformation tools« such as notches, endscrapers and denticulates. The industry is also very crude and is manufactured on casual flakes. The projectile inserts disappear in the Lower Mesolithic, which could be related to changes in hunting techniques or limited big game hunting. Furthermore, there is a temporal hiatus between phases VI and VII (Perlès 1990, 26; 1999, 315).

The Mesolithic industries are poor in microliths in accordance with other Mediterranean Mesolithic sites, while the toolkits are very crudely fashioned. In lithic phase VIII, dated to the second half of the 9th millennium BP, there aren't great differences with the previous phase, while it is a very rich phase in terms of the material found. During the Upper Mesolithic period, fishing activities increase together with the microliths that reappear. Perlès suggests that there is a relation between these tools and the fishing activities which may be related to the preparation of the fishing equipment, the capturing of the fish and storage activities. There are also more trapezes and no microburins (Perlès 1990, 46; 1999, 316–317).

In lithic phase IX of the Final Mesolithic period, dated to the end of the 9th millennium BP, is a very poor occupation phase. The lithic assemblage is similar to the Early Mesolithic period and is characterised by the presence of laterally retouched pieces, notches, endscrapers and denticulates. A novelty of this phase are the few transverse arrow-heads with concave sides, while non geometric microliths are almost absent (Perlès 1990, 84; 1999, 317).

Lithic phase X of the Initial Neolithic is not a rich lithic industry. The basic toolkit maintains to some extent the Mesolithic traditions, while some elements belong to the Neolithic tradition. This phase includes a crude flake-based industry, as well as pressure-flaked blades and elongated trapezes. This phase is dominated by obsidian which is more frequently used than the local raw materials (Perlès 1990, 104; 1999, 317).

From the Early Neolithic onwards, Perlès adopts a new acronym FLP N for the Neolithic lithic phases (Franchthi lithic phase) and restarts from 0 (Perlès 2004, 2). Franchthi lithic phase N I is the first phase of the Early Neolithic and is the richest lithic phase of the Neolithic and is ascribed in the subsistence economy of the Neolithic. Local raw materials play an important role in the diverse toolkit of this phase, while microlithic tools are also well-represented (Perlès 2004, 69).

Lithic phase N II belongs to the Middle Neolithic and is marked by changes in raw materials as well as in the structure and the typology of the lithics. Exogenous raw materials now dominate the lithic assemblage, while among the characteristic elements is the abundance of blonde flint, glossed cutting tools longitudinally used, which are resharpened and recycled, and of oblique edge armatures with bifacial pressure retouch (Perlès 2004, 89). Even though there are certain changes between the Early and Middle Neolithic tool industry mainly regarding raw materials, according to Perlès the techniques and methods used remain the same and the two periods are characterised by homogeneity (Perlès 2004, 161).

Lithic phase N III (equivalent to FCP 3) belongs to the Late Neolithic is not particularly distinct, as it is a very poor one (in contrast with the pottery material), while it is completely absent at the Paralia (Perlès 2004, 104).

Lithic phase N IV corresponds to the second half of the Late Neolithic while it is separated from the previous phase by a hiatus. It corresponds to FCP 4 and is characterised by percussion refittings and large obsidian flakes with multiple retouch. The material is less abundant and diverse than the Early and the Middle Neolithic, while it is further differentiated in the preparation methods of tool making (Perlès 2004, 107–114).

Lastly, lithic phase N V is inscribed in the Final Neolithic period. During this time, the lithic assemblage is characterised by the almost exclusive use of obsidian and the long parallel pressure retouch (Perlès 2004, 117).

2.6.8. Subsistence Remains

In this section, I will summarise the subsistence remains. Botanical and zooarchaeological studies at Franchthi provide valuable insights into the environmental and cultural evolution spanning from the Upper Palaeolithic to the Late Neolithic periods. Hansen's (1991) botanical analysis on material from the cave trenches describes the transition from the exploitation of wild plant species to the cultivation of domesticated plants. This progression is evident in the diverse botanical assemblage, reflecting human agricultural practices over time. Zooarchaeological investigations across the cave trenches reveal changing faunal compositions, from early exploitation of Pleistocene fauna to the intensification of domestic animal husbandry during the Neolithic.

2.6.8.1. Botanical Remains

Botanical information is available from the study of Hansen (1991) and covering trenches FAS, FAN, H1A and H1B and all time periods from the Upper Palaeolithic to the end of the Late Neolithic levels. The botanical assemblage of Franchthi includes a wide range of plant remains, reflecting the site's various periods of occupation, from the exploitation of wild species to the cultivation of domesticated plants (Hansen 1991, 155). Thus, botanical information solely comes from the cave deposits that represent the flora of the vicinity of the cave.

Botanical zone Ia and Ib (contemporary with lithic phase I and II) predominantly contained *Lithospermum arvense*, as well as two species of the Boraginaceae family *Alkanna* sp., *Anchusa* sp. and small amounts of wood charcoal. Botanical evidence from zone 1, dated to the Upper Palaeolithic, indicates a cold and dry environment (Hansen 1991, 104–106; 2001, 410–411).

Plant remains occur sporadically in zones II and II/III. In botanical zone II, there are the same species as before but also *Medicago* sp., *Lens* sp., *Vicia ervilia*, *Vicia lathyrus*, *Pistacia* sp., *Vitis silvestris*, *Avena* sp., *Prunus amygdalus*, *Pyrus amygdaloformis*, *Hordeum* sp., *Galium* sp., *Liliacae* and others. This phase is contemporary to lithic phase IV and V. This zone reflects warmer and more moist conditions during the end of the Pleistocene as arboreal species (such as almond, pistachio and pear) increase. This trend further continues into zone III (Hansen 1991, 109–117).

The archaeobotanical evidence of interzone II/III is relatively poor, while it indicates the presence of *Alkanna* sp., *Lithospermum arvense*, *Lithospermum officinale*, *Pistacia* sp., *Prunus amygdalus*, *Pyrus amygdaloformis*, *Lens* sp., *Avena* sp., *Hordeum* sp. This botanical zone corresponds to lithic phase VI and to the molluscan zone I (Hansen 1991, 116).

Botanical zone III presents a drastic increase in the quantity and variety of the botanical species present, while *Pistacia* sp., *Prunus amygdalus* and *Pyrus amygdaloformis* dominate the assemblage. Furthermore, legumes and wild cereals (oat and wild barley) are regularly present. This zone corresponds to lithic phase VII and to molluscan zone II (Hansen 1991, 119–127).

In botanical zone IV and V, the uncarbonized seeds of *Lithospermum arvense* appear again. There are also smaller quantities of *Pistacia* sp., *Lens* sp., *Prunus amygdalus*, *Hordeum* sp., *Avena* sp. present in the archaeobotanical record. These phases are contemporary with lithic phase VIII and IX and molluscan zone II and III. Zone IV is characterised by an increase in gromwell (*Lithospermum*) and in alkanet, while all other species decrease. The same is the case with the plant remains of zone V and interzone V/VI when the density of plant remains is markedly reduced (Hansen 1991, 129–135). A similar drop is also present in other archaeological datasets such as in lithics (Perlès 1990, 46), shells (Shackleton 1988) and animal remains (Payne 1973, 59–62) indicating a decrease in the intensity of occupation at the cave during this period.

The early Neolithic deposits include botanical zone VI is notably characterised by the almost complete absence of botanic remains after the sporadic evidence of occupation of zone V. Domestic two-rowed barley (*Hordeum distichum*) and emmer wheat (*Triticum dicoccum*) first appear in stratum X2 which corresponds to an Initial Neolithic occupation. These domesticated species appear alongside weed taxa (Boraginaceae and *Teucrium* sp.) that are often associated with agricultural fields. Furthermore, wild oats, which constitutes staple food during the Mesolithic, are completely absent during the Initial Neolithic, while the average size of lentils increases (Hansen 1991, 140; 2001, 414). The botanical evidence from zone VII, corresponding to the Middle to the Final Neolithic period, show an increase in numbers, especially of domesticated species such as emmer wheat, barley, lentils. In addition, einkorn wheat appears for the first time and several wild species, found in earlier phases, reappear such as pistachio and almond (Hansen 1991, 145).

2.6.8.2. Zooarchaeological Remains

Zooarchaeological information on terrestrial fauna comes from the study of the Palaeolithic and Mesolithic layers of trench H1A (Payne 1975) and more recently from the Final Mesolithic and Initial Neolithic layers of trenches FAN and FAS (Munro/Stiner 2015) and covering trenches FAS, FAN, H1A and H1B and all time periods from the Upper Palaeolithic (25,000 BP) to the end of the Late Neolithic levels (5,000 BP). As for the archaeoichtyological evidence, material from the late Upper Palaeolithic to the end of the Neolithic of Franchthi has been studied from trench FAS (Rose 1995).

Payne distinguishes five faunal phases for the Palaeolithic and the Mesolithic (Payne 1973, 59). The first macrofaunal remains of the Palaeolithic are dominated by equids (*Equus hydruntinus*) and cervids (*Cervus elaphus*) (faunal phase A), while there are also remains of tortoise (*Testudo*), hare (*Lepus*), pig (*Sus*) and birds (*Aves*) (Payne 1975, 122; 1982, 133). This phase is contemporary to lithic phase II and III. During faunal phase B, coeval to lithic phase IV, faunal evidence is characterised by a larger variety and includes large bovids (*Bos*), caprids (ibex or chamois), cervids and equids. Fox (*Vulpes*) also appears halfway through this phase. During faunal phase C, coeval to lithic phase V and VI, cervids become dominant. In addition, there is a reduction in large bovids and equids, and an increase in caprids and suids. The sporadic appearance of goat (*Capra*) also takes place (Payne 1973, 59; 1975, 122). Fish appear for the first time in layers dated to the 11,000 BP. During the Upper Palaeolithic and the Lower Mesolithic eel (*Anguilla anguilla*), sea bream (Sparidae) and grey mullet (Mugilidae) dominate the assemblage. Among the secondary species are the conger eel (*Conger conger*), the sea bass (*Dicentrarchus labrax*) and the shade fish (*Umbrina cirrosa*). This evidence indicates that a brackish water environment was exploited (Rose 1995, 21).

During the Lower Mesolithic and the second half of the 10th millennium BP, faunal phase D1 is marked by a change in the faunal assemblages of the site: Pleistocene species such as equids and caprids disappear, while the cervids become more abundant. Large bovid are scarce. Hare and birds are also present, while there are also some small fish. During the Upper Mesolithic, faunal phase D2 is characterised by the appearance of large quantities of bluefin tuna vertebrae (*Thunnus thynnus*) indicating large-scale fishing activities, and secondly by the presence of red deer and boar (Payne 1973, 59; 1975, 122). The remains of groupers (*Epinephelus* sp.) and barracuda (*Sphyraena sphyraena*) are also present. Fish taxonomy points towards the exploitation of a fully marine environment during this time (Rose 1995, 22–23). During the Final Mesolithic the red deer constitutes the vast majority of the ungulates, while caprines are rare (Munro/Stiner 2015, 597). During the Final Mesolithic period large-scale fishing ceases (Perlès 1999, 317).

Across the Final Mesolithic and the Initial Neolithic period, an abrupt shift from a wild to a domestic animal economy takes place. Red deer (*Cervus elaphus*), dominant in the Final Mesolithic, is replaced by caprines, which is accompanied by a decline in small game. The Final Mesolithic-Initial Neolithic period also includes cattle (*Bos* sp.) and pig (*Sus scrofa*). The Initial Neolithic (coeval to lithic phase X) is marked by the presence of caprines, mostly sheep (*Ovis aries*), which dominate the Initial Neolithic caprine assemblage, and to a lesser extent goat (*Capra hircus*). Also, during the Initial Neolithic, pigs are rare by rise during the Early Neolithic. In addition, the faunal assemblages across all the phases of the Neolithic are predominantly composed of domestic animals, primarily sheep (*Ovis aries*) and goats (*Capra hircus*), while cattle (*Bos taurus*) and pigs (*Sus scrofa*) are present in lower frequencies. In addition, fishing activities rapidly decline during this period (Munro/Stiner 2015, 597–599; 2020, 5; Payne 1973, 66), while intensive fishing activities are further attested again during the Late Neolithic period (Rose in prep.; Perlès/Vanhaeren 2010, 305).

In general, the fish identified during the Neolithic period are primarily sea breams (Sparids), followed by tunny (*Thunnus thunnus*). Grouper, grey mullet and barracuda are also present but in small amounts (Rose 1995, 24). Fox (*Vulpes vulpes*) are the more common wild carnivore species, followed by wild cat fragments (*Felis silvestris*), which represent an uncommon find. Wild land mammals are mainly represented by hare (*Lepus europaeus*) and red deer (*Cervus elaphus*). Cattle are rare throughout the Neolithic. Between the Early and the Middle Neolithic there is a slight increase in the age of adult caprines, a tendency which remains in the sequence that follows (Munro/Stiner 2020, 5–10).

Payne further distinguishes four more faunal phases (E to G) for the Middle to the Late Neolithic period although subsequent studies do not follow this distinction. The Middle Neolithic resembles the Early Neolithic faunal assemblage while the faunal assemblage is fairly diverse. Sheep and goat are more dominant, followed by cow and red deer to a smaller extent (Payne 1973, 62–66; 1975, 129–131; Munro/Stiner 2020, 9–10). Fish and pigs increase slightly, while the number of dogs and canid chewing marks both in the cave and the Paralia increase. Furthermore, the Late Neolithic is distinguished in Early Late Neolithic 1 and 2. During the Early Late Neolithic 1, faunal diversity is quite low. Overall, sheep and goat continue to be the most abundant species, followed by pig. Red deer and cow are very scarce. Fish decline abruptly during this phase. The Early Late Neolithic 2 period resembles the previous period, although evidence of burning increases. This is

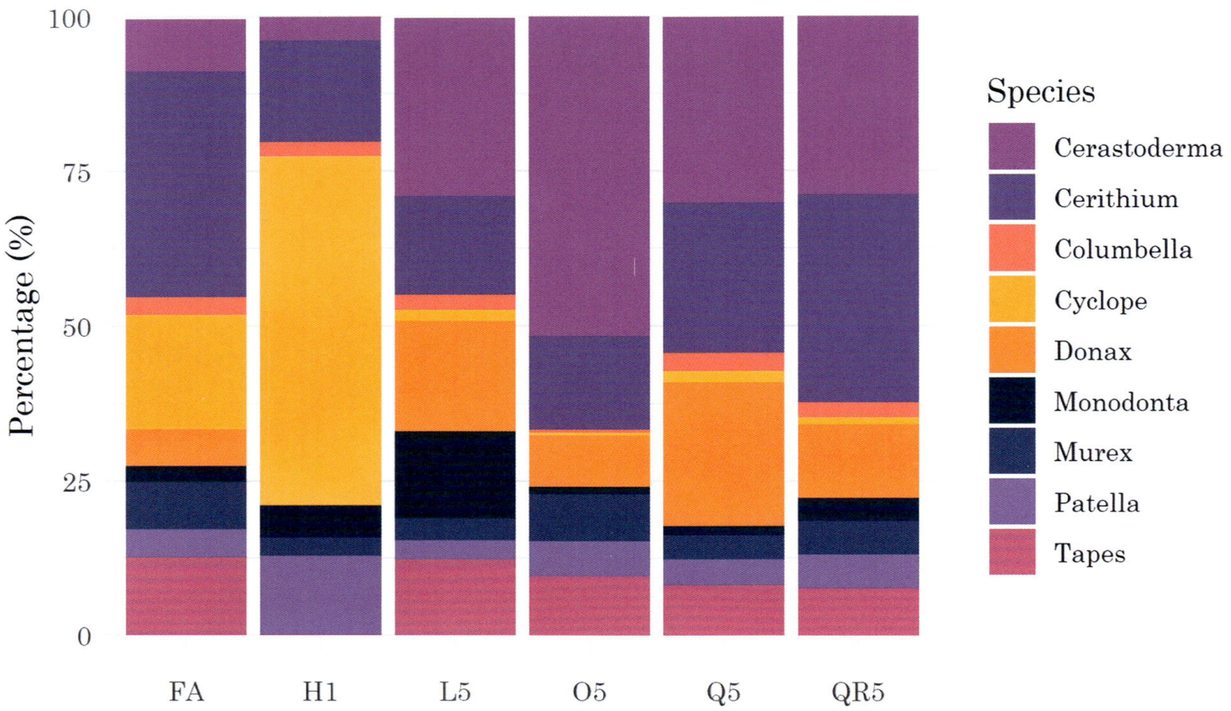

Fig. 6 Stacked bar chart of species dominance in each trench according to Shackleton (1988, Appendix C). Here the percentages of each unit were averaged and presented by grouped trenches.

associated with a number of hearths which are related to fish vertebrae. These remains belong to sea breams and tunny. During the Final Neolithic period, taxonomic diversity increases, as ungulates increase alongside the evidence of red deer hunting. Suid size also increases while cattle, although generally rare, slightly increase in numbers (Munro/Stiner 2020, 9–10; Rose in prep.).

2.6.9. Shell Assemblage

In this section, I will present the published work on the site as it is known so far from the fascicles volumes and various articles. My own contribution on the site material is presented in the following chapters. The assemblage of Franchthi is one of the earliest systematically studied and published material in the research history of Greece, dating to the Upper Palaeolithic and Mesolithic period.

During the earlier phases of the site (coeval to lithic phase I–IV), the molluscan remains are scarce: Small *Bittium* seashells found in the cave are considered natural deposits, as Shackleton (1988, 11–13) found no evidence of human introduction. These shells define Shackleton's molluscan zone 0. The first anthropogenic molluscan remains appear during the Upper Palaeolithic period at around 11,000 BP. Molluscan zone I (11,000–9,400 BP/9,050–7,450 BC) is coeval to lithic phase VI and is called the *Patella* zone. During this time, the assemblage consists almost exclusively of limpets (*Patella*), turbans (*Monodonta*) and tops (*Gibbula*) species (**fig. 6**). These species are all rock-dwellers. Molluscan zone II belongs to the Mesolithic period (9,400–8,500 BP). It is called the *Cyclope* zone and it is coeval to lithic phase VII and VIII (Shackleton 1988, 12–13. 38–39). *Cyclope neritea* dominates the molluscan assemblage, while some smaller specimens may belong to *Cyclope pellucida* (Perlès/Vanhaeren 2010, 298). There are also some ceriths (*Cerithium vulgatum*) and lagoon cockles (*Cerastoderma glaucum*). The small gastropod *Cyclope neritea* tolerates brackish waters and temperature fluctuations. Molluscan zone III is coeval to lithic phase IX (8,500–6,900 BP/6,550–4,950 BC) and is almost exclusively dominated by *Cerithium vulgatum*. Lastly, molluscan zone IV (6,900–5,000 BP) is more diverse and includes mostly bivalves such

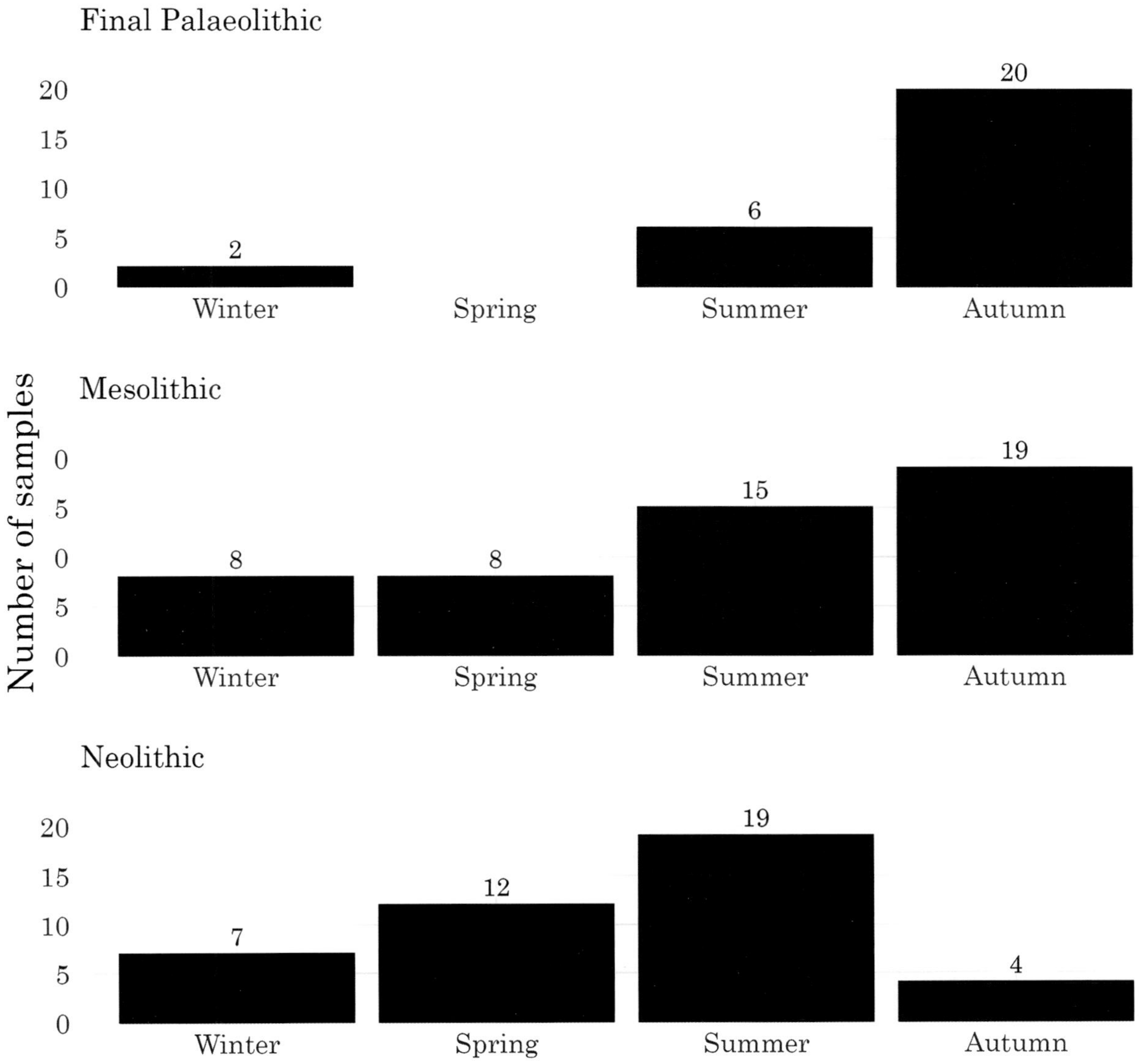

Fig. 7 Season of collection results of the study by Deith and Shackleton (1988) by chronological period.

as *Cerastoderma glaucum*, *Tapes decussatus*, *Donax trunculus* and *Donacilla cornea*. These species inhabit muddy and sandy substrates (Shackleton 1988, 12–13. 38–39). The *Cyclope neritea* from the site bears evidence of intentional burning, something that has been associated with their use for aesthetic and/or symbolic reasons and not with their consumption (Perlès/Vanhaeren 2010, 307–308).

An isotopic study on the marine molluscs of Franchthi was carried out by Deith and Shackleton (1988). They analysed 120 *Monodonta*, *Cerastoderma* and *Cerithium* shells from all the chronological periods of the site (**fig. 7**). This study concluded that there is a shift from a summer and autumn collection bias during the Palaeolithic and the Mesolithic towards a summer and autumn bias during the Neolithic period (1988, 154). More specifically, during the Final Palaeolithic period, Deith and Shackleton contend that the site was visited primarily during autumn and secondly during summer (1988, tab. 6). In addition, they identify two winter-collected shells, however they exclude the possibility of winter visits of the cave during this period (1988, 151). The analysis of the Mesolithic group of shells showed a stronger summer and autumn preference and thus they suggest that during that time the cave was used as a summer/autumn camp with occasional visits in winter and spring. Furthermore, during the Early and the Middle Neolithic period, Deith and Shackleton conclude that most of the shells were collected during spring and summer, with few exceptions (1988, 153–154 tab. 6). Lastly, a small number of Late Neolithic shells showed a widespread seasonal distribution, while a summer preference is more evident during the Final Neolithic period (1988, tab. 6).

Regarding the terrestrial molluscs, during the earlier phases of the site (coeval to lithic phase I–III), species include *Helix* sp., *Pupillidae* and *Helicella* sp. Later, during the 12th millennium BP (coeval to lithic phase V), the species *Helix figulina* increases in numbers which becomes further abundant during the next phase (lithic phase VI, molluscan zone I). Massive amounts of this land snail form dense shell middens (Perlès 2010, 118; Farrand 2000, 54–58. 84). During the next period (lithic phase VII, molluscan zone II), *Helix figulina* decreases in numbers, while during lithic phase VIII and IX, there are no important anthropogenic deposits (Perlès 1999, tab. 29.1). These land snail deposits found in trenches H1A and FAS of stratum T, are in most cases found broken and burnt, as well as mixed with other finds such as animal bones, seeds, shells suggesting the intensive consumption of the snails during that period. Overall, land snails are consistently present during the Palaeolithic and the Mesolithic, with their abundance being more prevalent during phase 5 (c. 11,000 BP) of the Palaeolithic and phase 7 (c. 9,500–9,000 BP) of the Mesolithic (Perlès/Vanhaeren 2010, 305).

The stable isotope composition of the *Helix figulina* from Late Pleistocene-Holocene (13,000 to ~10,500 cal BP) was analysed in a study by Colonese et al. and compared to modern specimens from the area. $\delta^{18}O$ values showed lower temperatures in the late glacial shells, while carbon isotope results yielded values that indicated the presence of C_3 vegetation as the main source of carbon (Colonese et al. 2013, 70).

Chapter 3: Methods and Materials

3.1. Background of the Methods

3.1.1. Sclerochronology

The term derives from the Greek compound words of *scleros*, hard, *chronos*, time and *logos*, word, explanation, reason. Thus, etymology itself describes it as the study of the hard parts of organisms as they evolve through time. According to the scientific definition of the 1st International Sclerochronology Conference from 2007,

> »sclerochronology is the study of physical and chemical variations in the accretionary hard tissues of organisms and the temporal context in which they formed. Sclerochronology focuses primarily on growth patterns reflecting annual, monthly, fortnightly, tidal, daily, and sub-daily increments of time entrained by a host of environmental and astronomical pacemakers. Familiar examples include daily banding in reef coral skeletons or annual growth rings in mollusk shells. Sclerochronology is analogous to dendrochronology, the study of annual rings in trees, and equally seeks to deduce organismal life history traits as well as to reconstruct records of environmental and climatic change through space and time.« (Oschmann 2009, 1).

Among other definitions, Andrus defines sclerochronology as »the study of information preserved in the accretionary growth of skeletal materials, often including geochemical analyses, to construct time-series data profiles« (Andrus 2011, 2893).

Sclerochronology has been a known technique for more than 70 years for interpreting past environmental conditions and is widely used by archaeologists, geologists, ecologists and marine scientists (Perhada et al. 2021). Davenport was the first to point out the potential of shells as climate indicators after conducting a study on *Pecten* shells (Davenport 1938, 514–515). In the following decades, other researches followed their example such as Epstein et al. (1953), Clark (1968) or Berry and Barker who also described the annual growth lines of fossil shells as well as the relation of the shell growth increments and the tidal cycles (Berry/Barker 1968, 938–939). Emiliani et al. (1964) were among the first to produce Sea Surface Temperature (SST) records from archaeological limpets. Sclerochronology is also applied on a number of different organisms of the past such as benthic or planktonic foraminifera (e. g. Lea et al. 2000), cold water or shallow water corals (Watanabe et al. 2019), fish otoliths (Coulson et al. 2014; Vignon 2015), etc.

3.1.1.1. Shells as Environmental Archives

Shells function as biogeochemical archives by recording in their chemical composition a substantial amount of information. This process takes place when the shells grow and the molluscs precipitate their calcium carbonate by sequentially depositing new layers of shell (Gröcke/Gillikin 2008, 266). The environmental information that we can deduct from this process depends on the organisms rate of growth, lifespan, growth stops during the year due to several parameters such as external stress or reproduction as shell growth may often not be continuous (Leng/Lewis 2016, 295). By identifying and characterising shell growth increments, they become temporal milestones, »geochronometers« or »geological calendars« (Wells 1963, 948; Pannella/MacClintock 1968, 69). The hard parts of molluscs as well as other marine organisms are mainly composed of calcium carbonate ($CaCO_3$), which can take the crystalline structure of calcite, aragonite, amorphous $CaCO_3$ and seldom vaterite, with the most common forms being the first two (Spann et al. 2010; Jacob et al. 2011; De la Pierre et al. 2014; San et al. 2023). In the case of *Patella* shells, the carbonate crystal in the apex consists of aragonite (orthorhombic crystal structure) and the extremities of calcite (trigonal structure)[1] (Fenger et al. 2007, 4–5). As the organism precipitates its calcium carbonate, various ions are incorporated into it. The elements that are incorporated make the shells useful as proxies for various environmental characteristics present in the carbonates. Variations in certain chemical elements of which they are composed such as magnesium, strontium and barium or their $\delta^{18}O$ composition can reflect changes in different environmental phenomena and parameters (Schöne/Gillikin 2013, 2).

In order to analyse the shell growth increments, I employed two methods: stable isotope analysis and trace element analysis. The two methods will be exhibited in the following subchapters.

3.1.2. Stable Isotopes

Oxygen isotope analysis is a method that allows for the estimation of past water temperatures. The analysis of the stable isotope ratio ($^{18}O/^{16}O$, expressed as $\delta^{18}O$) is a widely used indicator of the temperature or a range of temperatures. There is an empirically tested relationship between temperature and the ratio of the stable oxygen isotopes ^{18}O and ^{16}O. This ratio in water is temperature-dependent primarily due to evaporation, where lighter isotopes ^{16}O are more likely to evaporate leaving behind water enriched in the heavier isotope ^{18}O. When a mollusc forms its shell, it fractionates these isotopes often by preferring lighter isotopes. This biological fractionation is influenced by the ambient temperature, resulting in more ^{18}O being incorporated into the shell at colder temperatures. The technique is based on the measurement of the abundance ratio between two of the three isotopes of oxygen. The most abundant oxygen isotopes in nature are the light oxygen atom ^{16}O (~99 %) and the heavier ^{18}O atom (~0.2 %). The isotopes have similar chemical properties due to their electronic structure, but because of the differences in mass between their nuclei, they behave slightly differently in physico-chemical reactions and have different physical properties. Separation or fractionation of isotopes will occur during processes such as evaporation or condensation and the extent of fractionation is dependent on temperature (Allaby 2013).

3.1.2.1. Palaeothermometers

The applications of stable isotope analysis are numerous. Mollusc shells can also serve as palaeothermometers or geothermometers. By analysing archaeological shells, palaeoenvironmental data can be produced to reconstruct past climates as it allows for the estimation of past sea surface temperatures (SST) at a seasonal resolution. Furthermore, in contrast with other palaeoenvironmental proxies, stable isotope analysis of mollusc shells opens a window to local seasonal patterns and local climates as they are derived from largely sessile animals reflecting conditions at a single location (West et al. 2018, 192). What is more, the information derived from these archives, provides direct chronological and stratigraphic relationships between human activities and the environmental background, as the shell proxies belong to anthropogenic deposits (West et al. 2018, 195; Kwiecien et al. 2022, 14).

The use of stable isotopes is based on the observation that molluscs can precipitate their shells in isotopic equilibrium with the surrounding water (Urey 1947; Epstein et al. 1951; 1953; Craig 1957; Grossman/Ku 1986). In this way, chemical signatures from the surrounding habitat conditions are incorporated into the shell structure (Wefer/Berger 1991; Dettman et al. 1999). When the temperature decreases, this ^{18}O selectivity becomes more pronounced, and a greater proportion of the heavier oxygen atom is concentrated in the shell. Thus, the ratio of the two isotopes becomes negative when the ambient temperature increases, while it acquires more positive values when the same decreases. Therefore, the values of $\delta^{18}O$ indicative of the marine environment where the mollusc precipitate their shell, are dependent of the temperature which can be recon-

1 I tested the preservation of the calcium carbonate crystals by conducting Raman spectroscopy analysis on an archaeological shell sample (see section 3.1.5).

structed according to the following equation by Hays and Grossman (1991):

$$T_{\delta 18_O}(^{\circ}C) = 15.7 - 4.36 * (\delta^{18}O_{shell} - \delta^{18}O_{water}) + 0.12 * (\delta^{18}O_{shell} - \delta^{18}O_{water})^2$$

To employ this equation, the isotope composition of the water or the salinity is needed, which is unknown for ancient environments (Schöne/Gillikin 2013, 2). Water $\delta^{18}O$ is related to salinity in a near-linear way, thus in marine settings water $\delta^{18}O$ and salinity are considered to be constant in the life of an organism (Culleton et al. 2009, 1355). For this reason, in cases where seawater samples are unavailable, some researchers (e. g. Prendergast/Schöne 2017) derive $\delta^{18}O$ water values from water salinity using an equation by Pierre (1999).

Before employing oxygen isotope analysis on archaeological samples, it is important to test the technique on modern samples of the same species and from the same or similar localities. Calibration studies on modern limpet specimens detected a consistent offset in shell $\delta^{18}O$ values from equilibrium with surrounding water (Shackleton 1973; Fenger et al. 2007; Ferguson et al. 2011; Gutiérrez-Zugasti et al. 2017; Prendergast/Schöne 2017). This offset is attributed to physiological or »vital effects« of the molluscs which can disrupt isotopic equilibrium and is also species-dependent. Its regularity has deemed it applicable to palaeotemperature equations serving as a »correction factor« to calibrate the archaeological shell results based on the modern shell results (Prendergast/Schöne 2017, 34).

Interpretation of the isotope composition of the shells can further be influenced by that of the seawater ($\delta^{18}O$). Many factors have an impact on isotope composition of the seawater such as global ice melting, precipitation, evaporation, freshwater input (Rohling 2013). Surface water salinity and isotope composition of the water are highly correlated and thus salinity is also an important factor on a local level (Ravelo/Hillaire-Marcel 2007, 742). Physiology and ontogeny of molluscs have a major influence on the shell record (Schöne 2008). Thus, when evaluating and interpreting isotope data, shell growth rate and growth trends should be considered, as under certain conditions growth cessation may occur during which environmental conditions are not recorded in the shell increments (Gutiérrez-Zugasti et al. 2017, 4).

Beyond accounting for growth patterns, a critical element in this approach involves accurately identifying the annual temperature minimum and maximum values reflected in the shell's seasonal growth record. In the case of stable isotope analysis this requires a year-round seasonal record which allows the identification of the annual minima and maxima. In this thesis, seasonal extremes were identified with the use of LIBS as it will be demonstrated in the following chapter sections.

European limpet shells of the *Patella* spp. have also been established as reliable proxies using isotopic analysis of oxygen and carbon isotope ratios ($\delta^{18}O$ and $\delta^{13}C$) combined with sclerochronology. A number of studies on modern specimens have been conducted in preparation for the analysis of archaeological samples: Fenger et al. conclude that $\delta^{18}O$ of *Patella* shells are reliable proxies for climatic reconstructions as oxygen isotope variations of modern samples analysed revealed sinusoidal patterns that matched observed (satellite) SST variations by taking into consideration a systematic offset (Fenger et al. 2007). A similar modern study was undertaken by Gutiérrez-Zugasti et al. further validating the applicability of the method on limpet shells (*Patella vulgata*), while also pointing out that offsets vary among different species and localities (Gutiérrez-Zugasti et al. 2017, 4). Other researchers have used limpet shells to study climatic anomalies during Mediaeval times (Surge/Barrett 2012) as well as climate variability during the Roman Warm Period (Wang et al. 2012). Ferguson et al. studied fossil and modern *Patella* shells from the strait of Gibraltar reconstructing SST pronounced seasonality patterns in the area during the last glacial period (Ferguson et al. 2011), while Colonese et al. (2012) investigated SST fluctuations and season of exploitation patterns using modern and archaeological *Nacella* spp. limpets from Tierra del Fuego in South America.

3.1.2.2. Determination of Season of Collection

Oxygen isotope analysis on mollusc shells and the determination of the season of collection, is not only a reliable palaeoenvironmental proxy, but can also be employed in archaeology to reconstruct subsistence strategies and seasonal settlement patterns of coastal communities (e. g. Mannino et al. 2003; Rick et al. 2006; Colonese et al. 2009; Burchell et al. 2013). The sampling and interpretation of stable isotope data from archaeological mollusc shells, allows the determination of the season of collection, and therefore the investigation of seasonality[2] patterns.

The »season of death« of a mollusc shell refers to the time when the mollusc was collected by humans and the environmental conditions of that season are embedded in the last increment of growth. This estimation can be done through the study of the growth increments via microscopic and/or geochemical methods. Microscopic analytical methods involve the detailed analysis at a microstructural level of annual growth patterns visible as translucent and opaque growth lines visible under transmitted light which is also combined with stable isotope analysis. Cross-sections are created perpendicular to the shell growth which are then polished and stained to facilitate their study under a reflected-light microscope. Their digital images enable the detailed measurement of the growth increments (Hallmann et al. 2009; Milano et al. 2017). For example, Hallmann et al. (2009) demonstrated how life-history traits of the species *S. gigantea* from Alaska and British Columbia can be interpreted to precisely identify growth patterns and thus accurately determine the season of death and environmental conditions, using high-resolution macro photographs of cross-sections of shells and oxygen isotope analyses. This study emphasised the need to understand the growth history of shells on a microscopical level prior to the analysis with the use of isotopes for seasonality purposes (Hallmann et al. 2009, 2363).

In addition, geochemical analysis involves the oxygen isotope analysis of the shell carbonates to infer seasonal estimates (Mannino et al. 2003; Leng/Lewis 2016; West et al. 2018). The oxygen isotope ratio analysis at the edge of the shell needs to be interpreted in relation with the annual range of values (Bailey et al. 1983, 390). However, a single sample of the final growth increment is not sufficient to estimate the season of death, but it is able to provide information about water temperature at the specific point in time that it was collected (Kennett/Voorhies 1996, 694). Sequential sampling is needed in order to understand pattern changes of the isotope signals, while in order to determine the seasonality of collection it is important that the $\delta^{18}O$ values of the shell record reflect year-round growth and thus temperatures (Thomas 2015a, 136). Furthermore, sufficient growth of the shell is required in order to minimise time-averaging[3] of the isotope ratios, and the growth rate of the shells should be taken into account in order to best sample the carbonate material (Mannino et al. 2003, 668–669). For example, in the study by Mannino et al. 2003, investigated the employed sampling strategies and species selection using long and short series of $\delta^{18}O$ samples from archaeological and modern specimens of *Phorcus lineatus* shells from Culverwell. Initially, the modern species study revealed the presence of seasonal variation in the shell growth which could be correlated with seasonal temperature fluctuations. Having established the suitability of the methods, the study proceeded to the analysis of the archaeological samples and the extraction of seasonality of collection results (Mannino et al. 2003).

2 The term »seasonality« lacks a clear definition across different disciplines and can refer to the annual cycle of temperature, the annual cycle of precipitation as well as the annual cycle of human activities. Seasonality of temperature and rainfall together make climate seasonality (Kwiecien et al. 2022, 2). In this work, by seasonality I refer to the annual cycle of temperature as it can be reconstructed in the calcium carbonate of the shells.

3 By time-averaging I refer to the mixing of different time intervals or seasons of shell growth (Goodwin et al. 2004). By definition, time-averaging is the process by which events that happened at different times appear to be synchronous (Flessa 2001, 292).

3.1.2.3. Isotope Sampling Preparation Workflow

Different ways of calcium carbonate powder extraction were used on the shell samples. In the case of the modern samples, both the drilling and the micromilling techniques were employed. All shells were already cut into two halves alongside the maximum growth axis for the needs of LIBS scanning. Certain modern samples were sampled via drilling with no further prior preparation. Edge samples were milled from the extremities and sequential holes were followed by drilling on the shell section with the use of a hand drill and the carbonate powder was collected. The sequential drilling results were not satisfactory enough and for this reason, the micromilling carbonate powder extraction method was followed for the modern samples that were henceforth analysed.

3.1.2.3.1. Testing with a Minimal Toolkit

Initially, the first four shells from Franchthi (FR-P_20210515_A to D) were sampled for isotopes after scanning them with LIBS. As more sophisticated equipment was not yet available, the samples were drilled sequentially alongside the section as part of a preliminary experiment. Approximately ten carbonate samples were extracted from each of the four shells with the use of a DREMEL hand drill with no prior sample preparation. This was an attempt to test the different carbonate powder extraction methods, however the results were quite rudimentary compared to the elemental line scans (see **fig. A.26**) as the resolution of the drilled isotope results was very low. However, the results of the drilling sampling were deemed to be of very low resolution and the hand-drilling method was deemed inefficient (see section 4.1.3).

3.1.2.3.2. Micromilling

A more elaborate sampling for isotope analyses was also carried out on a number of modern shell specimens (n = 14). For micromilling[4], shell halves were mounted on plexiglass cubes with a plastic epoxy (Multipower 3, GlueTec) and then covered with metal epoxy resin (WIKO 05), which protects the shell from fracturing under the pressure of the saw. The next step was to cut the shells into 3 mm-thick sections, after straightening the previously created shell section. The cross-sections were cut along the maximum growth axis with a low-speed precision saw (Buehler, Isomet 1000) and then positioned on glass slides using a metal epoxy resin. When ready, the sections were ground on glass plates using 800 and 1200 grit powder and finally polished with 1 µm aluminium oxide suspension (Al_2O_3) on a BuehlerMasterTex cloth. In between each grinding and polishing step, the sections were rinsed with water and placed in an ultrasonic bath to remove any grains of powder and then air-dried (Hallmann et al. 2009, 2354).

3.1.2.3.3. Archaeological Samples

A different approach was used for the archaeological samples. Archaeological shells were not sampled via micromilling for bureaucratic (the necessary permits for the creation of sample cross-sections had not been obtained) and logistic reasons, as I aimed at analysing larger sample assemblages with fewer powder samples per specimen. Specific areas of the shell halves were sampled for isotopes based on the elemental results by hand micromilling using a Minimo dental drill with a cylindrical diamond coated dental drill bit (1 mm diameter) mounted on a Wild Heerbrugg M3 binocular stereo microscope. Only the minima and the maxima growth increments were sampled which had been marked with the use of LIBS (see section 3.1.3.3). The micromilling technique was conducted parallel to the growth increments and perpendicular to the direction of growth allowing for high-resolution sampling.

After collecting the calcium carbonate powder, the samples were weighed (50–120 µg samples) using a Mettler Toledo XP26 microbalance. The weighted powder was then placed into exetainer tubes and then given for analysis. The instrument used for the analysis of the samples was an Thermo Finnigan MAT 253 gas source isotope ratio mass spectrometer in continuous flow mode coupled to a GasBench II located at the Johannes Gutenberg University in Mainz.

4 Samples were prepared and analysed at the Institute of Geosciences at the Johannes Gutenberg University in Mainz.

3.1.3. Laser-Induced Breakdown Spectroscopy

3.1.3.1. Basic Principles

Laser-Induced Breakdown Spectroscopy (LIBS), also known as Laser-Induced Plasma Spectroscopy (LIPS), is an elemental analysis technique or a spectrochemical analysis technique used to determine the elemental composition of materials. LIBS exploits the light emitted by a laser-produced plasma to obtain the elemental composition profile of a sample. The technique can be generally employed on gaseous, liquid, or solid samples. LIBS can be used to analyse a broad variety of objects such as metal, glass, stone, pottery, paint, geological or biological samples, polymers or archaeological objects in order to characterise them qualitatively and quantitatively (Cremers/Radziemski 2006a; 2006b; Anglos/Detalle 2014, 532–533). In this work, LIBS is applied in order to analyse and assess the elemental shell records.

To briefly describe the technique, a laser pulse is propagated through beam-guiding optics and focused via a lens on the sample. The pulse interacts with the surface of the shell section, ablates the material and generates a plasma that lasts for a few microseconds. A percentage of the plasma-emitted light is then collected and transferred through optic fibres to the spectrometer (**fig. 8**). A diffraction grating inside the spectrometer separates the light into single component wavelengths. These wavelengths are recorded and quantified by the detector that produces the spectral data presenting one or multiple spectral peaks, each corresponding to a chemical element. The spectral data are processed by the computer software which determines the intensity of the emission lines of each element in the sample. Each laser pulse produces a spectrum, and multiple spectra may be averaged to obtain more accurate results. Subsequent spectral analysis indicates the elemental content of the material as each characteristic atomic-emission peak in the LIBS spectrum allows the identification of the material (Cremers/Radziemski 2006b, 5–6; Anglos/Detalle 2014, 533; Giakoumaki et al. 2007, 750).

LIBS is capable of rapid measurement and analysis speed and provides results in real-time, while minimal or no sample preparation is required, compared with other elemental spectroscopic techniques such as Laser Ablation Inductively Coupled Plasma Mass Spectrometry (LA-ICP-MS), Secondary Ion Mass Spectrometry (SIMS), X-Ray Fluorescence (μ-XRF).

Moreover, LIBS can be used by non-specialised users since it is a straightforward and simple analytical technique. Furthermore, successive spectral data from the same point can provide depth profile analysis since every laser pulse penetrates deeper into the sample. Lastly, LIBS allows for the identification of lighter elements such as lithium, fluorine, sodium or magnesium (Anglos 2001, 187A; Anglos/Miller 2006, 363; Nevin et al. 2012, 340).

Generally, LIBS is classified as a micro-invasive technique, because the mass removed by the laser pulse is in the sub-microgram range and the dimensions of the craters created during ablation are much less than 1 mm in diameter, rendering them barely visible to the naked eye. Typically, the spots or craters created by LIBS have a diameter ranging between 50 and 200 μm, depending on the focusing beam and optics (Giakoumaki et al. 2007, 752). Therefore, LIBS is an effective tool for the analysis of archaeological or cultural heritage objects.

Nonetheless, it is possible to further decrease the spot diameter to 10 μm, enabling the analysis of small-size features and minimising the destructive effect on the sample (Menut et al. 2003, 6065). In the case of the LIBS instrument used in this work, SEM analysis showed that the craters created were 10–15 μm in diameter, which increased to 30–50 μm when the same spot was ablated multiple times (see section 3.1.5.2).

3.1.3.2. LIBS and Sclerochronology

LIBS spectroscopy can contribute significantly to sclerochronological studies as elemental mapping allows the chemical characterisation of the accretionary increments of mollusc shells. For this purpose, previous studies have employed LA-ICP-MS, ICP-

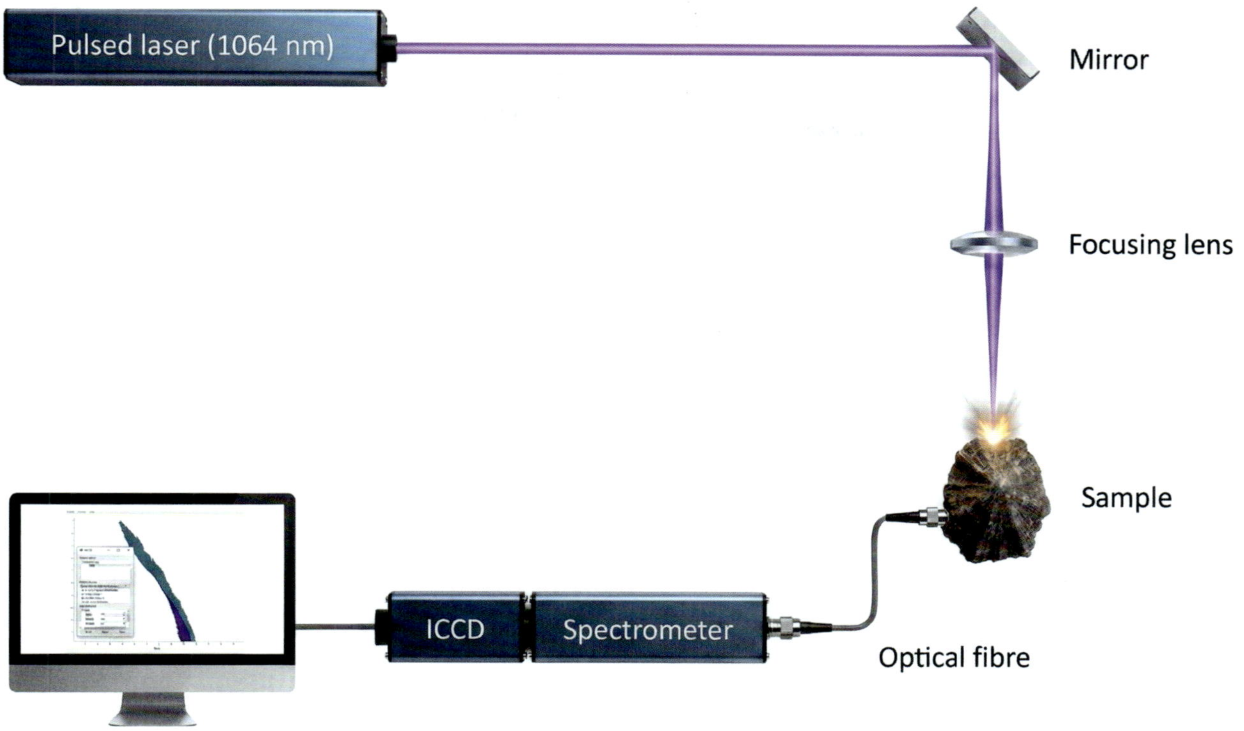

Fig. 8 Schematic illustration of LIBS instrumentation.

MS or Inductively coupled plasma atomic emission spectroscopy (ICP-AES) together with oxygen isotope analysis to investigate magnesium, manganese, strontium, barium and lead to calcium ratios seasonal and interannual variability in bivalve molluscs (e. g. Stecher et al. 1996; Vander Putten et al. 2000; Freitas et al. 2005; 2006; Elliot et al. 2009), in corals (Wei et al. 2004), foraminifera (Shen et al. 2007) and other materials. In this work, the Mg/Ca ratio was effectively used to analyse the shell specimens for seasonality, whereas the Sr/Ca ratio mapping did not yield meaningful results and thus it was not employed in further analyses (see section 4.1.1.3).

The application of these spectroscopy techniques on the samples can have various forms on the sample surface. It can be in the form of firings on single spots, while some studies produced 1-dimensional data series by ablating the samples in a linear way using either LA-ICP-MS (e. g. Batenburg et al. 2011; Marali et al. 2017) or LIBS (Cáceres et al. 2017) to create elemental profiles. Furthermore, 2-dimensional LIBS maps have been created in the past for compositional mapping of aluminium alloy to identify the distribution of a specific element and create an image of the surface (Kim et al. 1998). Regarding the elemental mapping of mollusc shells, 2D elemental maps of scallop shells were produced by Lu et al. (2015), who combined LIBS with Raman spectroscopy to study the shell growth by ablating the shells' external surface. More recently, 2-dimensional elemental maps of mollusc shells were produced with the use of LIBS (Hausmann et al. 2017; 2019a; 2019b), while Hausmann et al. (2019a) also proposed LIBS imaging as a pre-screening step for oxygen isotope sampling by targeting seasonal extremes and the subsequent reduction of the oxygen isotope samples per shell specimen, allowing the sampling of larger archaeological assemblages. An example of a Mg/Ca map produced in the framework of this study, can be seen in **figure 9**.

Mg/Ca ratios on mollusc shells with the use of LIBS were first successfully obtained by García-Escárzaga et al. (2015) demonstrating the capacity of *Phorcus lineatus* shells to serve as reliable palaeoenvironmental proxies. The Mg/Ca ratio of the shell apertures of monthly collected shells was measured and sequential elemental analysis was carried out in three shell samples.

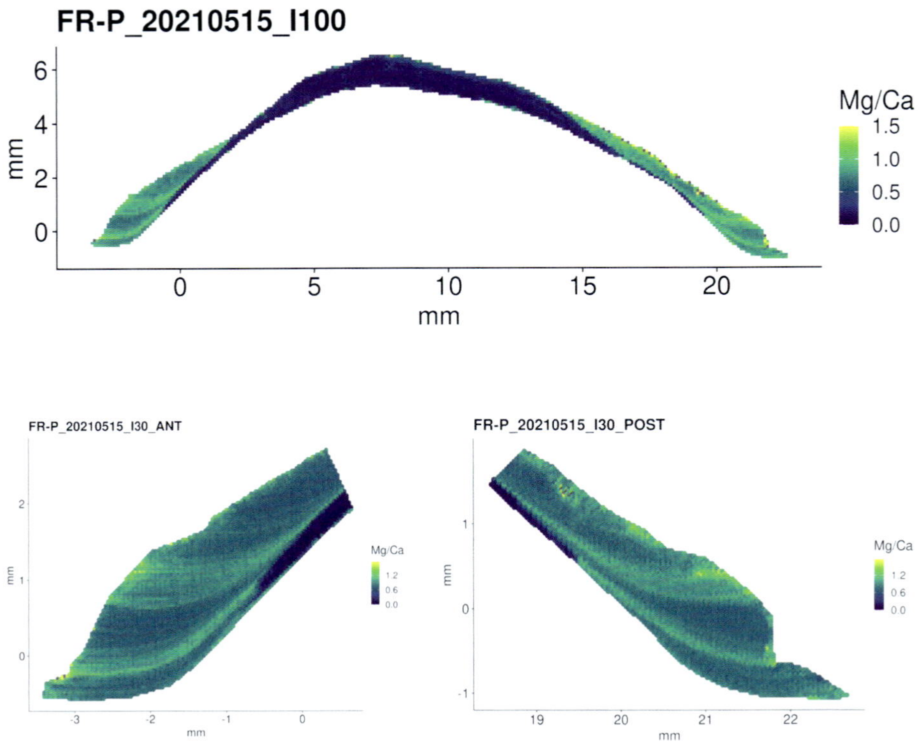

Fig. 9 Example of a Mg/Ca elemental map of a limpet shell.

3.1.3.3. Micro-LIBS Instrument

The specific micro-LIBS instrument[5] that was employed in this study synchronises three main components: the laser system, the spectrometer, and the motorised XYZ-stage (sample holder) that operate together in order to create 2D elemental maps of shells. The components synchronisation is accomplished using a custom-made electronic controller (Biomimetic PC).

The micro-LIBS uses a pulsed DPSS (diode-pumped solid-state) and Q-switched Nd:YAG laser with a Gaussian unstable resonator, with a repetition rate of 100 Hz and an output energy of 1.7 mJ at 1,064 nm, producing 8 ns pulses. The laser beam is focused using a high-power objective lens with infrared coating and an effective focal length of 20 mm. The system is equipped with a CMOS camera aligned with the objective lens to image the shell surface. The camera shares focus with the laser beam, albeit with a slight offset. The spectrograph coupled with an Intensified Charge Coupled Device (ICCD) detector with a 1800 l/mm grating, is installed in the micro-LIBS system. This setup provides a resolution of 0.06 nm/pixel resolution and a spectral range of approximately ~65 nm. An internal delay generator is used to set the delay time to 0.5 µs and the integration time to 1 µs. To generate the Mg/Ca elemental maps, the Mg II emission line at 279.553 nm and the Ca II emission line at 315.887 nm (**fig. 10**) were selected as the analytical wavelengths (after Hausmann et al. 2023).

Shell analysis requires initially cutting the sample into two sections along the major growth axis using a diamond-coated blade on a slow-speed saw (Buehler; IsoMet 1000). If the section is not properly cut and there are irregularities, the section is ground or polished using metallographic grit paper (P800, P1200, P2400). The shell section is placed into the sample chamber and firmly attached to the fast-moving XYZ stage (**fig. 11**). The attachment is secured using aluminium foil and adhesive (Blu Tac) to ensure it remains in place and can withstand the inertia caused by the movement of the stage.

5 The system was developed and manufactured by a group of researchers (Victor Piñón, Panagiotis Siozos, Andreas Lemonis) in collaboration with Niklas Hausmann at the Institute of Electronic Structure and Laser headed by Prof. Demetrios Anglos at the Foundation for Research and Technology-Hellas (IESL-FORTH) in Herakleion, Greece and was transferred to the Leibniz-Zentrum für Archäologie in Mainz. The system is extensively described in Hausmann et al. 2023.

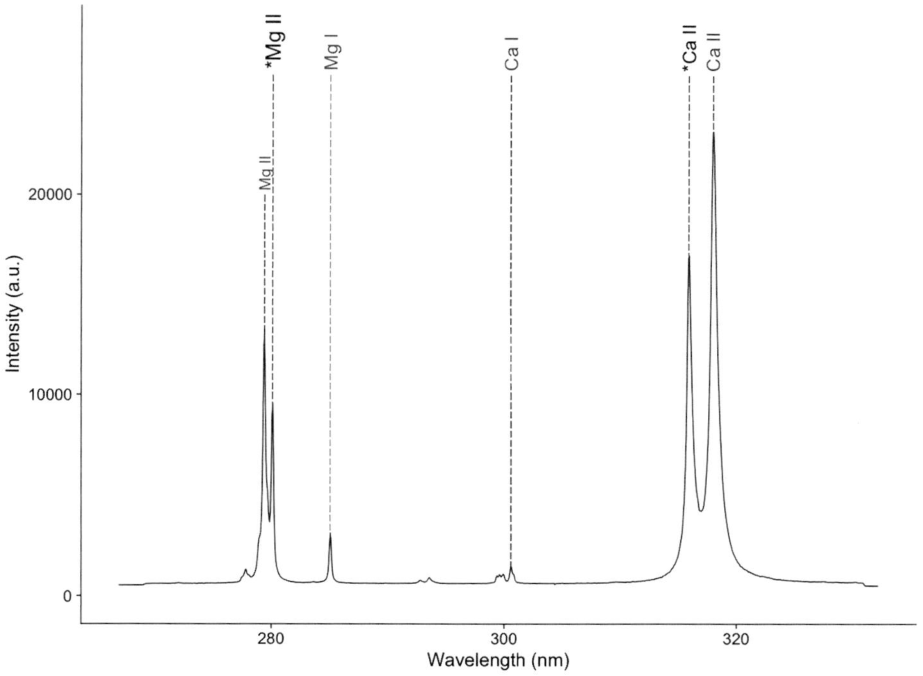

Fig. 10 LIBS spectrum of a shell sample. Asterisks indicate the Mg II (279.553 nm) and Ca II (315.887 nm) emission peaks. – (Reproduced from Hausmann et al. 2023, fig. 2, under CC BY 4.0).

3.1.3.4. LIBS Software and the Shiny App

After the chamber door is shut, the scanning process (sample movement, auto-focus, laser firing, spectra acquisition, etc.) is carried out via the custom-made Python software developed by Biomimetic PC. The sample position is adjusted to produce focused images of the shell section on the imaging camera (**fig. 11**). Afterwards, the perimeter of the shell section is defined using a point-and-click software tool. Three tilt points within the shell section are selected to calibrate the sample plane on the Z-axis. Based on the defined perimeter and tilt points, the sampling resolution is set (typically 100 or 30 µm), and a grid of points is automatically generated, with the distance between grid points equal to the sampling resolution. This process ensures accurate imaging and analysis by precisely defining the area and resolution for sampling the shell section.

After the end of the LIBS spectra acquisition process of the sample, the acquired spectra are processed and visualised using the Spectral Analysis window of the software (**fig. 12**). The software calculates the intensity ratio of the two predetermined emission lines of magnesium and calcium and the ratio's average value on several spectra acquired at the same sample point. It also calculates the average value, standard deviation, and relative standard deviation for each individual sample spot, with the option to discard additional spectra as cleaning shots. A Comma Separated Values (CSV) file is generated to store all sampled data from the sample scan. It includes the XYZ relative coordinates of the sample with the calculated mean ratio, standard deviation, and relative standard deviation for each sample spot. This data can be plotted separately as either a 2D elemental map or as a line graph.

The software provides additional features in order to analyse more efficiently the shell samples. For example, the software enables the execution of a cleaning step by discarding certain spectra from each spot. By discarding usually the first three out of ten spectra per point, the seven remaining spectra per spot are averaged and saved in the CSV file. Additionally, after scanning the sample, specific points of the elemental shell records can be selected and marked for further measurements with different sampling methods. For example, annual minima and maxima when detected can be selected and marked with the laser using higher energy via a point-and-click interface. The markings are created by ablating the sample with two perpendicular lines of a length of 1 mm and a 30 µm step among the points, with the number of pulses set at 100. These marked crosses created on the sample surface are visible by the naked eye and can be used to accurately choose the position of the extract carbonate material sampling for oxygen isotope analysis.

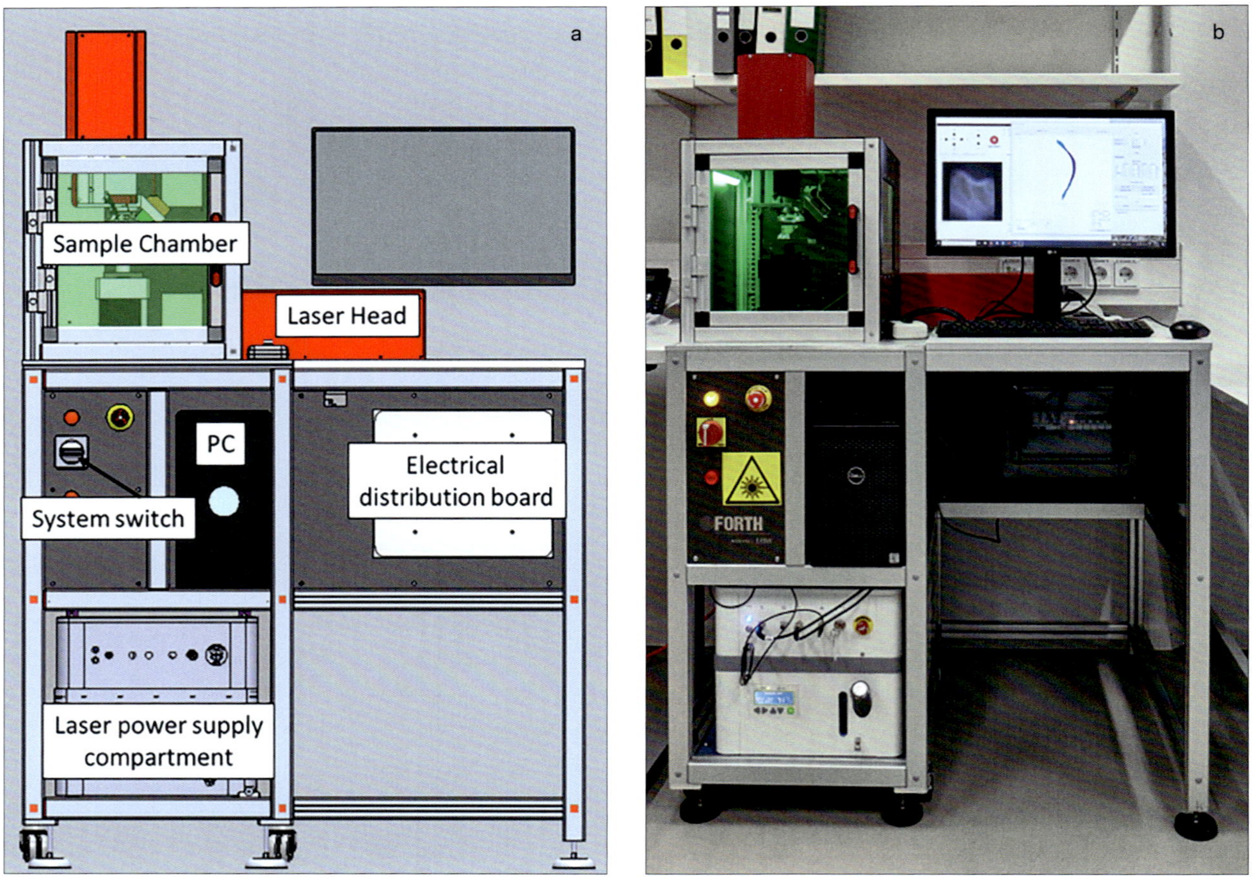

Fig. 11 Figure (**a**) and photograph (**b**) of the LIBS setup. – (a reproduced with permission from V. Piñón; b photo S. Steidl, LEIZA).

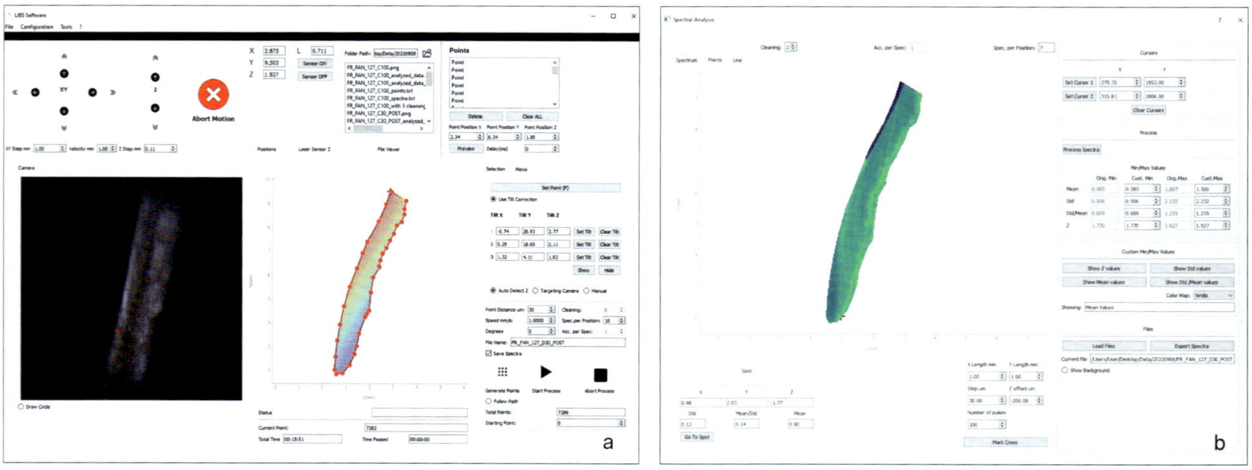

Fig. 12 Screenshots of the main window of the LIBS software, where the perimeter of a section of a limpet shell has been defined (**a**), and the Spectral Analysis window (**b**).

After the scanning process, the CSV files produced by the software are processed via the openly accessible Shiny Shell Shiny App. The application[6] is an interactive web application that executes R code and allows the visualisation of the data, turning the CSV files into elemental maps. The CSV files are uploaded and can be modified according to the needs of each user. The application allows the user to select the geometry type of the points displayed (tiles or points) and the x and y axes can be changed (horizontally, vertically or flipped). With the use of bars the colour options tab allows the colour filtering of the extremely high or low Mg/Ca ratio and the deviation filtering tab similarly allows the filtering of the standard deviation and the relative standard deviation values. Furthermore, the colour scale, the title of the plot and its font size can be modified. In the case of line scans, the application enables the visualisation of the line scans into linear graphs where the error bars can be displayed. A smoothed trend line can be displayed, which averages the data given a specific reference window. Lastly, the plots generated can be downloaded via the download plot tab and saved as PNG files. All the elemental map plots in this piece of work were generated with the use of the Shiny Shell Shiny App (**fig. 13**).

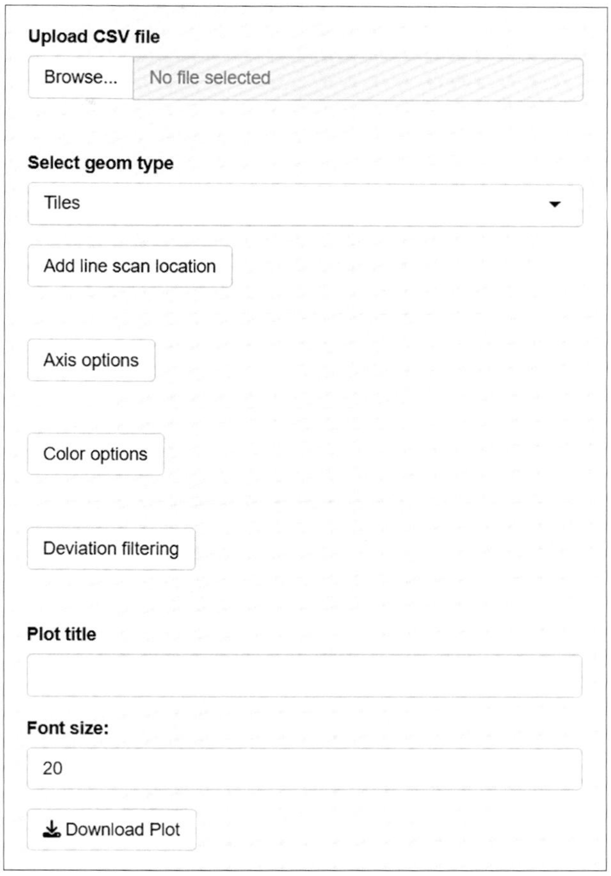

Fig. 13 Screenshot of the Shiny Shell Shiny App.

3.1.4. Amino Acid Racemisation Analysis

Amino Acid Racemisation (AAR) is used as a geochronological tool and is widely accepted as a reliable dating technique for Quaternary research (Kaufman/Manley 1998, 987). It is a relative dating technique that is quick, effective, and allows the low-cost analysis of large sample numbers. It has been proven to provide reliable age estimation results for a number of carbonate-based biominerals, as the radiocarbon dating technique has its limits in various cases (Barker 1958, 261–263; Cubizolle et al. 2007, 188–189). The AAR dating of shells has a long history in research and has been known since the 1970s, however it was not widely applied because of the interference of external factors, especially high temperature that can significantly alter the results (e. g. Masters/Bada 1977; Wehmiller 1993; Parfitt et al. 2005; Ortiz et al. 2009; Demarchi et al. 2011). Nonetheless, in recent years, increased attention has been set on the protected closed system intra-crystalline fraction of the proteins of the biominerals (Penkman/Kaufman 2013; Penkman 2017), which allows for the recovery of well-protected intra-crystalline amino acids and the secure dating of the samples.

The AAR method has already been successfully applied to enamel (Dickinson 2018; 2019), shells (Penkman et al. 2008; Demarchi 2011; Ortiz et al. 2015), foraminifera (West et al. 2019; Wheeler 2022), corals (Hendy et al. 2012; Wehmiller et al. 2012), and

6 The application was created by N. Hausmann and can be found online: https://nikhausmann.shinyapps.io/Shiny_Shell/.

ostrich eggshell (Crisp 2013). The technique is most commonly used to date Pleistocene deposits dating up to ~2,500,000, depending on the latitude (e. g. Refsnider et al. 2013), as well as samples that are dated to the Holocene (Demarchi 2011, 123; e. g. Kosnik et al. 2017). For molluscan samples, it can be used on materials dating up to 150,000 years (Wehmiller/Miller 2000, 191).

3.1.4.1. Principles of AAR for Dating

More specifically, amino acids are the building blocks of proteins and form an integral part of all biominerals, including shells. The proteins of interest are located in the intra-crystalline structure, more protected from the outer environment from contamination or leaching. The inter-crystalline fraction susceptible to leaching, contamination or changes in the chemical environment can be an unreliable source of amino acids for dating. For this reason, an »intra-crystalline« protected from contamination fraction of proteins, functioning as a closed system, has recently been isolated, leading to improvements in AAR by reducing the unpredictability of the outcomes (Towe 1980; Brooks 1990; Sykes et al. 1995).

The majority of the amino acids have an asymmetric carbon atom at their centre (a chiral centre), resulting in each amino acid having two enantiomeric isomers. The L = laevorotatory (left) and D = dextrorotatory (right) assignment is used to distinguish the two types of chiral molecules, and the two pairs are called enantiomers (Demarchi/Collins 2014, 3). The AAR technique is based on the ascertainment that, in nature, amino acids exist in the L enantiomer form in living organisms. The L and D isomers are chemically identical and upon the death of the organism, amino acids slowly degrade and are changed to their D enantiomer form. The extent of racemisation is usually expressed by the D/L value. Eventually, equilibrium is reached, which for most amino acids equals 1 (Bada 1972). The amino acids give different carbon ages of materials. As different amino acids racemize at different rates, they can be used for different periods. The D/L ratio for all amino acids is used to define the relative age of carbonate material (Demarchi et al. 2011, 115–116; Demarchi/Collins 2014, 3).

The rate of the racemisation chemical reaction increases as the temperature increases, as dictated by the Arrhenius Equation (Laidler 1984). In AAR, the D/L value corresponds to a sample's thermal age; for example, D/L is equal to 0 at time zero and equal to 1 in a thermally old sample at equilibrium. This means that the technique assumes that the initial D/L is zero and that racemisation is post-mortem (Crisp 2013, 41). AAR geochronology can also be used to generate numerical chronologies in addition to relative ones, but protein diagenesis needs to be precisely defined and calibrated for a particular geographical area (Demarchi et al. 2011, 122; Penkman et al. 2011, 446).

3.1.4.2. Workflow and Laboratory Preparation

We conducted analyses for twelve amino acids on 66 archaeological shells using the AAR technique at the NEaar Laboratory (North East Amino Acid Racemization) located at the Chemistry department of the University of York[7]. In order to analyse the intra-crystalline proteins, oxidative pre-treatment of the samples is performed in order to remove the inter-crystalline and any exogenous proteins. The isolation of intra-crystalline proteins is a common practise in many biominerals, such as mollusc shells, and it has been shown that it can provide more reliable data and reduce variability for AAR (Sykes et al. 1995; Penkman et al. 2008; Demarchi et al. 2013). Penkman et al. (2008, 22) showed that exposure of powered mollusc shells to concentrated NaOCl for 48 h removes the organic matrix between mineral crystals and isolates the intra-crystalline protein fraction.

The analysis is typically achieved using the Reverse Phase High Pressure Liquid Chromatography (RP-HPLC) using fluorescence detection based on the method of Kaufman and Manley (1998). Run time takes approximately 115 minutes (Penkman

7 Laboratory work was conducted together with Niklas Hausmann under the guidance and supervision of the North East Amino Acid Racemization Laboratory (NEaar) of Prof. Kirsty Penkman and the laboratory members Samantha Presslee, Lucy Wheeler and Martina Conti.

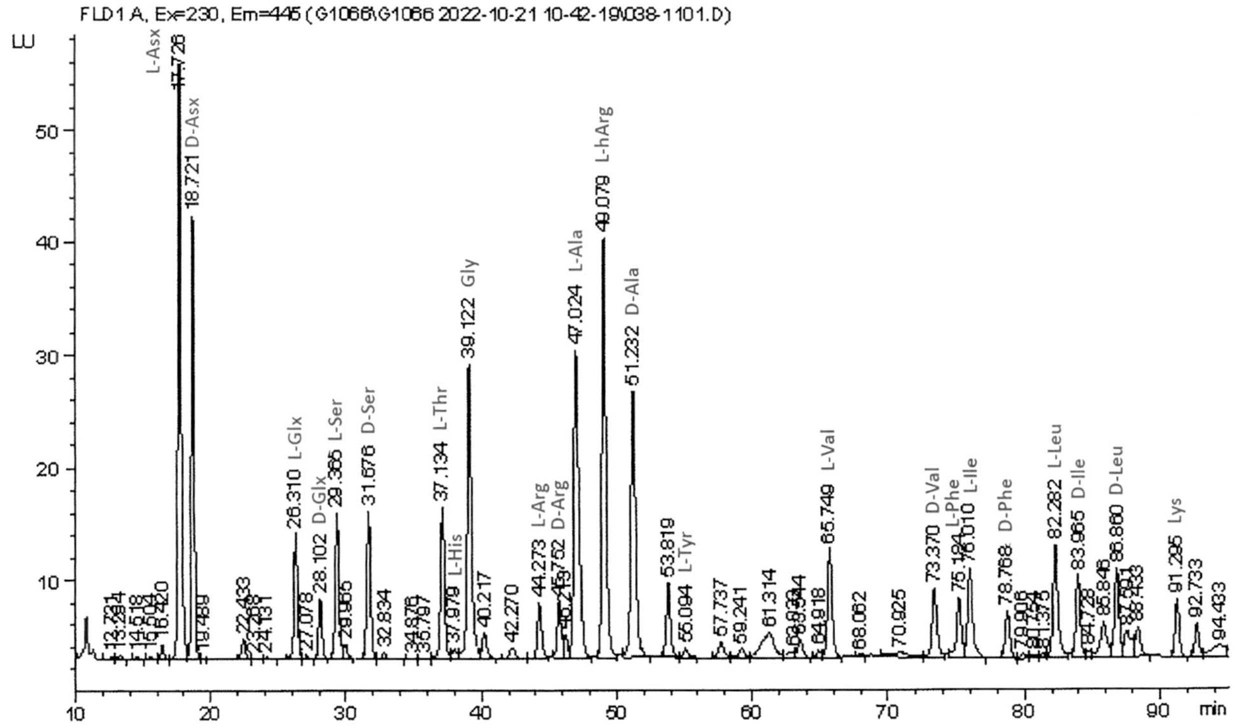

Fig. 14 Example chromatogram illustrating amino acid peaks (annotated in blue) in shell sample FAN_163_A.

2005), and amino acids elute in the following order: L-Asx, D-Asx, L-Glx, D-Glx, L-Ser, D-Ser, L-Thr, Gly (co-eluting with D-Thr, L-His and D-His), L-Arg, D-Arg, L-Ala, L-hArg, D-Ala, L-Tyr, L-Val, L-Met, D-Met, D-Val, L-Phe, L-Ile, D-Phe, L-Leu, D-alle, D-Leu (**tab. A.1**).

Sample extraction took place according to the AA protocol for calcium carbonate biominerals of the NEaar Laboratory. After selecting 66 samples among the 180 archaeological shell samples (**tab. A.2**), which were divided into two batches, small pieces of material were chipped off from each specimen by using a pliers. The technique requires a small mass of sample (~10 mg). Subsequently, the samples are crushed and turned into powder with a mortar and a pestle. The samples are then bleached (with 12 % Sodium hypochlorite, NaOCl) in order to isolate the intra-crystalline fraction and then separated into »Frees« fractions for the analysis of the amino acids unbound in protein chains and »Hyds« for hydrolysed fractions for the analysis of all the amino acids present in the sample. The Frees are demineralised by using 2 molar Hydrochloric Acid and then the samples are dried overnight in a centrifugal evaporator. As for the Hyds, the samples are demineralised and hydrolysed by using 7 molar Hydrochloric Acid, then are flushed with nitrogen and subsequently placed in the oven at 110 °C to break all protein-bound amino acids for analysis. After 24 hours in the oven, the samples are dried under vacuum in a centrifugal evaporator to remove the acid. It has been found that this hydrolysis time releases the maximum concentration of amino acids and induces minimum racemisation (Kaufman/Manley 1998). Subsequently, all the samples Frees and Hyds are rehydrated by using a rehydration fluid that contains an internal standard (L-homo-arginine) which enables the quantification of the concentrations of the amino acids in the sample.

The results of the analysis are called chromatograms (**fig. 14**), where the different amino acids are shown as distinct peaks at their designated racemisation times. The amino acid peaks in the chromatograms are characterised by recognising the patterns and retention times in which they appear and comparing them with standards that are frequently run between the samples. Finally, the chromatograms are digitised with the use of Excel and the Peakachoose add-on and calculations are made to produce the results for each amino acid.

3.1.5. Validating Methodologies

3.1.5.1. Introduction

In order to demonstrate the suitability of the analytical methods employed in this thesis (LIBS and stable isotope analysis), I conducted background experiments. Firstly, regarding LIBS elemental mapping, I conducted SEM imaging analysis on the ablated shell surface using different experimental parameters. This enabled the assessment of the impact of LIBS on the shell crystal fabric as well as to validate the best sampling parameters for optimal analysis. Secondly, with the use of Raman spectroscopy I tested the crystal polymorph preservation of the shell samples. The different crystal structure of the two polymorphs of calcium carbonate (calcite and aragonite) were examined in order to accurately characterise the different areas of the sample. One of the oldest archaeological shells of the shell assemblage was inspected and no recalcification of the aragonite structure was detected. Furthermore, LIBS mapping indicates the location and extent of the two different calcium carbonate structures of limpets shells, and with the use of Raman I was able to specifically verify this assumption on the specific shell samples. By confirming the location and the preservation of the crystal polymorphs, it enables the secure interpretation of stable isotope results as well as the use of palaeotemperature equations by Hays and Grossman (1991). The two different sets of validating experiments that demonstrate the appropriateness of the methods employed for this type of samples, are presented extensively below.

3.1.5.2. SEM Analysis

SEM experiments[8] were conducted in order to assess the impact of LIBS analysis on the shell surface. Shots with different parameters were made in order to test the influence of the focus condition, the energy influence, the influence of different numbers of pulses, of grids with different steps between the points, and the ablation of the edges. Overall, SEM imaging showed that cracks and fissures created on the shell material can vary, even when the firing conditions are identical and this is due to the underlying differences of shell's crystal fabric structure.

The first experiment I conducted was about the influence of the different number of pulses. More specifically, SEM analysis showed that the craters created when firing with 1 pulse were barely noticeable and with 2 or 3 pulses very small, while the signal recorded was very faint. In addition, spots of 10–15 µm in diameter were created when ablating the surface, which gradually increased to 30–50 µm when repeatedly ablating the same spot (**fig. A.1**).

In order to evaluate the influence of the focus condition, I ablated the shell section at different distances from the optimal position (Z = 0) where the sample is into focus. Measuring the distance on the Z axis, I fired with the laser three times moving vertically with 10 pulses per spot at 100, 80, 60, 40, 20, 0, -20, -40, -60, -80, -100 µm distance from the optimal focus position. Each spot was 200 µm apart. This way I created three parallel lines of eleven spots each, where the columns include the same distance spots (**fig. A.2**). The spectral signal became weaker as the laser moved further away from the optimal plane, while the craters created became smaller when the distance was larger at Z = 100, 80 and 60.

Another experiment that was carried out to test the capacities of the system was regarding the energy influence. In this case I fired again with 10 pulses per spot with increasing attenuator values that reflect the amount of energy. Firings with the same amount of energy were performed three times. The attenuator percentages of 5 %, 10 %, 15 %, 20 %, 25 %, 30 % correspond to 0.83, 1.38, 2.45, 3.97, 5.96 and 8.31 mJ accordingly. Energy was measured with the use of an oscilloscope, while the energy that is regularly used when handling the system is at 10 %. 10 % is found

8 SEM imaging was conducted at the University of Crete by the technician A. Manousaki. Experiments on the sample with LIBS were carried out together with V. Piñón. Prior to the analysis, the shell sample (FR-P_C) was cleaned and then sputter-coated with a 3 nm-thick gold film, which is necessary to obtain high quality images for non-conductive materials. The sample was examined using a JSM-6390 LV Scanning Electron Microscope (JEOL Ltd.).

to be the optimal attenuator value as at 5 % the spots created by the laser are not sufficient enough, while all the firings above produced excessively large spots and destroyed the shell surface at a larger extent.

A fourth experiment was made to investigate how different maps with different steps and different number of pulses per position are imprinted on the shell surface. In this case, twelve maps or grids of spots were fired using 1 pulse per position in four maps with 100, 50, 30 and 15 µm steps, four maps with 5 pulses per position with the same distances and four with 10 pulses per position. 1 pulse maps yielded very faint spots, some of which I was not able to locate on the shell.

As shown on the images, at 15 µm distance (**fig. 15g–h**), in the case of the two grids of 4 × 4 spots with 5 and 10 pulses, the craters become all merged into almost one bigger crater. At 30 µm distance (**fig. 15e–f**), the different separate craters are intertwined, forming a larger ablated rectangular surface. At 50 µm (**fig. 15c–d**), the 5 × 5 grids are discernible, however the spots are still disordered. The more clearly defined grids are at 100 µm distance between the spots, whereas at the 5 pulse grids (**fig. 15a**) the spots are weaker.

The 30 and 100 µm resolutions are most frequently used in our analyses and the effect of the laser pulses is seen on **figures 15f** and **15b**. Samples are first sampled at a 100 µm resolution which provides a fast and efficient coverage of the whole surface and afterwards a more detailed scan at 30 µm resolution at the extremities where the calcitic parts of the shell are located and where I can extract the seasonality information that is of interest. At the grid of the 10 pulse spots at 30 µm distance between the spots, the whole surface of the inside the grid, as seen in the SEM image, is ablated and this provides the highest resolution of the shell samples.

The fifth and last experiment conducted to test the system was to evaluate the impact of the laser on the edge values. I created with the laser 10 lines on the edge of the shell each with different numbers of points, 5 on the interior and 5 on the exterior side of the shell (**fig. 16**). In the SEM images, I found fewer spots from what I counted before firing, as some of the points were positioned on purpose outside of the shell in order to ablate the edge. The spectra acquired showed that the signal deteriorates abruptly when the laser pulse is at the edge as it gets out of focus and there are no intermediary quality signals.

Overall, from all five different experiments I can conclude that 10 pulses per spot with an energy of 10 % (1.38 mJ) are the sufficient parameters for the measurement of the calcitic shells. The shell sample needs to be positioned at the optimal position (Z = 0) from the objective in order to obtain clearer results, while on a flat surface. Edge shots provide noisy signals that cannot be of use and are not taken into consideration when interpreting the elemental maps.

3.1.5.3. Raman Spectroscopy Evaluation of the Shell Carbonate

Sample FAS_206_C was analysed with Raman spectroscopy in order to test the preservation of its calcium carbonate (**tab. 2**). The specific shell was selected because it belongs to the oldest shells that were analysed and is dated to the Final Palaeolithic period of the site. Eight measurements were performed on the sample; four measurements/spots on the ablated with LIBS section and four on the non-ablated. On the ablated with LIBS half, spots 1 and 2 are located on different areas of the calcitic part and spots 3 and 4 on the aragonitic part. On the non-ablated half, spots 5 and 6 are on the calcitic area of the shell, while spots 7 and 8 are on the aragonitic area (**tab. 2; fig. A.5**).

Raman spectroscopy is a rapid non-destructive molecular analytical technique where no sample pretreatment is required. Raman is based on the principle that every molecule or material has a distinct chemical »fingerprint« (or characteristic spectral pattern) and Raman spectra provide information about the molecular structure of samples. As a Raman spectrum corresponds to its molecular fingerprint, it can be used to quickly identify and characterise the material. Raman is a light scattering technique, where a molecule scatters incident light from a high intensity laser light source. The scattered light has a different frequency from that of incident light and is used to construct a Raman spectrum. A Raman spectrum consists of a number

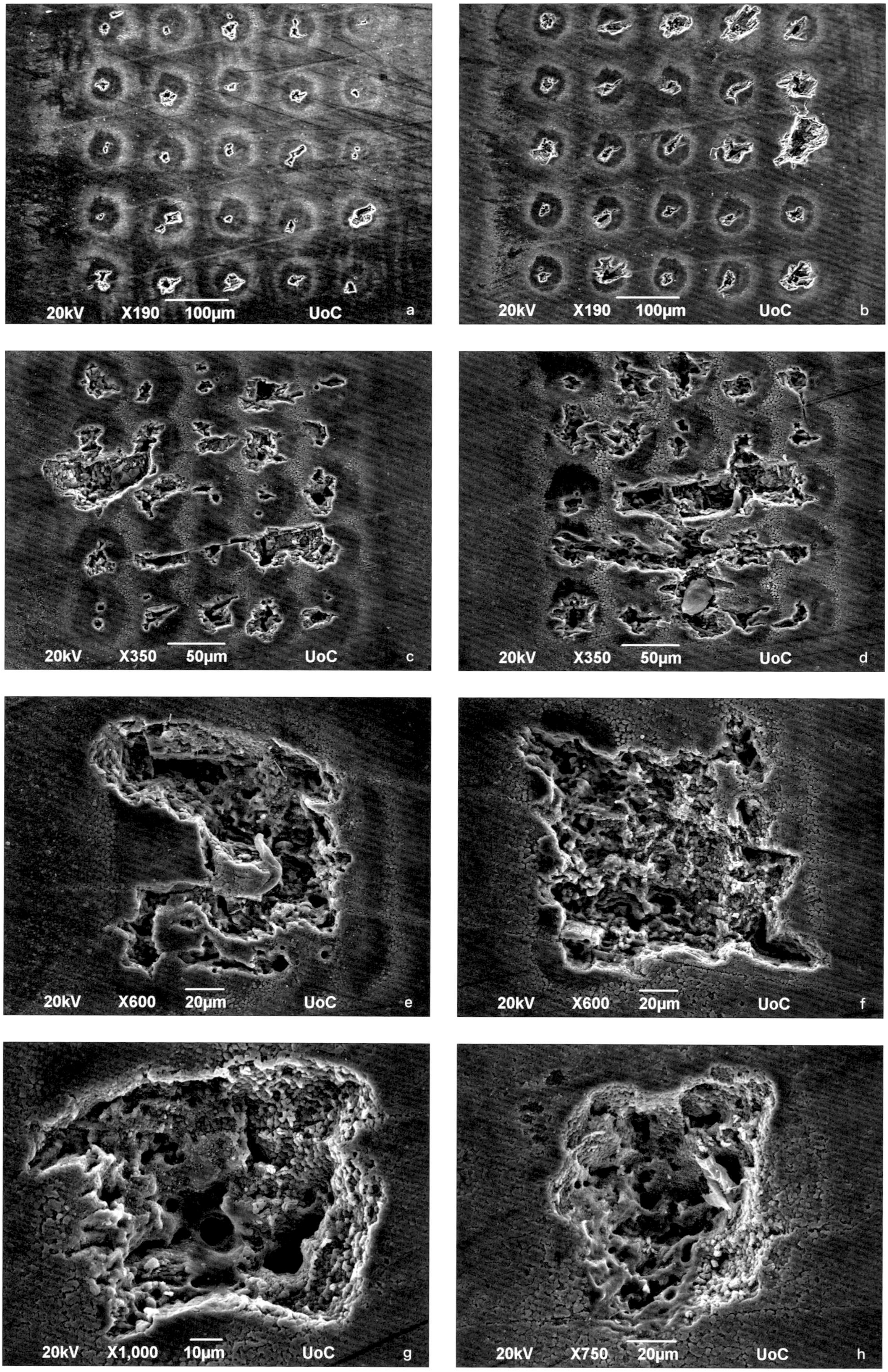
20kV X190 100µm UoC a
20kV X190 100µm UoC b
20kV X350 50µm UoC c
20kV X350 50µm UoC d
20kV X600 20µm UoC e
20kV X600 20µm UoC f
20kV X1,000 10µm UoC g
20kV X750 20µm UoC h

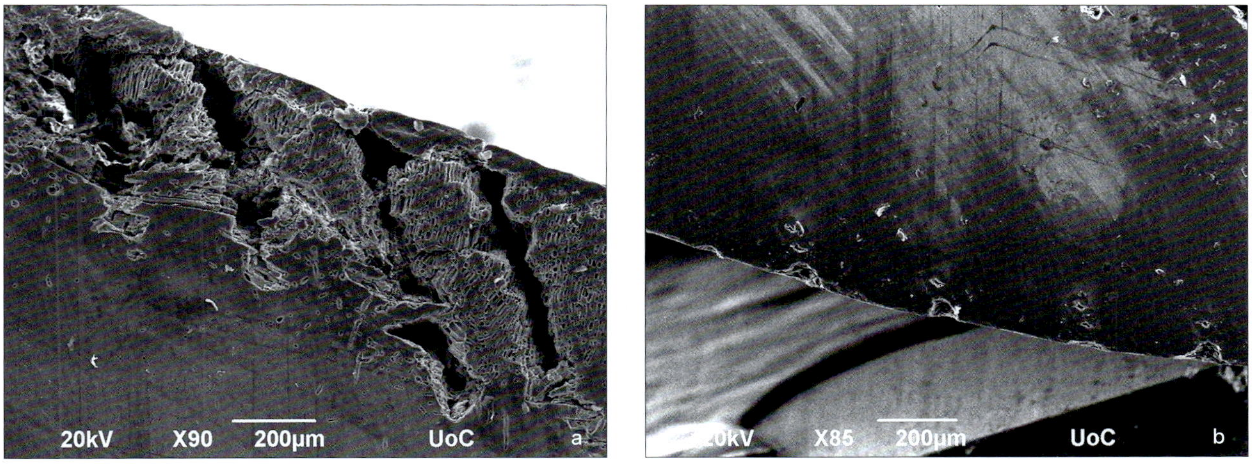

Fig. 16 SEM imaging of line scans fired towards the exterior (**a**) and the interior edge (**b**).

Tab. 2 Locations of the Raman measurements on the shell sample.

Raman spots	Shell half	Shell area	Shell crystal
Spot 1	Ablated	Posterior	Calcite
Spot 2	Ablated	Posterior	Calcite
Spot 3	Ablated	Apex	Aragonite
Spot 4	Ablated	Apex	Aragonite
Spot 5	Non-ablated	Anterior	Calcite
Spot 6	Non-ablated	Posterior	Calcite
Spot 7	Non-ablated	Apex	Aragonite
Spot 8	Non-ablated	Apex	Aragonite

of peaks, showing the intensity and wavelength position of the Raman scattered light. Each peak corresponds to a specific molecular bond vibration and is presented as an intensity-versus-wavelength shift (Smith/Dent 2005; Larkin 2011, 1–2).

The measurements were performed using the same parameters for all the samples: A continuous wave (CW) laser emitting at 785 nm with a power of 50 mW. The acquisition time was 20 seconds and the number of scans accumulated was 2. Reference spectra for calcite and aragonite were retrieved from the RRUFF database (www.rruff.net, see Lafuente et al. 2015).

The spectra produced positive results confirming the expected preservation of the shell's calcium carbonate crystal polymorphs at their corresponding locations on the shell section. Four main bands at 156, 281, 711 and 1,085 cm^{-1} contributed to the identification of calcite and four main bands at 154, 206, 705 and 1,085 cm^{-1} contributed to the identification of aragonite (**tab. 3**). Calcitic (1, 2, 5 & 6) and aragonitic (3, 4, 7 & 8) spots produced spectra that were very similar to the standard spectra with matching peaks (**figs 17–18; A.6–A.13**), while no differences were noticed among the spectra of similar spots on the ablated and the non ablated halves. Minor weak peaks at 352, 903 and 926 cm^{-1} in the spectra of the spots 2, 5 and 6, represent noise and are due to the scattering of light from other sources of light in the room where the measurement was carried out and are not

Fig. 15 SEM imaging of grids of spots: **a** 5 pulse spots at 100 µm distance. – **b** 10 pulse spots at 100 µm distance. – **c** 5 pulse spots at 50 µm distance. – **d** 10 pulse spots at 50 µm distance. – **e** 5 pulse spots at 30 µm distance. – **f** 10 pulse spots at 30 µm distance. – **g** 5 pulse spots at 15 µm distance. – **h** 10 pulse spots at 15 µm distance.

Raman band (cm^{-1}) observed (spot 1)	Raman band (cm^{-1}) calcite (de la Pierre 2014)	Vibration assignment
156	155	Lattice mode
281	276	Lattice mode
711	711	$\delta_{in\ plane}\ CO_3^-$
1,085	1,086	$\nu_{sym} CO_3^-$
Raman band (cm^{-1}) observed (spot 7)	**Raman band (cm^{-1}) aragonite (Frech et al. 1980)**	
154	152	Lattice mode
206	206	Lattice mode
705	707	$\delta_{in\ plane}\ CO_3^-$
1,085	1,086	$\nu_{sym} CO_3^-$

Tab. 3 Raman band assignments of calcite and aragonite.

related to the nature of the material. The same is the case for the thin negative peak at 541 cm^{-1} of spot 6.

Overall, Raman spectroscopy on the archaeological shell FAS_206_C confirmed previous assumptions on the presumed nature and location of the calcium carbonate crystal structures. No recalcification of the aragonitic structure was detected in the sampled areas, while it was confirmed that the low Mg/Ca ratios of the shell section in the area of the apex as indicated by LIBS mapping, correspond to aragonite structures and the higher Mg/Ca ratios to calcite structures.

3.1.6. Combined Methodologies

In this PhD thesis, all of the above methods are employed together to analyse effectively the shell specimens in order to best extract seasonal and environmental data. First, from the mollusc shells, the growth rate of the organisms needs to be taken into account and in this case, the methods employed are borrowed from sclerochronology studies. The growth of the accretionary parts is assessed with the use of LIBS and the generation of the 2D elemental maps. With laser ablation I was able to decipher the season of death of the mollusc organisms, when sufficient shell growth is present and subsequently I can target the minima and maxima annual Mg/Ca values. Carbonate powder is subsequently extracted from these targeted locations on the shell section and the carbonate samples undergo oxygen isotope analysis. In order to extract palaeotemperature data, water samples were collected from the same location as the archaeological site under study and also analysed for oxygen isotopes. These values were subsequently applied in palaeotemperature equations to approximate temperature data. With the use of the AAR technique, a more refined and detailed chronology was attempted to be established among samples of different contexts in order to better understand the intra-site chronological correlations.

The benefits of these combined methodologies are numerous. Molluscs shells being largely sessile animals are advantageous when studying site-specific climatic variability, while in the case of *Patella* shells, the fact that they are relatively short-lived specimens makes them advantageous for studying short-term climatic events. Furthermore, their immediate connection with the archaeological record directly links the palaeoenvironmental and palaeoclimatic data with the broader or more specific archaeological context of the site under investigation, thus connecting the events with past human activities (Leng/Lewis 2016, 298; West et al. 2018, 191).

Sample sizes and resolutions are also important factors that are efficiently tackled by this combined approach. Sclerochronology studies are often based on extensive isotope sampling of specimens which demands an increased budget to perform the analysis which restrains the number of samples. However, increased cost and time investment does not conform with the desirable highest resolution sampling strategy which is essential in order to restrain from time-averaging and to have more accurate results (West et al. 2018, 197). In this study, high-resolution sampling is effectuated via LIBS as the shells are

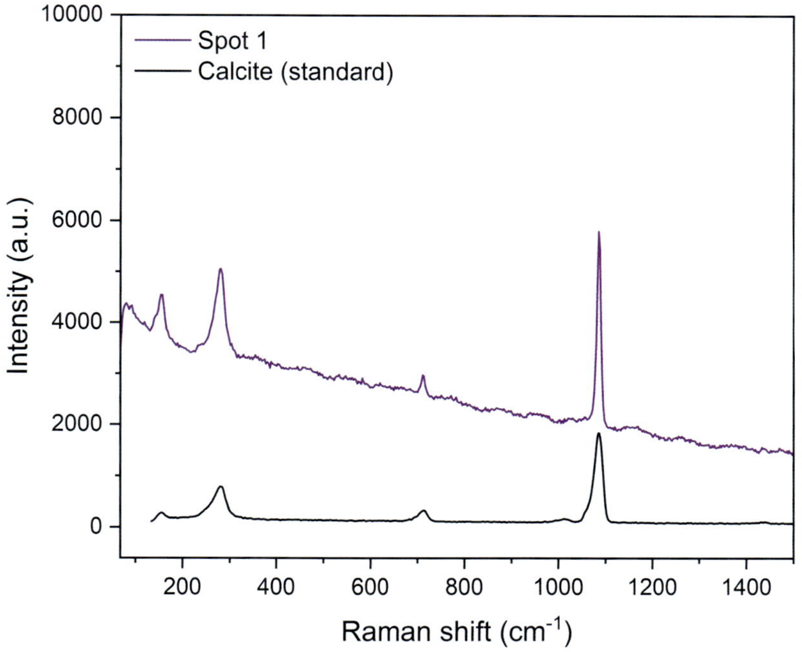

Fig. 17 Raman spectrum of calcite spot 1 without background correction compared with standard calcite spectrum (black line).

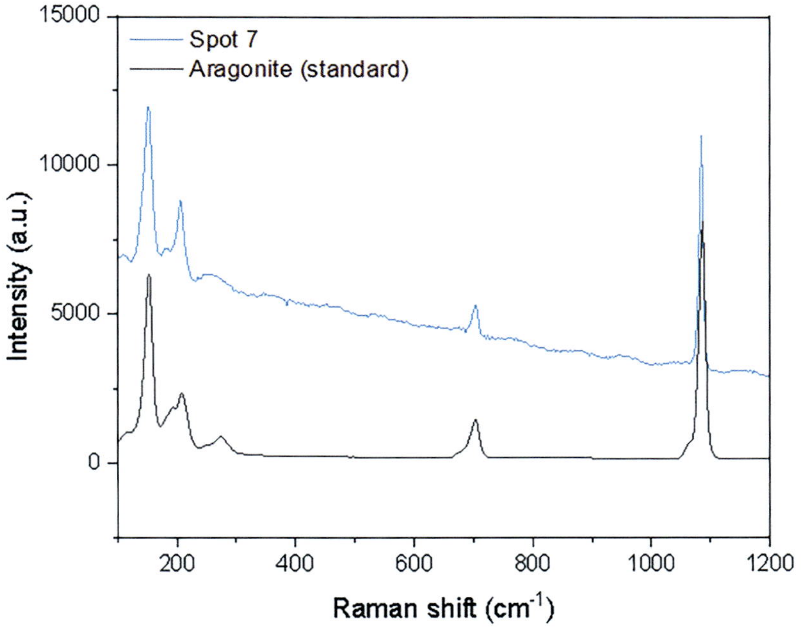

Fig. 18 Raman spectrum of aragonite spot 7 without background correction compared with standard aragonite spectrum (black line).

scanned at 100 µm resolution and if Mg/Ca variability is detected, further scans are performed at 30 µm resolution on the extremities. When a LIBS setup is available for analysis, apart from maintenance costs, the cost of running samples is near to zero. The small number of samples for oxygen isotope analysis per shell enables the analysis of larger archaeological assemblages.

To conclude, this combined proposed method offers the opportunity to study past seasonal and palaeoenvironmental conditions by analysing large assemblages in a cost-effective manner and study short-lived events in the past. Increased resolution within the samples together with large sample assemblages can significantly enhance the scope of the archaeological material and thus the interpretations of past coastal exploitation and past human activity (Hausmann et al. 2019a, 7). High-resolution elemental mapping combined with micro-scale sampling of marine calcium carbonates for stable isotope analysis allows for the increased representativity of the shell data across a site or a time period and therefore maximises the robustness of archaeological interpretations (Thomas 2015b, 161).

3.2. Modern Samples

3.2.1. Introduction

A modern reference collection of limpet shells has been gradually created for the needs of the calibration study of this thesis. In total, 131 live-collected specimens from across 22 different coastal locations were collected and analysed with the use of LIBS (**fig. 19**). In order to better understand and cross-calibrate data prior to the geochemical analyses of the archaeological samples, I conducted a pilot study on modern samples. This way, I was able to test the modern shell archive for potential problems under known conditions, before I tackled the archaeological samples. The collection of live shell specimens from a similar environmental background as the archaeological ones and with a known season of collection, can serve as control data and facilitate the interpretation of the climatic proxies. Additionally, water samples were taken in order to perform sea surface temperature (SST) calculations.

This modern work serves as the backdrop for the archaeological study of the Franchthi shells as well as of the rest of the archaeological shells that will be analysed in the future. The collection of limpets initially concentrated on shells from the surrounding shore of Franchthi and gradually included additional shells gathered from different areas throughout Greece. As more archaeological shell will be investigated by the SEAFRONT project in the future, live specimens were also gathered from the vicinity of

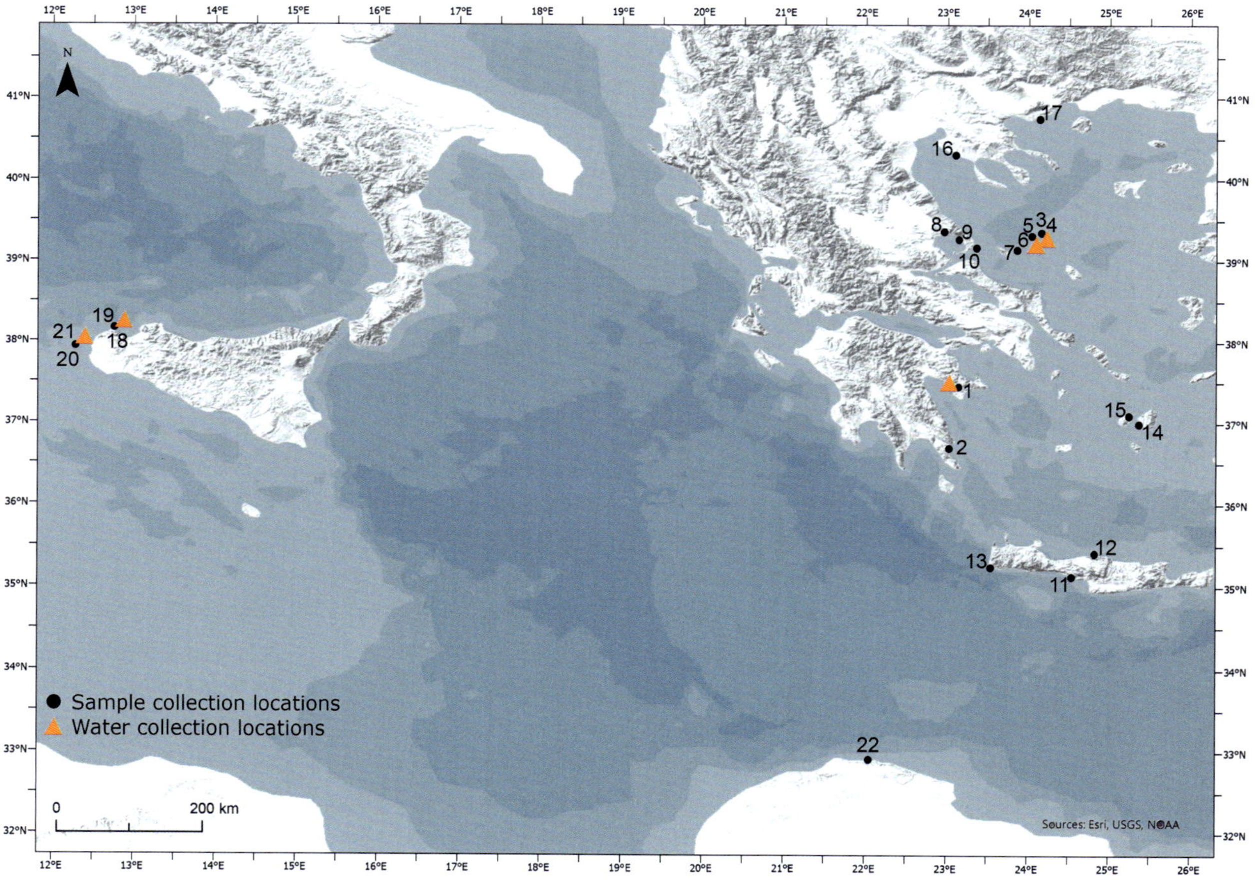

Fig. 19 Geographical distribution of modern shell collection locations. Numbers correspond to **tab. 4**. Triangles indicate the location of the water samples (see **tab. 5**).

	Country	Geographic region	Collection site	Archaeological site	Site code	Coordinates	Collection date	No. of limpets analysed
1	Greece	Peloponnese	Franchthi	Franchthi	FR-P	37.42280, 23.13084	15/05/2021	24
2	Greece	Peloponnese	Monemvasia		ML	36.68648, 23.03960	17/12/2021	5
3	Greece	Thessaly	Youra isl. (west)	Cyclop's Cave	Y-W	39.37678, 24.14985	17/09/2021	12
4	Greece	Thessaly	Youra isl. (south)	Cyclop's Cave	Y-S	39.36861, 24.14602	17/09/2021	8
5	Greece	Thessaly	Agios Petros isl.	Agios Petros	KP-AP	39.32127, 24.05445	17/09/2021	15
6	Greece	Thessaly	Kyra-Panagia isl.		KP-KO	39.31602, 24.06613	17/09/2021	4
7	Greece	Thessaly	Patitiri, Alonnisos isl.		PA	39.14285, 23.86363	18/09/2021	5
8	Greece	Thessaly	Plakes, Volos		V-P	39.34755, 22.97251	16/09/2021	3
9	Greece	Thessaly	Koropi, Pagasetic Gulf		KO-PG	39.28927, 23.14363	10/09/2021	5
10	Greece	Thessaly	Katiyorgis, Aegean Sea		V-KB	39.17255, 23.34283	17/09/2021	6
11	Greece	Crete	Ligres, Rethymno		RE-L	35.13481, 24.53191	06/08/2021	2
12	Greece	Crete	Glaros, Rethymno		RE-G	35.41310, 24.82722	05/08/2021	1
13	Greece	Crete	Kedrodasos, Chania		K-CH	35.26844, 23.55886	27/07/2021	1
14	Greece	Cyclades	Alyko, Naxos isl.		AN	36.98486, 25.38561	06/08/2021	3
15	Greece	Cyclades	Ysterni, Paros isl.		GYP	37.08777, 25.26442	14/08/2021	2
16	Greece	Macedonia	Nea Kallikrateia, Chalkidiki		NKC-M	40.30134, 23.08626	20/06/2021	4
17	Greece	Macedonia	Pyrgos Apollon-ias, Kavala		PARK	40.74020, 24.14294	01/09/2021	2
18	Italy	Sicily	Tonnara del Secco (east)	Grotta dell'Uzzo	TS-E	38.16794, 12.77000	16/05/2023	5
19	Italy	Sicily	Tonnara del Secco (west)	Grotta dell'Uzzo	TS-W	38.16694, 12.76886	16/05/2023	3
20	Italy	Favignana isl.	Faro Punta Sottile (east)	Grotta d'Oriente	FPS-E	37.93468, 12.27661	18/05/2023	4
21	Italy	Favignana isl.	Faro Punta Sottile (west)	Grotta d'Oriente	FPS-W	37.93521, 12.27333	18/05/2023	5
22	Libya	Cyrenaica	Sousa	Haua Fteah	MP6, M0, MBRP	32.90801, 22.04439	Spring, Sum-mer, Autumn	12
	Total							**131**

Tab. 4 Site provenance and collection dates of modern reference shells.

the archaeological sites of Cyclop's Cave (Greece), Haua Fteah (Libya), Grotta dell'Uzzo, and Grotta d'Oriente (Italy). Apart from the locations immediately associated with an archaeological site, the collection also expanded to locations of different environmental conditions as the aim was to cover different locations from a variety of environments.

Background work on modern specimens from the same geographic area includes studies that have already been conducted demonstrating the efficacy of using limpet shells as palaeoclimatic high-resolution proxies. Prendergast and Schöne (2017) conducted oxygen isotope analysis on 19 modern shells from eight Mediterranean locations and provided satisfactory results on the reliability of *Patella caerulea* shells as palaeothermometers. In more than 80 % of the cases, the proper season was interpreted, showing that this species is a suitable candidate for seasonal shellfish foraging research by using ancient shells (Prendergast/Schöne 2017, 44–45). As for the use of LIBS as a pre-screening method for stable isotope analysis, Hausmann et al. (2019) applied spatial mapping on shell specimens from Croatia, Greece, Israel, Libya, Malta, Tunisia and Turkey using LIBS and successfully aligned the Mg/Ca seasonally-resolved record derived from LIBS with $\delta^{18}O$ values and SST. The authors demonstrated how elemental mapping can be employed to evaluate the shell record prior to isotope analysis and identify potential physiological traits interfering with the interpretation of the elemental record (Hausmann et al. 2019, 6).

3.2.2. Description

In terms of geographic origin, the distribution of the shell samples collected and analysed differs. The majority of the modern shell specimens were gathered in Greece, with fewer shells collected in Italy, and Libya. Regarding the shells from Greece, two of the collection sites were located in the Peloponnese, eight in Thessaly, three in Crete, two in the Cyclades, and two in Macedonia. Unfortunately, there was no opportunity to collect specimens from the Ionian sea which has higher mean SST values in comparison with the Aegean Sea (Carapiperis 1952, 155). Limpet shells were also collected from Sicily during a fieldwork trip from the Trapani province and the island of Favignana. Modern shells from Sousa in Libya had been collected and analysed in the past (Prendergast/Schöne 2017) and were given to me by A. Prendergast.

Most of the samples analysed were gathered during autumn, followed by spring and summer-collected shells, while fewer were collected during the winter time (see **tab. 4**). This was due to travel restrictions in Greece in early 2021 because of the covid pandemic, which did not allow for collection excursions. Additionally, weather conditions during winter made the seashores less accessible.

Sites of collection related to archaeological sites included Franchthi, Youra where Cyclop's Cave is located, as well as the islands of Kyra-Panagia and Agios Petros. 24 limpets were collected from the area of Franchthi Cave, 20 from the island of Youra, 15 from the islands of Agios Petros and 4 from that of Kyra-Panagia. The last three sites are part of the National Marine Park of Alonnisos Northern Sporades, a large protected marine area in Central Greece. A survey permit was acquired by the Greek Ministry of the Environment and Energy beforehand in order to collect fauna from the area due to its protected status. The shells collected in Tonnara del Secco, Trapani, are meant to be modern analogues of the Grotta dell'Uzzo archaeological shells, while the shells from Faro Punta Sottile on Favignana of the Grotta d'Oriente. The shells from Sousa serve as analogues for the archaeological shells from the site of Haua Fteah.

The limpets were gathered from the rocky coast, on the intertidal zone, using a simple knife. The collection of limpets requires no prior experience or knowledge, but the use of a tool, such as a knife, is usually necessary to detach them from the rocks. It should be noted that, in all of the instances of live-specimen collection, only the largest adult specimens were collected and in no case were the molluscs depleted from a specific habitat or location. One or two specimens were detached from a colony of limpets depending on its size and the next sample was always taken further afar in order to exert the minimum amount of stress on the mollusc populations. In addition, all of the shells were measured using digital callipers (Sylvac 30-810-1506/50) prior to being sectioned and their length and width were recorded.

3.2.2.1. Water Samples

Seawater samples were also collected alongside shell specimens from neighbouring shores with archaeological sites from which shell samples were or it is intended to be analysed. Water samples for oxygen isotope analysis were collected from the locations of Franchthi Paralia, the southern dock of the island of Youra, Agios Petros islet in Greece, Tonnara del Secco East and West in Sicily and from Faro Punta Sottile West on the island of Favignana (**tab. 5**). One water sample was analysed from each location. The water samples were analysed in order to perform SST calculations and compare them with satellite instrumental data from their respective areas.

3.2.2.2. Mean Sea Surface Temperature in Greece

The factors controlling seasonal variability and the spatial distribution of SST are multiple and complex. Among these factors are air temperatures, precipitation, evaporation, wind speed and direction, freshwater discharges and exchange of water masses with the Black Sea and the Eastern Mediterranean. SST is mainly influenced by seasonal atmospheric conditions, while there is a month's delay in the influence of air temperature onto SST. SST decreases with increasing latitude throughout the year. The SST maximum occurs in August (> 24 °C) and the minimum in February (< 13 °C), with some exceptions occurring in January (Poulos et al. 1997, 235–236). The mean SST of the months September, October, November and December is respectively higher than that of the months July, June, May and April. Recently, SST in the Aegean has rapidly increased since the nineties, especially in the southern part of the basin, under the influence of Levantine water inputs (Skliris et al. 2011, 137). The mean SST from the locations of Franchthi, Youra, Tonnara del Secco and Favignana is represented in **figure 20**. Among the different locations, the Franchthi mean surface temperatures are constantly higher from the other locations. Moreover, in all the locations analysed, the higher standard deviation values are present during the autumn months.

Sea Surface Salinity (SSS) also varies seasonally and fluctuates from < 31.0 psu to > 39.0 psu. Salinity distribution also has a north-south trend with less saline waters towards the north (Poulos et al. 1997, 237).

3.2.2.3. Other Archaeological Sites Included in the Modern Study

In this section, I will briefly provide the archaeological context for the sites of Grotta dell'Uzzo, Grotta d'Oriente and Haua Fteah. These sites are not described in the archaeological background as the focus in that chapter is the Aegean Sea.

	Country	Collection site	Archaeological site	Coordinates	Collection date
1	Greece	Franchthi Paralia	Franchthi	37.42280, 23.13084	15/05/2021
2	Greece	Youra isl. (south)	Cyclops Cave	39.36861, 24.14602	17/09/2021
3	Greece	Agios Petros isl.	Agios Petros	39.32127, 24.05445	17/09/2021
4	Italy	Tonnara del Secco (east)	Grotta dell'Uzzo	38.16794, 12.77000	16/05/2023
5	Italy	Tonnara del Secco (west)	Grotta dell'Uzzo	38.16694, 12.76886	16/05/2023
6	Italy	Faro Punta Sottile (west)	Grotta d'Oriente	37.93521, 12.27333	18/05/2023

Tab. 5 Provenance and collection dates of seawater samples.

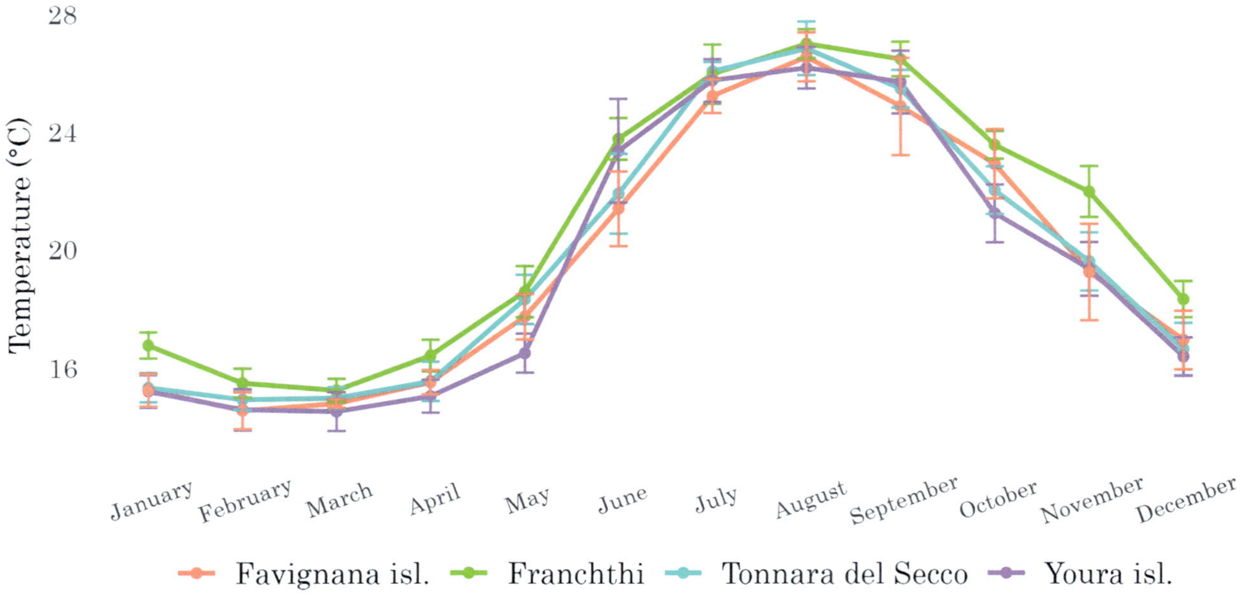

Fig. 20 Averaged monthly Sea Surface Temperature (SST) instrumental data from 2011–2021 from the locations of Franchthi, Youra isl., Tonnara del Secco and Favignana isl. Error bars represent standard deviation values. – (Data retrieved from the Copernicus Marine Service, copernicus.eu).

Fig. 21 View of Grotta dell'Uzzo from the north. – (Photo N. Hausmann, SEAFRONT project).

Fig. 22 Aerial views in the vicinity of Grotta d'Oriente: **a** View from north-east. – **b–c** View from north-west. – **d** Entrance of the cave. – (Photos N. Hausmann, SEAFRONT project).

3.2.2.3.1. Grotta dell'Uzzo

The Grotta dell'Uzzo (**fig. 21**) is a cave located in north-west Sicily that was first recognised in 1973 and excavated between 1975 and 1983 and later during 2004 (Mannino 1973; Segre/Piperno 1975; Piperno/Tusa 1976; Cassoli et al. 1987; Di Salvo et al. 2012a). It is considered an important site in the Central Mediterranean due to its long sequence which covers Mesolithic and the Neolithic occupation phases as well as a transitional period between the two (Piperno et al. 1980a; 1980b). Similarly to Franchthi, Grotta dell'Uzzo witnesses the transition from the Mesolithic lifeways towards agricultural subsistence (Cassoli et al. 1987, 815). Well-preserved stratigraphic sequences have been found at Grotta dell'Uzzo, especially in trenches A and F. Dating analyses on both wood-charcoal and marine shells place the earliest Neolithic occupation during the early 6th millennium BC (5,983–5,714 cal BC), and thus among the earliest sites in the Italian peninsula (Mannino et al. 2006, 25). Human presence is attested at the Grotta dell'Uzzo since the end of the Pleistocene and the beginning of the Holocene when hunter-gatherers were frequently visiting the cave (Mannino et al. 2006, 25–28). 13 Mesolithic burials of individuals have been found up to date at the cave leading to the conclusion that the site was used as a burial site during the Late Mesolithic (Di Salvo et al. 2012a; Borgognini Tarli 1993).

3.2.2.3.2. Grotta d'Oriente

The site of Grotta d'Oriente (**fig. 22**) is located on the island of Favignana at the Ègadi archipelago. The cave is located on the north-eastern slope of Montagna Grossa at 50 m a. s. l. The site was excavated in 1972 and later again in 2005. The stratigraphy of

Fig. 23 The beach near Franchthi.

the site spans from the Late Upper Pleistocene to the Middle Holocene. Pleistocene fauna has been recovered as well as artefacts and fauna from the Mesolithic, the Neolithic and later periods (Mannino 1972; 2002; Mannino et al. 2012; 2014). Two burials have been found, Oriente A dating to the Late Upper Palaeolithic or Early Mesolithic and Oriente B dating to the Mesolithic (Mannino 2002, 17–19). Dating analyses on material from the recent excavations have shown that layers belong to the late Upper Palaeolithic (~14,200 cal BP), the Early Mesolithic (~9,700 and 9,600 cal BP), and Late Mesolithic or Early Neolithic (Mannino et al. 2012, 4; Colonese et al. 2018, 2). Among the most common molluscs found in the Mesolithic deposits are *Patella* and *Phorcus* species alongside limited fish remains (Mannino/Thomas 2004, 24). Early Mesolithic faunal remains include a variety of mammals (e. g. *Equus hydruntinus*, *Bos primigenius*, *Cervus elaphus*), while subsistence evidence broadens during the later stages of the Mesolithic. Remains of domestic (*Ovis*/*Capra*), birds and marine fish belong to the Late Mesolithic/Early Neolithic layers (Mannino/Thomas 2004, 34–39).

3.2.2.3.3. Haua Fteah

The archaeological site of Haua Fteah is a semi-circular rock-shelter located in north-eastern Libya at the foot of the Gebel el Akhdar (»Green Mountain«) massif. It is approximately situated 1 km inland from the Mediterranean sea at 60 m a. s. l. The site may have been used already since 150,000 years ago, but was mainly occupied from the Marine Isotope Stage (MIS) 5 until modern times, covering a particularly long chronological sequence. Remains of two anatomically modern humans were dated at 73–65,000 years ago. The site was first excavated in the 1950s and more recently between 2007 and 2015. The Epipalaeolithic phase of occupation is divided between the Oranian and the Libyco-Capsian phase. This phase coincides with the deglaciation phase following the last Glacial Maximum. The Late Pleistocene to Early Holocene Capsian assemblage contained among others backed bladelets, microliths and bone and ostrich eggshell artefacts. The Neolithic layers include pottery, beads and painted pebbled, pressure-flaked stone artefacts as well as remains of domesticated sheep and goats (McBurney 1960, 199–

Fig. 24 **a** The collection location on the western coast of Youra. – **b** Limpets on the rocks. – **c** The steep western shores. – **d** The southern dock.

205; 1967; Barker et al. 2010; 2012; Douka et al. 2014; Farr et al. 2014; Jacobs et al. 2017).

3.2.2.4. Fieldwork and Description of Collection Locations

For the collection of the modern shell specimens, a number of fieldwork excursions were carried out. Of larger importance were the modern limpet specimens collected from the locations of Franchthi, Youra, Grotta dell'Uzzo and Grotta d'Oriente from where archaeological specimens were or will be analysed. In this section, the collection locations are described as well as the fieldwork trips to the sites of interest.

3.2.2.4.1. Franchthi

In May 2021, I visited the site of Franchthi. The road leads to Lambayana beach where submerged archaeological remains have also been found (see section 2.6.4; Beck/Koutsoumba 2015; 2016; 2017; Beck et al. 2021). A path that runs along a rocky shoreline, leads to the cave of Franchthi. While most trenches of the cave are covered up today, the deep F/A trench is still open with annotations of the occupation phases, summarising the palimpsest of the thousand years of human occupation of the site. The Paralia trenches are not visible today as they are covered by vegetation.

In front of the cave and the Paralia trenches lies a relatively small sheltered pebble beach. Limpets were collected from the few small rock boulders of the beach (**fig. 23**). Finding them was fairly easy and formed small groups of various sizes on the rocks. The adult specimens seemed relatively small- to medium-sized, and were detached from the rocks with the use of a simple knife.

3.2.2.4.2. Youra

In September 2021, I travelled to the islands of the northern Sporades complex in Central Greece and more specifically to the island of Youra (**fig. 24**), where Cyclops's Cave is located, as well as from the island of Kyra-Panagia, where the archaeological site of Agios Petros lies. These two small islands belong

Fig. 25 **a** Islet of Agios Petros. – **b** The collection location on the islet of Agios Petros seen from north-east. – **c** The collection location of Kokkino on the southern gulf of Kyra Panagia seen from north-west.

to the National Marine Park of Alonnisos Northern Sporades, the largest protected marine area in Europe (approximately 2,220 km^2). Their modern name of *Erimónisa*, meaning desert islands, is a witness to the fact that nowadays they are uninhabited with the exception of a monk on Kyra-Panagia. Given the protected status of the Marine Park, official permits to access the area and collect specimens were taken by the Greek Ministry of the Environment and Energy. Access to these areas was possible with the boat of the Managing Agency of the Marine Park, responsible for the protection and surveillance of the area.

I collected limpets from the western and the southern dock of Youra (**fig. 24a. d**). The shore was rough and rugged and the limpets I collected from the hollow parts of the rocks, relatively higher from the sea surface.

Afterwards, I left for the island Kyra-Panagia where the islet of Agios Petros is located in the homonymous bay (**fig. 25**). The island of Kyra-Panagia has two large bays, Planitis in the north and Agios Petros in the south, two natural harbours offering anchorage for ships. The islet of Agios Petros is fairly small and the trenches of the excavation are not

Fig. 26 The collection location of Tonnara del Secco West seen from the south.

visible because of the vegetation. Limpets were also collected from the rocky shore of Agios Petros on the south-west end of the islet and the beach of Kokkino in the southern bay of Kyra-Panagia.

3.2.2.4.3. Grotta dell'Uzzo

In May 2023, I travelled to Palermo and north-western Sicily to visit the Mesolithic-Neolithic sites of Grotta dell'Uzzo (Trapani province) and Grotta d'Oriente (Favignana island) and collect modern shell specimens. First, I ventured to the peninsula of San Vito Lo Capo, west of Palermo as the Grotta dell'Uzzo is part of the Zingaro Nature Reserve and collection is not allowed in the immediate area. The Zingaro, a nature reserve since 1981, is Sicily's first protected area. In the past, it was home to small farming villages, while today it is a popular tourist destination.

Modern limpets were collected from the Tonnara del Secco location, in front of an abandoned building dedicated to the catching and processing of tunny (**fig. 26**). The two locations from where specimens were collected, east and west, are very close to each other. The eastern location constitutes a small »pool« near the shore, while the western location was on the rocky shore.

3.2.2.4.4. Grotta d'Oriente

On the last day of fieldwork, I drove to Trapani and via a ferry I arrived at the island of Favignana which is part of the Egadi Islands, off the north-western coast of Sicily. Grotta d'Oriente is only accessible with the use of climbing equipment and for this reason the visit to the cave was done through the lens of a drone.

As the area nearby the cave is steep and not accessible, limpet specimens were collected from the location of Faro Punta Sottile on the western side of the Favignana (**figs 27–28**). More specifically, they were collected from the small protected gulf east of the small lighthouse called Punta Sottile. The two locations called east and west simply correspond to their position in the gulf.

Fig. 27 The collection location Faro Punta Sottile East seen from the south.

Fig. 28 The collection location Faro Punta Sottile West seen from the north.

3.3. Archaeological Samples

3.3.1. Introduction

In this section, the 180 archaeological samples selected for analysis are presented in detail. The criteria for sample selection, including the focus on secure, dated contexts and a focus on specific excavation units and chronological periods are presented. Detailed observations on sample preservation, abrasion, and evidence of burning were recorded, alongside the provisional species identification. This background work constituted a foundation for the elemental and isotope analyses (see Chapter 4) that were carried out afterwards on the shell assemblage.

3.3.2. Description

The selection of the archaeological samples took place at the Argolid Ephorate of Antiquities in Nafplio in July 2021, where the finds from the Franchthi excavations are stored. As the shell assemblage of the excavation has not been studied in its entirety, and the analytical methods employed that were going to be employed are destructive, I selected samples from excavation units that had been previously studied and published by J. Shackleton (1988) in the monograph »Marine Molluscan Remains from Franchthi Cave« . All of the shells that were selected belong to secure and dated contexts (**tab. A.3**). I also selected examples from a wide range of time periods, with a focus on the Early Neolithic. I wanted to include most of the trenches that were part of Shackleton's publication: trenches FAS and FAN from the cave and trenches L5, L5NE, O5, Q5N and Q5S from the Paralia site (**tab. 6**). Initially, 534 samples were selected for analysis, however the permit application was rejected by the Greek archaeological service due to the high amount of samples. Finally, a permit to analyse approximately one third of the requested archaeological material was granted and a total of 180 shell samples were selected which belonged to 62 excavation units (**tab. 7; figs A.14–A.25**). The selection of the excavation contexts was further based on the availability of the samples according to the non-exhaustion principle of the archaeological record as the techniques employed are destructive for the samples. The non-exhaustion principle consisted of preserving a certain fraction of the available specimens per excavation unit and not selecting samples from units that contained only one specimen.

Broad chronological period	Chronological period	Date range (cal BC)	Trenches cave	Trenches Paralia
Palaeolithic	Final Palaeolithic	10,700-9,000	FA	
Mesolithic	Mesolithic	9,000-7,000	FA	
Neolithic	Mesolithic/Initial Neolithic	7,000-6,600	FA	
	Initial Neolithic	6,800-6,600	FA	
	Early Neolithic	6,600-5,800		L5, O5, Q5
	Middle Neolithic	5,800-5,300	FA	
	Late Neolithic	5,300-4,500	FA	
	Final Neolithic	4,500-3,700	FA	

Tab. 6 Summary of the chronological periods considered in this study, showing the broader periods, calibrated date ranges, and the corresponding trenches from which the shell material was obtained. – (Douka et al. 2015; Perlès et al. 2013; Stiner/Munro 2011; De Vivo et al. 2001; Farrand 2000; Perlès 1987; 1999; 2004; Vitelli 1993; 1999).

3.3.2.1. Sample Distributions

More specifically, as far as the spatial distribution of the samples is concerned out of the 62 excavation units that were sampled 33.9 % of the units belong to the FAS trench, 25.8 % to the FAN trench[9], 3.2 % to the L5 trench, 11.3 % to the L5NE trench, 1.6 % to the O5 trench, 12.9 % to the Q5N trench, and 11.3 % to the Q5S trench (**fig. 29**). If we combine the trenches into the two areas, 59.7 % of the units originate from the cave contexts, and the remaining 40.3 % belong to the Paralia site. Therefore, there is a good statistical representation of both areas of Franchthi in terms of the spatial coverage.

As for the percentage of the shell samples per trench, 28.9 % of the samples belong to the FAS trench, 25.6 % to the Q5S trench, 26.7 % to the FAN trench, 10.6 % to the Q5N trench, 1.1 % to the L5 trench, 6.7 % to the L5NE trench, 0.6 % to the O5 trench. These differences in the number of the samples selected were dictated by their abundance or scarcity of the shell record preserved to date, as mentioned before.

	Site area	Trench	Excavation unit number	Number of excavation units	Number of samples
1	Cave	FAN	123-124, 127, 129, 131, 132, 134, 144, 148, 152, 156, 162-163, 172, 174-175	16	48
2	Cave	FAS	74, 83, 91, 122, 124, 128, 139, 142, 145, 148, 151, 155-156, 174, 182, 186, 189, 195, 203, 206, 207	21	52
3	Paralia	L5	33, 37	2	2
4	Paralia	L5NE	43, 45, 48, 54-56, 64	7	12
5	Paralia	O5	81	1	1
6	Paralia	Q5N	23, 25-26, 45, 47, 55-56, 65	8	19
7	Paralia	Q5S	185, 187-189, 191, 222, 230	7	46
	Total			62	180

Tab. 7 Summary of site areas, trenches, excavation unit numbers, and corresponding sample counts.

	Chronological period	Number of excavation units	Number of samples
1	Final Palaeolithic	3	12
3	Lower Mesolithic	5	8
4	Upper Mesolithic	3	16
5	Final Mesolithic	3	5
6	Final Mesolithic/Initial Neolithic	3	8
7	Initial Neolithic	5	5
8	Early Neolithic	25	80
9	Early Neolithic/Middle Neolithic	3	3
10	Middle Neolithic	9	35
11	Late Neolithic	2	5
12	Final Neolithic	1	3
	Total	62	180

Tab. 8 Summary of excavation unit numbers and sample counts across chronological periods.

9 Trench FA is divided into FA North (FAN) and FA South (FAS).

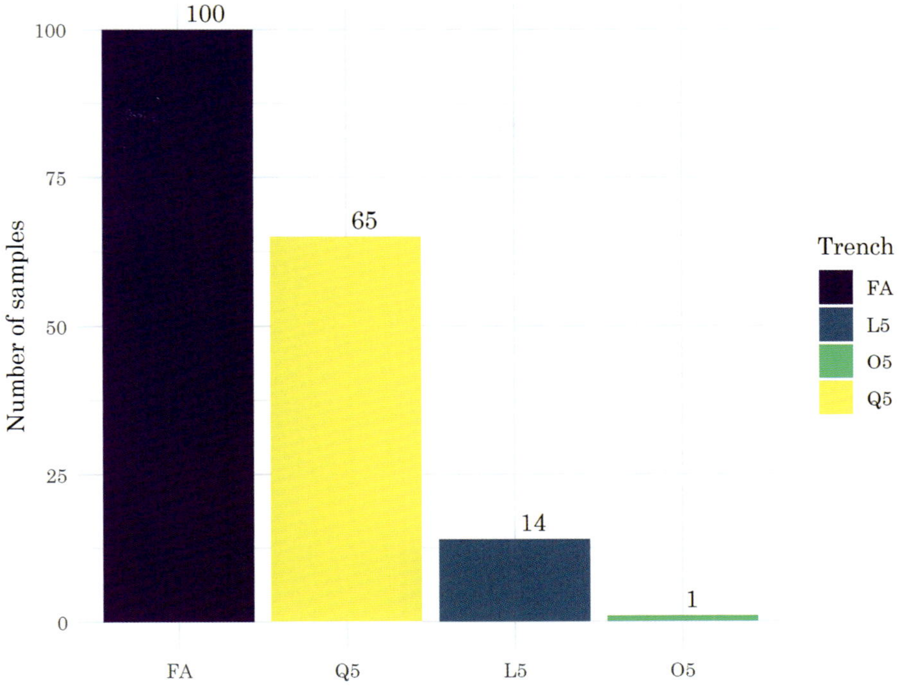

Fig. 29 Bar plot with the number of the archaeological samples analysed per trench (grouped).

Early Neolithic units make up 40.3 % of the excavation units, Middle Neolithic units make up 14.5 %, Initial Neolithic units make up 8.1 %, and Lower Mesolithic units make up another 8.1 %. 4.8 % corresponds to the Upper Mesolithic, 4.8 % to the Final Palaeolithic, 4.8 % to the Final Mesolithic, 4.8 % to the transitional stage between the Final Mesolithic and the Initial Neolithic period, and 4.8 % to the Early and Middle Neolithic periods. 3.2 % and 1.61 %, respectively, of the excavation units are dated to the Late Neolithic and Final Neolithic periods (**tab. 8**). The general distribution of the shell samples reflects the distribution of the excavation units, although the percentages are slightly different.

Regarding the temporal distribution of the samples, the greatest number of samples (44.4 %) belongs to the Early Neolithic. 6.6 % of the sampled shells are dated to the Final Palaeolithic and 16.1 % to the Mesolithic phases of the site that are all located in the cave, while the rest (72.8 %) are dated to the Neolithic phases and belong to both the cave and the Paralia site (**fig. 30**). Evidently, an emphasis in the selection of specimens was placed on the Mesolithic and the Early and Middle Neolithic phases since the investigation of the Early Neolithic is a central theme of this thesis. In this manner, I analysed larger datasets from these specific chronological periods in order to have more concrete statistical results.

The shells selected for this work correspond to Shackleton's »Molluscan zones« I, II, III and IV as *Patella* shells at Franchthi first appear at Molluscan zone I or the *Patella* zone according to Shackleton (1988, 12). This phase is equivalent to the Franchthi General Phase (FGP) 6 and the Final Palaeolithic period (Perlès 2018, 9–10) (see section 2.6.6). For this reason, my work does not contribute to the history of the earlier phases of the site. The shells sampled here start from stratum U/V (Farrand 2000, 53–54).

3.3.2.2. Sample Observations

Even though the assemblage had been previously studied, detailed morphological and depositional and post-depositional descriptive observations are not reported in the publication by Shackleton (1988). Shackleton reports only summarised information on morphological characteristics of the studied material, as for example the average length per species (Shackleton 1988, 23). In order to best describe the sub-assemblage selected in this work, prior to the analysis of the 180 archaeological shell specimens, the dimensions (length, width and height) and weight of all the specimens were measured with digital callipers (**tab. A.3**). Further observations were also recorded regarding the degree of preservation, abrasion and traces of burning. These descriptive

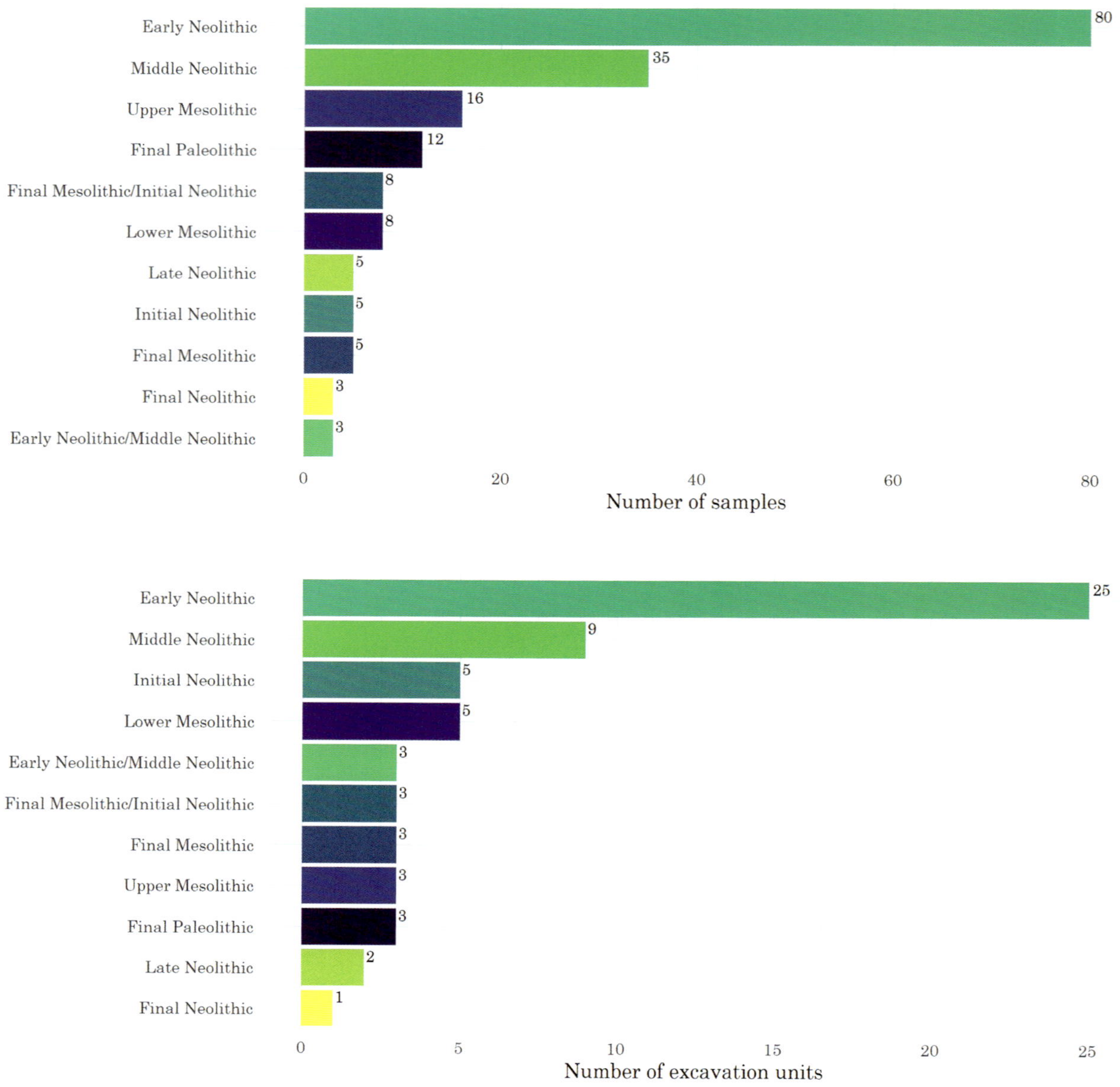

Fig. 30 Two bar plots displaying the number of samples and the number of excavation units by chronological unit.

observations and classifications are fundamental to all subsequent analyses in order to better understand and interpret the molluscan assemblage in the specific temporal and spatial context and also make them more comparable with other assemblages (Claassen 1998, 53; Veropoulidou 2011, 17).

As all of the specimens selected for elemental and isotope analysis belonged to the *Patella* genus, species identification was carried out following a morphological approach. From the 180 specimens, 133 were identified as *P. caerulea* (73.8 %), 2 as *P. rustica* (1.1 %) and 2 as *P. ulyssiponensis* (1.1 %). For the rest of the 43 samples (23.8 %), species identification could not be conclusive (**fig. 31**). Nonetheless, a number of works on *Patella* species contend that identification of the Mediterranean limpet species is only possible through a genetic approach due to the high degree of shape and colour variability as well as the overlapping morphological traits (Mauro et al. 2003; Sá-Pinto et al. 2005; Petraccioli et al. 2010; Sanna et al. 2012). For this reason, species attribution in this case is provisional, lacking a firm identification approach. This also extends to the modern assemblage which has not shown any species specific differences.

Furthermore, regarding taphonomic processes, the degree of preservation records the entirety of

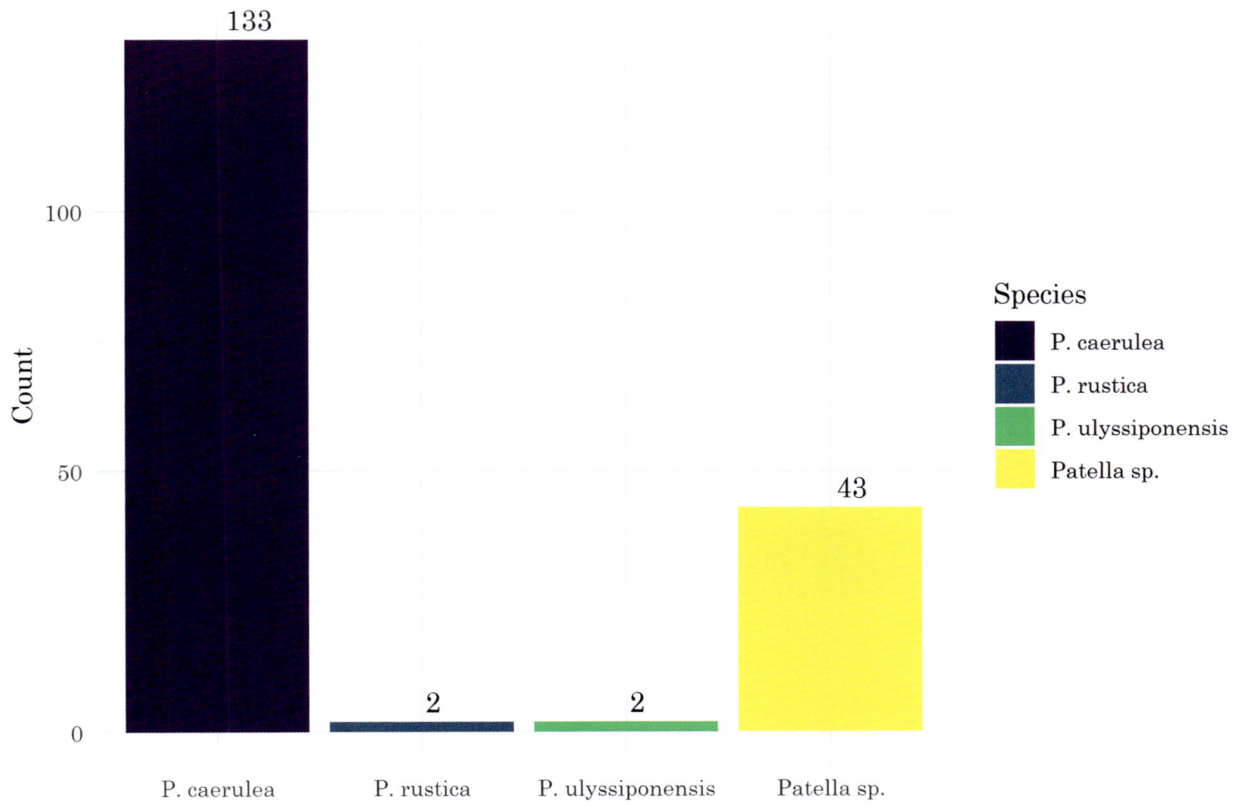

Fig. 31 *Patella* species representation.

the shell preservation on a tripartite scale based on whether they were preserved whole or mostly whole, if the apex and more than half are preserved and if the sample is a fragment. During the sample selection process, it was ascertained that no fragments of the same specimen were selected for analysis. For this reason and also because the specimens are gastropods, the Minimum Number of Individuals (MNI) and the Number of Identified Specimens (NISP) in this work are identical and equal 180. Regarding the degree of post-depositional abrasion, samples were classified into three categories: low, medium and high degree of abrasion. By low I refer to a state of preservation similar to a fresh or modern specimen and by high to a state where anatomical features are severely altered. As for the evidence of burning, the impact of fire was evaluated on a tripartite scale of being scorched, grey or black, while the absence of the impact of fire and the potentially scorched cases were also recorded (**figs 32–33**).

After examining the 180 specimens, 55 % of the specimens were preserved whole or almost whole (also characterised as intact specimens), 32.7 % preserved the apex and more than half of the shell, while 12.2 % of the specimens were fragments. The mean length of the first two categories (whole or almost whole specimens and preserving the apex and more than half) is 30.2 mm, the mean width 25.2 mm and the mean height 8.5 mm. The mean weight of this group of samples was 2.4 g. The mean length of the fragments is 27.3 mm, the mean width 22.7 mm, the mean height 6.3 mm and the mean weight 1.3 g.

As to the abrasion degree, the majority of the assemblage illustrated medium degree of abrasion. More specifically, 18.8 % of the samples were characterised by low abrasion, 48.8 % by medium abrasion and 32.2 % by high degree of abrasion. Lastly, the majority of the samples did not present any evidence of fire (84.4 %). 8.8 % of the samples were characterised as potentially scorched and 5.5 % as scorched. Only one sample was characterised as grey (0.5 %) and one as black from burning (0.5 %). These observations will be taken into account in Chapter 4 in order to evaluate their potential effect on the elemental and isotopic record of the samples.

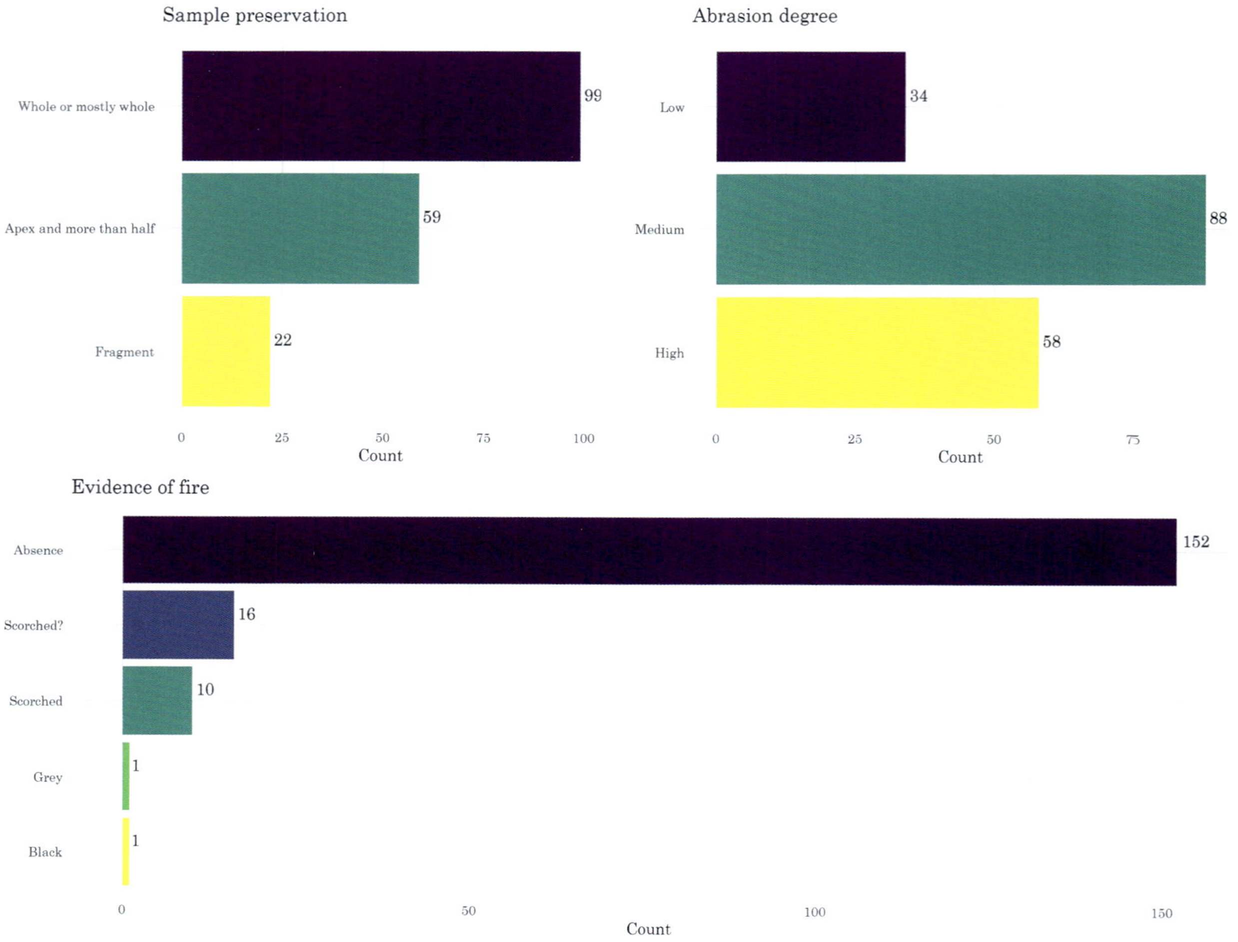

Fig. 32 Three bar plots illustrating observations recorded regarding sample preservation, abrasion degree and evidence of fire by number of samples.

3.3.3. Summary

The selection of the material aimed to ensure a comprehensive representation of the shell assemblage from Franchthi Cave and the Paralia site, with a particular emphasis on the Early Neolithic period. The morphometric, depositional and post-depositional characteristics of the samples across different chronological periods were recorded in detail. The degree of preservation of the samples showed that most shells were well-preserved with limited evidence of burning. The detailed recording of the shell assemblage serves as a basis for the subsequent elemental and isotope analyses of the samples.

Fig. 33 Examples of the different states of preservation of the archaeological samples: **16** Burnt specimen (FAN_131_A). – **127** Highly abraded specimen (Q5N_55_B). – **177–180** Fragmented specimens (Q5S_230_X-AA). – Remaining numbers: Specimens in a standard state of preservation.

Chapter 4: Results

4.1. Modern Samples

After successfully collecting 131 modern shell specimens from locations in Greece, Italy and Libya, the samples were approached in two different ways:

Firstly, all of the samples were scanned with LIBS and Mg/Ca data were acquired for each specimen. In total, 400 elemental maps and 132 line scans from the 131 modern samples were produced. Initially, a scan of the entire shell section was created at a 100 µm resolution and afterwards in the cases where it was deemed necessary higher resolution maps at 30 µm were created for the calcitic extremities in order to detect seasonal cyclicity patterns. Line scans efficiently show seasonal cyclicity trends and help identify the season of collection recorded in the last increment of growth. All maps and line scans were taken into account in order to evaluate the results of each shell specimen.

Secondly, sequential oxygen isotope analysis was carried out on representative samples from locations of importance in order to validate the employed LIBS methodology. Different ways of calcium carbonate sampling extraction were tested depending on the available means. In total, 374 calcium carbonate samples were extracted for isotope analysis from 20 shell specimens (on average = 17, minimum = 1) (see **tabs A.5–A.6**). Furthermore, isotope analyses were run on water samples from six different locations. Based on the water sample oxygen isotope results, the Hays and Grossman (1991) equation for SST reconstruction and instrumental SST data, site-specific empirical offsets were calculated.

The isotope results were compared with the elemental data and the efficiency of the combination of the two methods was evaluated. Also, the elemental mapping of a large number of modern shells prior to the mapping of archaeological shells, allowed me to familiarise myself with elemental data and seasonal cyclicity patterns from known seasonal contexts in order to be in a better position to interpret the archaeological data.

4.1.1. Main Results

Both elemental and isotope results that are presented in this section, produced from the modern reference samples, were satisfactory. Seasonal cyclicity was reflected in an interpretable way, while in most of the cases the season of collection reflected in the elemental maps was in agreement with the known season of collection. Isotope results were also in accordance with the elemental results, with sequential milling $\delta^{18}O$ values resembling largely the elemental line scans.

Generally, Mg/Ca ratios followed a predictable pattern with higher values (0.5–2.0) in the calcitic parts of the shell and lower (0.01–0.3) in the aragonitic parts at the apex. Towards the exterior of the shell on the M+3 layer, Mg/Ca ratios were sometimes significantly higher (over 2.0) and these were filtered out of the elemental maps based on the relative standard deviation as they did not offer any significant information (e. g. **fig. 35**) (Hausmann et al. 2023, 10). Summer periods are often shorter than those in winter periods, indicating that shells do not grow or grow at a slower rate during the warmer months of the year.

»Good« elemental results means that seasonal increments are well-defined and easy to identify as such (i. e. winter or summer increments). Winter increments tend to fluctuate between 0.5 and 0.6 ratios of Mg/Ca and summer increments between 0.75 and 0.9. Autumn and spring increments are identified in the intermediate values. When molluscs are more than 1 year old, seasonal cyclicity can be usually identified and interpreted as such. Young fast growing molluscs as well as older are more difficult to be evaluated, with the first not exhibiting seasonal cyclicity patterns and the latter slowing down their growth as they get older.

Generally, ambiguous results were produced in the cases where shell growth was not sufficient and seasonal cyclicity could not be detected. In these cas-

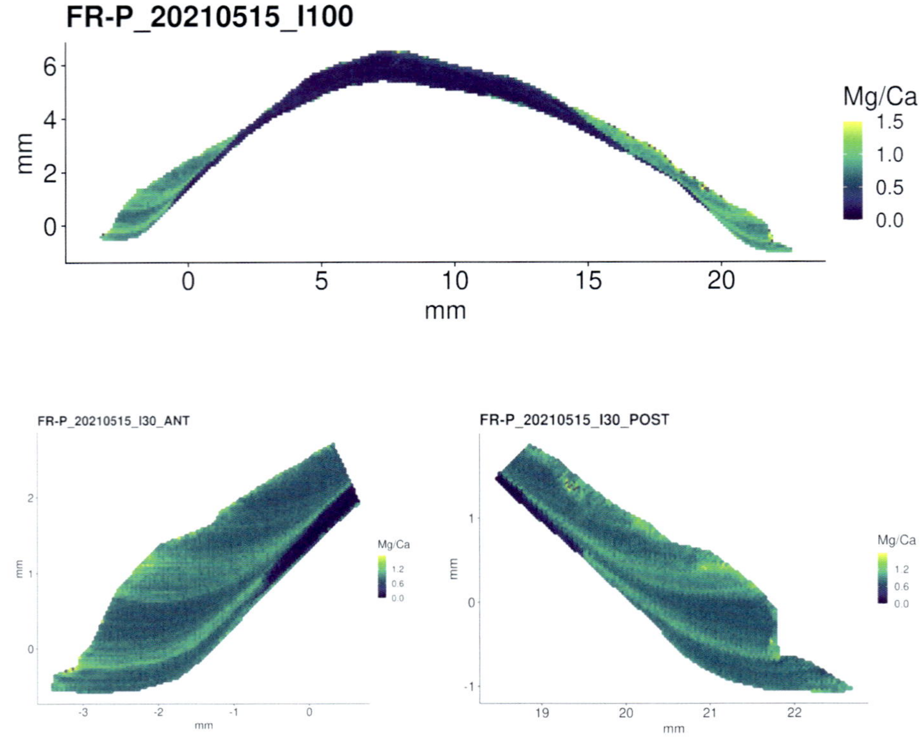

Fig. 34 A sample from Franchthi exhibiting well-defined seasonal cyclicity: Sample FR-P_I, 100 µm scan of the whole shell section and 30 µm scans of both calcitic extremities.

es, the shells were considered uninterpretable. Also, a small proportion of the shell specimens exhibited more than a year's growth.

In the following section, elemental and isotope results will be presented by location of collection in a concise manner. First, I will present the 92 modern shell specimens that are associated with current (Franchthi), past (Sousa), or future (Youra, Kyra Panagia, Tonnara del Secco and Favignana isl.) studies of archaeological shells from nearby collection sites.

4.1.1.1. Franchthi

In the case of the analysed shells from Franchthi, Mg/Ca ratios followed the standard patterns. Seasonal cyclicity in this large set of shells (n = 24) was not always well-defined as seasonal alternation was not straightforward (FR-P_D, G, L, N, S) and Mg/Ca ratios often did not distinctly show seasonal patterns. This can be due to the shells' young age as 33 % of the shells recorded less than one year in their growth increments (FR-P_D, J, L, M, P, Q, U, V) and another 33 % presented growth patterns that could not be interpreted in terms of seasonal cyclicity (FR-P_B, N, O, R, S, T, W, X). Sample FR-P_I constitutes a »good« example that recorded three winters and two summers (**fig. 34**), where the different seasons are well-resolved and distinguishable, whereas this is not the case in sample FR-P_S (**fig. 35**). The latter does not exhibit any seasonal cyclicity patterns and elemental data are homogeneous not allowing for any meaningful characterisation.

Furthermore, the Franchthi modern shells collected from the seashore close to the archaeological site, were generally relatively small specimens (average length: 25.0 mm, average width: 20.7 mm), while the specimens that did not exhibit any seasonal cyclicity patterns were even smaller in average (average length: 22.6 mm, average width: 19.2 mm). A further instance of this is sample FR-P_X (see **Appendix I**), which is the smallest specimen of this group of shells (height: 18.5 mm, length: 16.1 mm). Indeed, elemental data from this shell do not reflect any cyclicity. Apart from their size, the shells from Franchthi are relatively young with an average age of 1.5 years, while growing relatively fast when considering their size.

In terms of the identification of the season of collection, in the case of the modern scanned shells from Franchthi (n = 24), 70 % of the shells reflected the actual season of collection (spring) and 13 % the preceding. Only two samples, characterised as

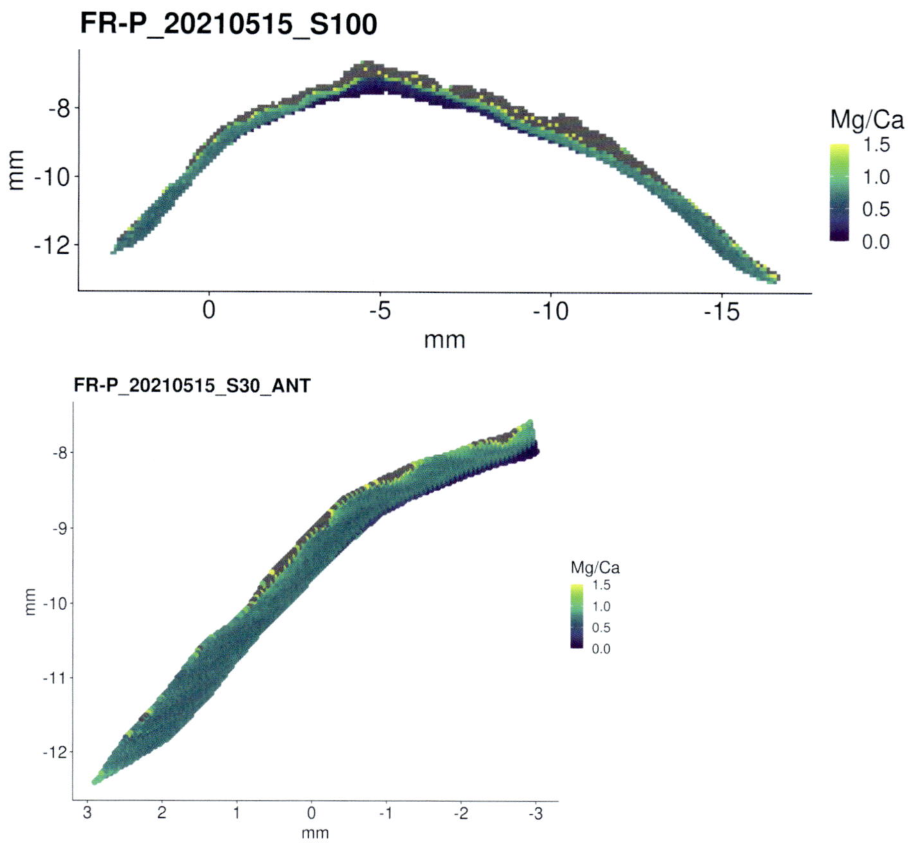

Fig. 35 A sample from Franchthi not exhibiting seasonal cyclicity: Sample FR-P_S, 100 µm scan of the whole shell section and 30 µm scan of the calcitic anterior part.

too young to be trustworthy, showed the subsequent season (specimens FR-P_N and O) and in the case of two other samples no determination about the seasonality could be made (specimens FR-P_W and X).

Regarding the isotope analysis, ten modern shells from Franchthi Paralia (FR-P_A to J) were analysed using oxygen isotopes. As explained in Chapter 3, these shells were analysed using two different methods: drilling and micromilling.

4.1.1.1.1. Testing with a Minimal Toolkit

First, the specimens from Franchthi (FR-P_A to D) sampled via drilling, produced quite rudimentary results (**fig. A.26**). In all four shell samples, edge values between the isotope and the elemental results were similar and (with the exception of sample C) seasonal curves resemble broadly.

4.1.1.1.2. Micromilling

Subsequently, four modern samples from Franchthi were sampled for isotopes via micromilling (FR-P_C, G, H and I) and produced notably more refined results (**fig. 36**). The overall data trends of the isotopes match those of the elemental data and the known season of capture (spring) is well-reflected on both sets of data. Time-averaging is present in the isotope micromilling results, especially in samples C and G. This is due to the nature of the sample as the sampled surface was very small (approximately 3.9, 3.45, 1.2 and 1.95 mm respectively), as well as my then limited experience with the micromilling method.

4.1.1.2. Youra

Elemental maps from the location of the island of Youra provided data that reflected a clearer picture of seasonal cyclicity compared to those from Franchthi. The 20 samples collected belong to two groups, south and west, as they were collected from two different areas (small docks) of the uninhabited island. To begin with, six of the shells from Youra exhibited an unusual calcitic distribution, namely six of the shells had very small calcitic layers (Y-W_D, E, F, G, I, Y-S_D), resulting in a very compacted picture of seasonal cyclicity. Seasonal variation is so condensed in some of the cases and becomes visible only at the 30 µm resolution maps (e. g. Y-W_E). Second, the high Mg/Ca increments related to summer periods

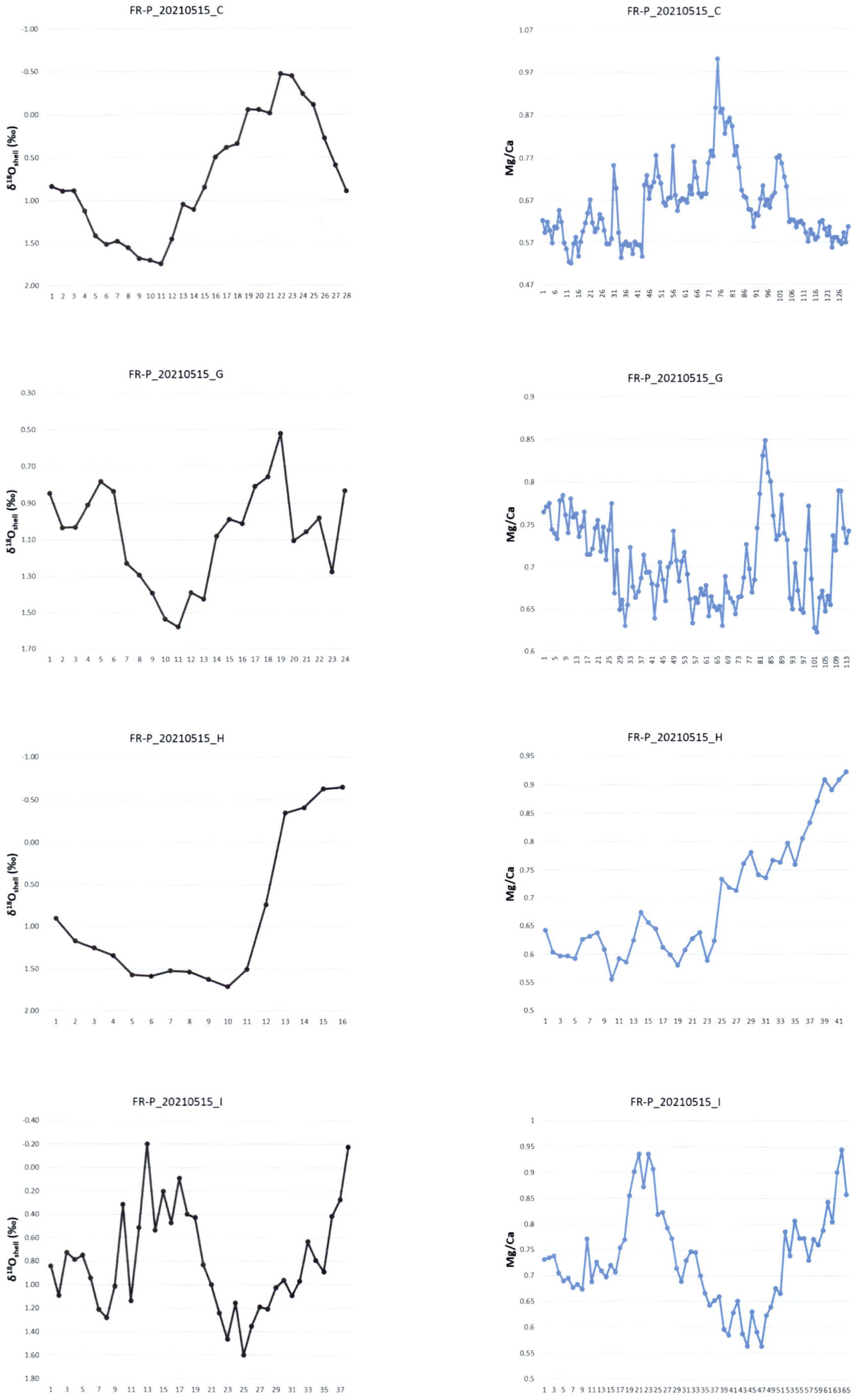

Fig. 36 Samples from the shell edge. Isotope results (micromilling) of samples FR-P_C and G–I compared to elemental line scans.

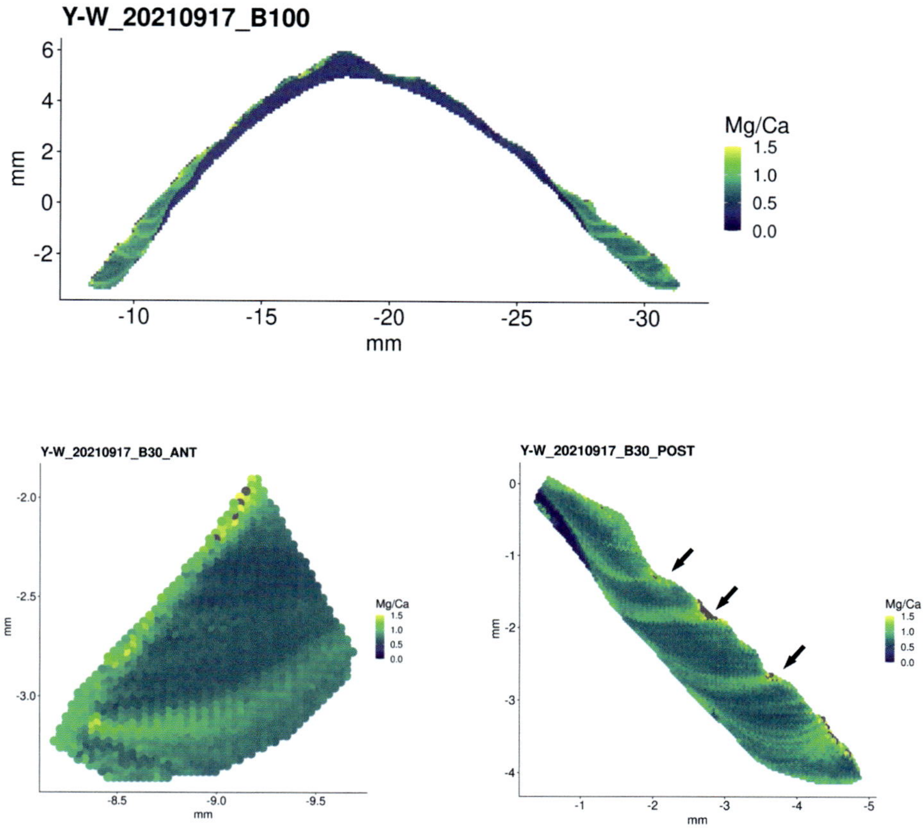

Fig. 37 A sample from Youra exhibiting well-defined seasonal cyclicity: Sample Y-W_B, 100 µm scan of the whole shell section and 30 µm scans of both calcitic extremities. Arrows show the position of the growth stops.

in this group of shells are shorter than those in winter periods, indicating that shells did not grow significantly during the warmer months compared to the colder months of the year during.

A long-lived and easy to interpret specimen from this location is sample Y-W_B, which recorded five summers and five winters (**fig. 37**). The summer increments of sample Y-W_B coincide with growth stops visible on the curved shape of the shell exterior at the calcitic parts.

In contrast, specimen Y-W_E is an example where the seasonal cyclicity cannot be easily interpreted because seasonal increments are too densely grouped (**fig. 38**). Interestingly, seasonal increments are not identical on the posterior and the anterior of the shell. On the anterior side of the shell four summers are approximately counted while on the posterior almost five or more. This may mean that the growth rate is relatively low and that the analytical resolution is not high enough and some years are too close to differentiate.

Overall, the shell specimens from Youra are slowly growing shells as well as older than other specimens in my collection. The average age of the Youra shells is 3.6 years, which is the highest of the modern groups of shells studied. Moreover, the average length of the Youra shells is 23.1 mm, while the average width is 18.3 mm. I mention this, in order to highlight the fact that age is not analogously related to size, as the Franchthi shells for example (with average age of 1.5 years) are larger (by 1.9 mm in length and 2.4 mm in width) than the shells from Youra (**tab. A.4**).

Regarding the identification of the season of collection, results were quite positive as 85 % of the shells from Youra (n = 20) indicated the actual season of collection (summer), as August and September are the warmest months of the year in that area (**fig. 39**). Lastly, no differences were noted between the shells from the southern and the western dock.

The fact that these limpets are among the oldest specimens of this study, could be due to the fact that Youra is an uninhabited island in an ecologically protected marine zone (the National Marine Park of Alonnisos Northern Sporades) where human intervention is nowadays minimal. As the fauna in the area is protected, no or minimal exploitation of the marine ecosystem and its resources takes place,

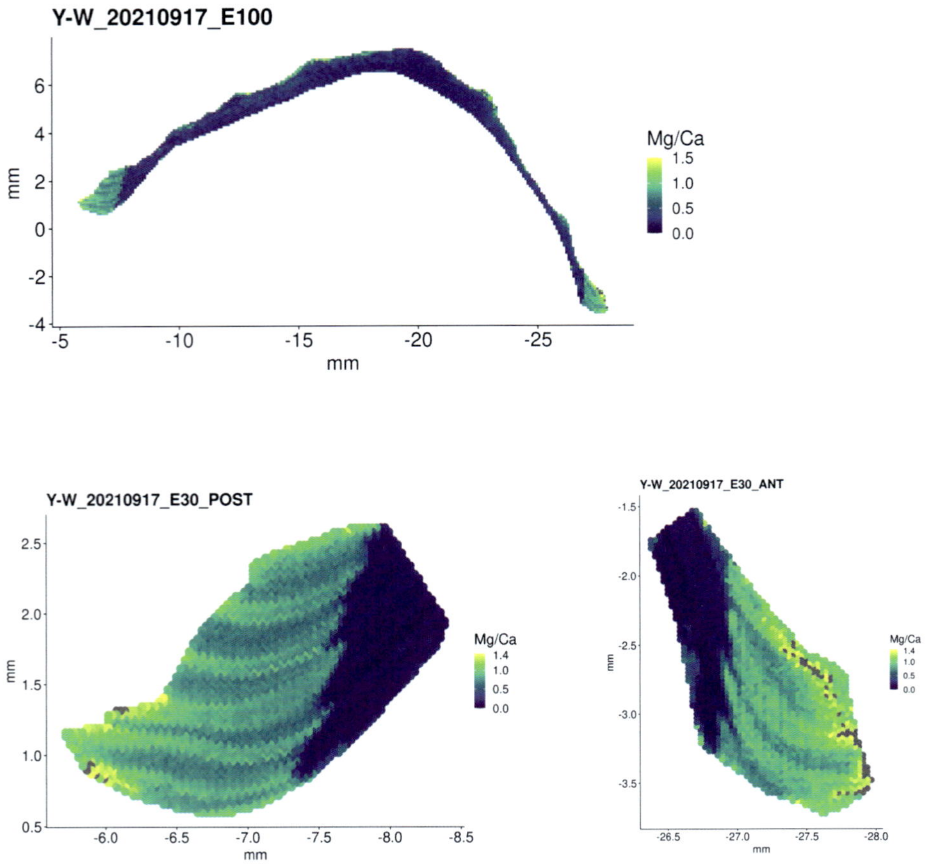

Fig. 38 A sample from Youra exhibiting unclear patterns of seasonal cyclicity: Sample Y-W_E, 100 µm scan of the whole shell section and 30 µm scans of both calcitic extremities.

allowing for populations to grow in an undisturbed manner. It may also mean that predation by other animals is minimal (e. g. Kowalewski et al. 2014).

Isotope results from the samples from Youra (**fig. 40**) are less refined when compared to their respective elemental line scans. This is due to the

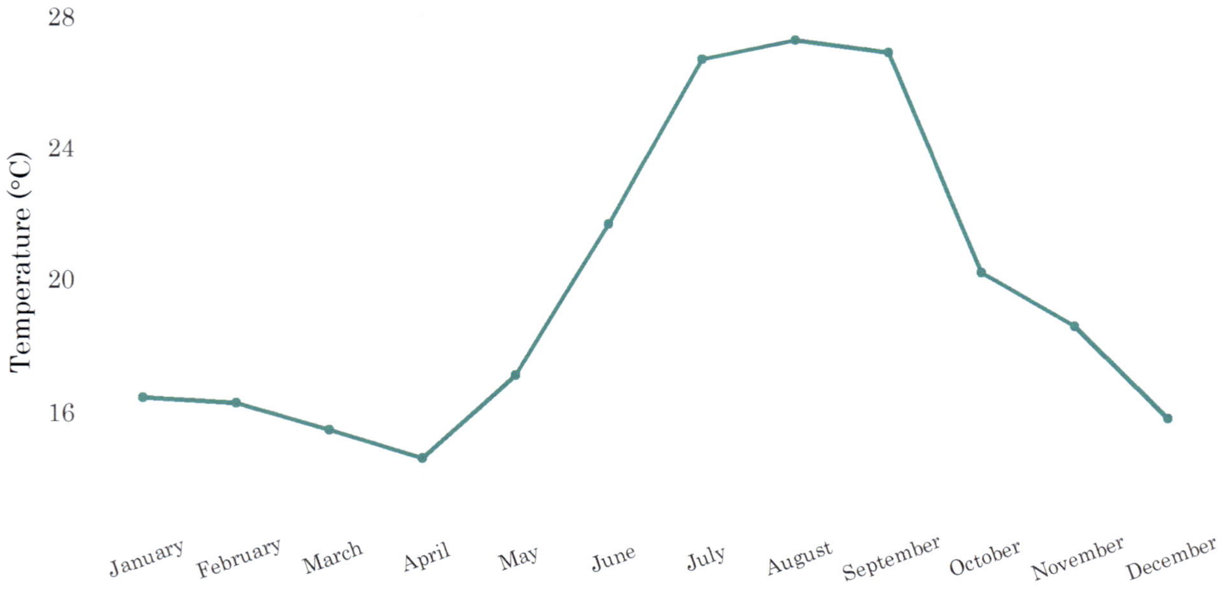

Fig. 39 Monthly Sea Surface Temperature (SST) instrumental data on the island of Youra from 2021. – (Data retrieved from the Copernicus Marine Service, copernicus.eu).

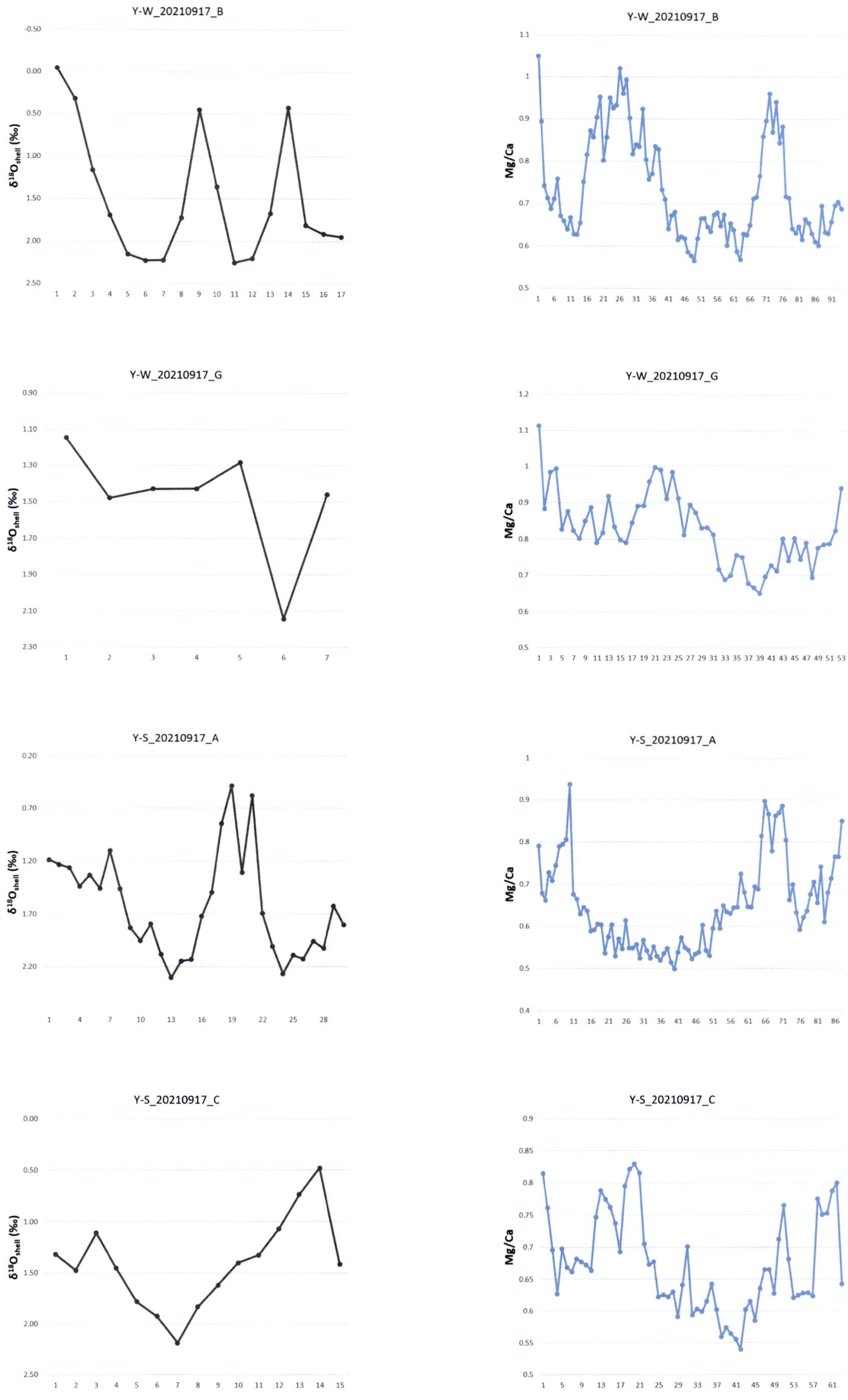

Fig. 40 Isotope micromilling results of samples from the island of Youra (Y-W_B and G, Y-S_A and C) compared to elemental line scans.

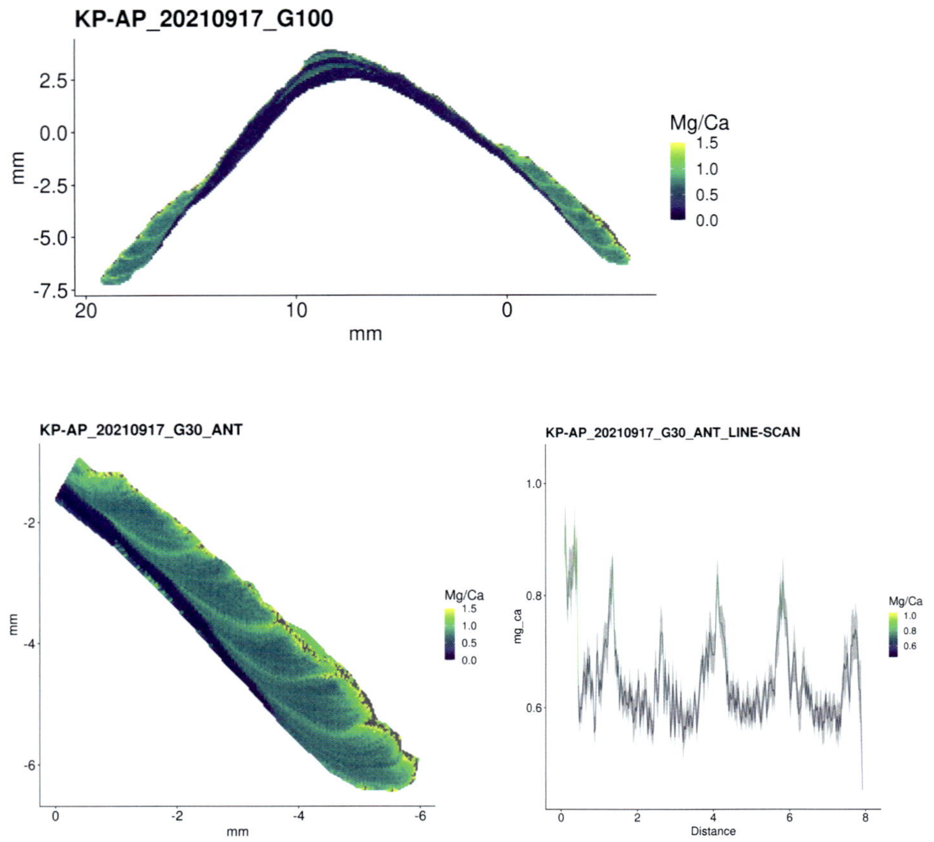

Fig. 41 An indicative example from Agios Petros, Kyra Panagia exhibiting well-defined seasonal cyclicity: Sample KP-AP_G, 100 µm scan of the whole shell section, 30 µm scan of the anterior side and a linear scan.

very restricted calcitic area of these shells where the sampling took place. Seasonally resolved increments were compacted together in an area of a few millimetres, rendering the micromilling sampling more challenging. Time-averaging is evident in these samples, especially in samples Y-S_C and Y-W_G, where the sampled surface of the latter had a length of only 1.5 mm. Isotope results of samples Y-W_B and Y-S_A, compare better with the elemental line scans, and demonstrate, even coarsely, seasonal cyclicity patterns. Additionally, all but one sample (Y-S_C) reflect in the $\delta^{18}O$ values high summer temperature that correspond to the mid-September warm temperatures.

4.1.1.3. Kyra Panagia

The modern shells (n = 15) from the tiny uninhabited islet of Agios Petros in the immediate vicinity of the island of Kyra Panagia all produced clear results (**fig. 41**). Seasonal cyclicity is very well-reflected on almost all of the elemental maps from this location (with the exception of specimen KP-AP_O). Summer periods are visible and abrupt, as they are characterised by thin and intense peaks on the linear scans and thin distinct lines on the elemental maps.

All shells clearly indicated the actual season of collection (summer) as they were collected in mid-September when the highest SSTs occur in the region. Among these, 13 % of the shells indicated an autumn or summer/autumn season of collection which is valid for this liminal time of the year and the rest of the samples indicated a straightforward summer season of capture. Specimens from this location were also relatively long-lived with a mean age of 4.3 years of age, while being also relatively large (mean length: 25.8 mm, mean width: 21 mm).

A second location from this area was the beach of Kokkino situated on the Agios Petros bay of the island of Kyra Panagia. The four shells collected from this location also exhibited seasonal cyclicity but were generally younger (mean age: 2.3 years) and smaller specimens (mean length: 22.8 mm, mean width: 19.8 mm) in comparison with the shells from Agios Petros (**fig. 42**). This could be due to the fact that the Kokkino location is near a dock often frequented by small vessels and thus the environment

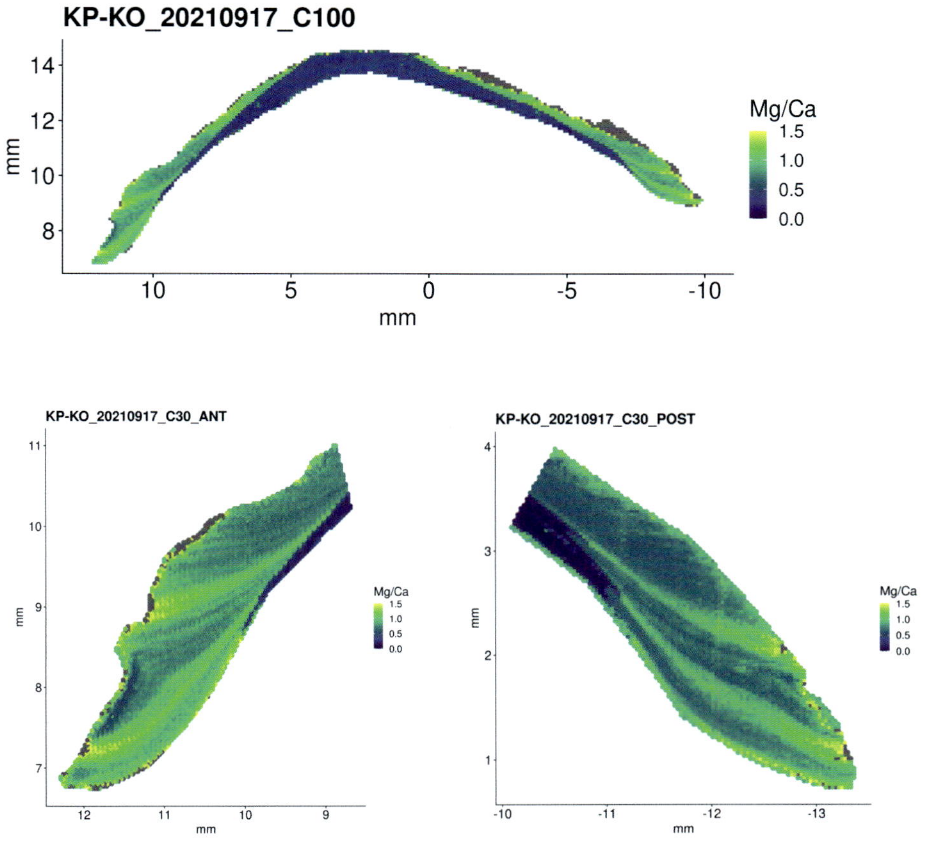

Fig. 42 A sample from Kokkino exhibiting well-defined seasonal variability: Sample KP-KO_C, 100 µm scan of the whole shell section and 30 µm scans of both calcitic extremities.

is more prone to the negative outcomes of human activity. Seasonal cyclicity was not as pronounced as the limpets from the nearby islet of Agios Petros. In general, elemental results from this location were not very consistent, as only half of them indicated the actual season of collection (summer), while the other half recorded seasonal variation with a delay as they reflected a spring/summer season of collection.

Furthermore, I scanned six shells with LIBS to test a different elemental ratio, that of strontium to calcium (specimens KP-KO_A, KP-KO_B, V-P_B, Y-S_A, Y-S_H, Y-W_B). For this purpose, the Sr/Ca elemental maps were generated by targeting the Sr II emission line at 407.771 nm and the Ca II emission line at approximately 373.690 nm. One of those was the specimen KP-KO_A from Kokkino (**fig. 43**). Unfortunately, no seasonal variation was detected when following this approach. Slightly higher Sr/Ca values were visible in the area of the apex, whereas no variation could be detected on the calcitic parts of the shell. For this reason, I decided not to proceed with further scanning of the shells for strontium.

4.1.1.4. Tonnara del Secco

Shell specimens from Tonnara del Secco in northwest Sicily provided seasonally-resolved elemental maps, however seasonality patterns were not always distinct. Seasonal increments were more extended and the summer periods were not pronounced or thin as in shells from other locations (**fig. 44**). From the eight shells that belong to this group, two of them did not reflect the actual season of collection (25 %, TS-E_E and TS-W_C), while some specimens produced ambiguous results, as is the case with specimen TS-E_E for instance. This is further enhanced by the fact that all of the shells are quite young (mean age: 1.1 years). Furthermore, the samples had a mean length of 37.2 mm and a mean width of 32.5 mm. Once again, size does not appear to be directly analogous to age.

A common feature of this group of shells, which were collected in mid-May, was that in many cases both the winter increments were broad, as were the summer increments, suggesting that summer growth cessation was not common. Moreover, higher Mg/Ca ratios on the very edge increments were vis-

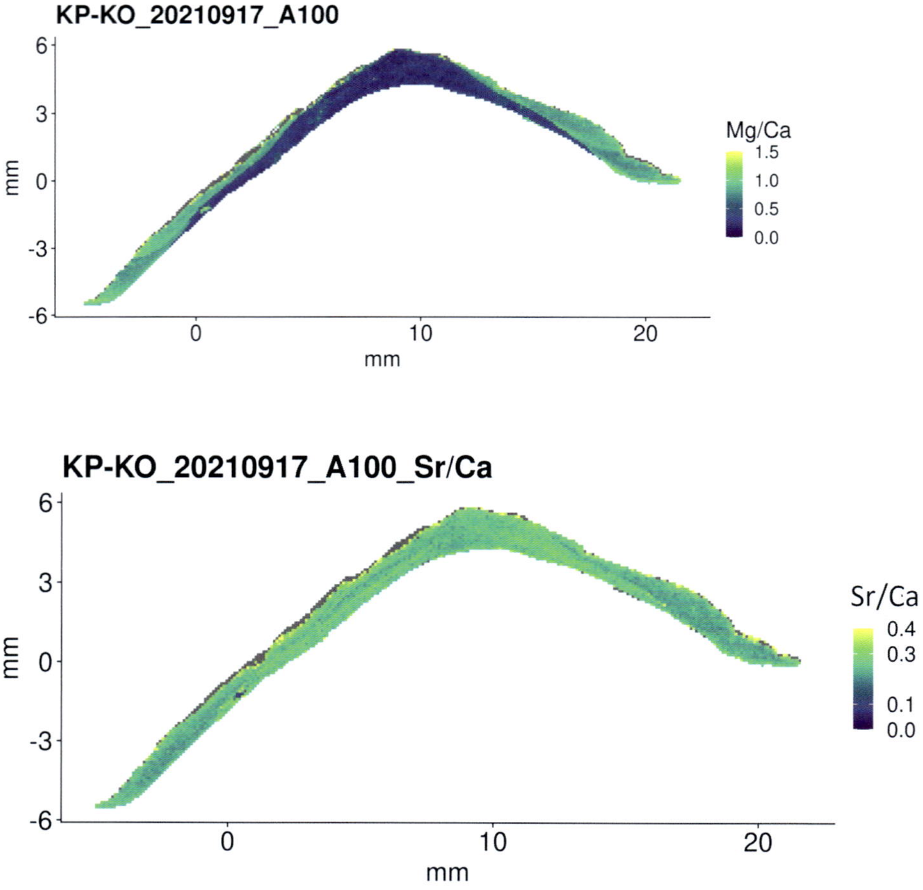

Fig. 43 A Mg/Ca 100 µm and a Sr/Ca 100 µm scan of specimen KP-KO_A from Kokkino where no seasonal cyclicity could be identified in the latter.

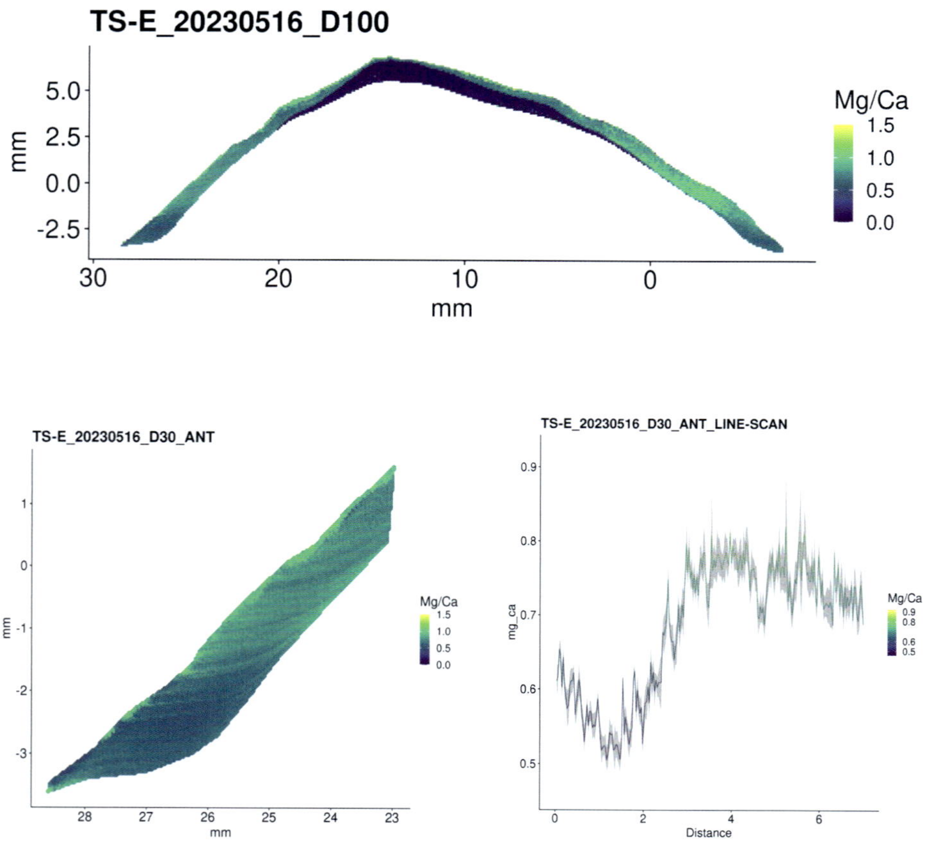

Fig. 44 A representative example from Tonnara del Secco with extensive increments of seasonal cyclicity: Sample TS-E_D, 100 µm scan of the whole shell section, 30 µm scan of the anterior part and the linear scan of the same part.

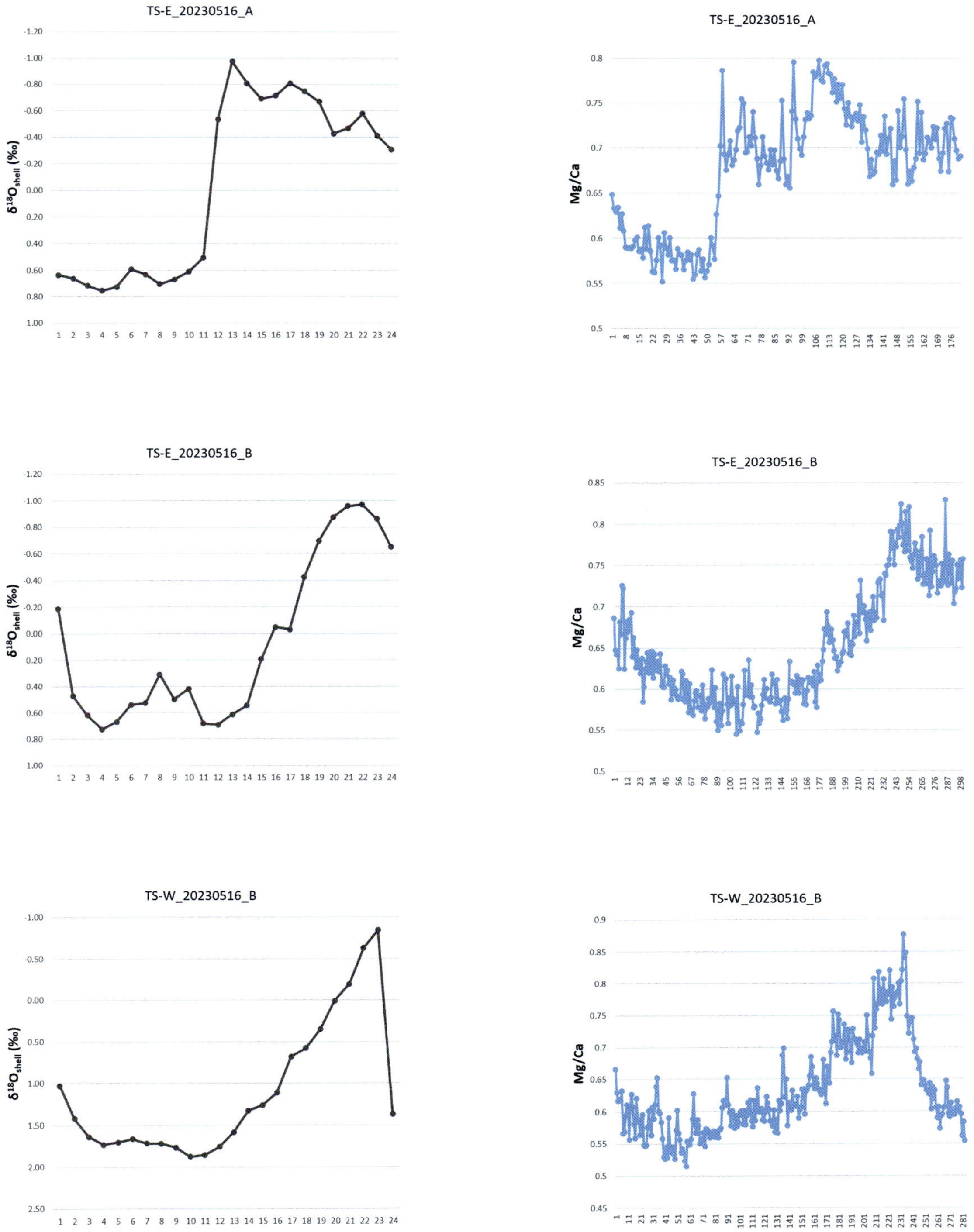

Fig. 45 Isotope micromilling results of samples from the Tonnara del Secco location in Sicily (TS-E_A and B, TS-W_B) compared to elemental line scans.

ible in some of the specimens or the higher values were reflected more clearly on the linear scans (TS-E_B, TS-E_D, TS-W_B).

Regarding the isotope results from this location, they compare very well with the elemental line scans. From the three samples that were analysed using micromilling, samples TS-E_B and TS-W_B reflected the expected rising spring temperatures (**fig. 45**). Edge values of sample TS-E_A are quite low indicating winter season of collection, which is not the case. This means that the last increment of growth was not accurately captured.

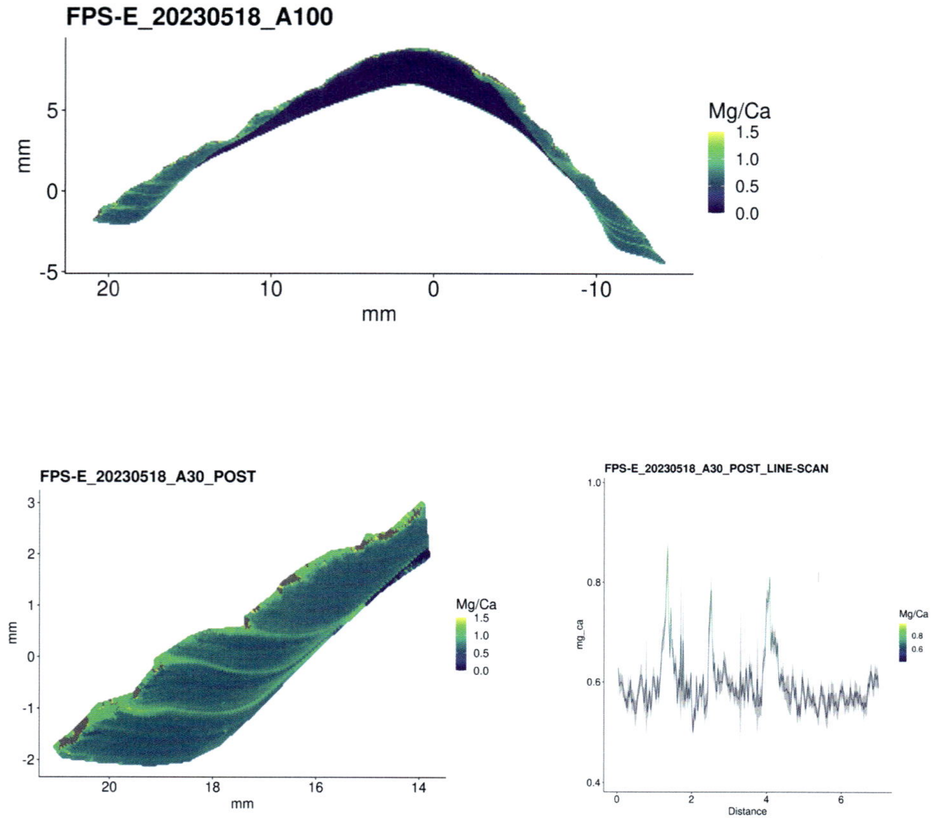

Fig. 46 A sample from Faro Punta Sottile East exhibiting well-defined seasonal cyclicity: Sample FPS-E_A, 100 µm scan of the whole shell section, 30 µm scan of the posterior calcitic edge and the linear scan of the same edge.

4.1.1.5. Faro Punta Sottile

Collection from Faro Punta Sottile on the island of Favignana took place on two different sides of the small homonym bay and for this reason samples are divided into samples from the east and the west side of the bay. All but one (FPS-E_B) of the specimens (n = 9) from Faro Punta Sottile gave seasonally interpretable results reflecting the actual season of collection (spring). The shells from this location recorded only a few years of growth (1–2 years), with the exception of sample FPS-E_A that was 4 years old, which recorded distinct short summer periods alternating with larger winter growth increments (**fig. 46**).

In addition, differences were noted among the shells of these two areas: In the case of samples from the east, seasonal increments of both winter and summer were quite broad (FPS-E_C and D) and easy to securely identify with the winter increments being still more extensive, a common feature in all of the analysed shells (**fig. 47**). Nonetheless, this was not the case in all of the five specimens (FPS-W_A to E) from the west area of the bay which were younger in age (mean age: 1.5 years) and their growth increments were more ambiguous. This group of shells are characterised by large winter increments that are visible on the elemental map, where the summer increments are often absent or very short. Even so, partial seasonal cyclicity is more pronounced on the linear scans and in all of them a warming trend is easily noticeable (e. g. FPS-W_C).

Oxygen isotope results of the Faro Punta Sottile group of samples compare well with the elemental data (**fig. 48**). In all of the cases isotope data agree with the elemental ones demonstrating similar seasonal cyclicity patterns. In samples FPS-E_C and D, the edge value is time-averaged as the isotope values are not as high as expected compared to those that follow. However, the oxygen isotope edge values of all three samples are not very different from each other (1.48, 1.85, 1.55 ‰). All three samples reflect the known season of collection (spring).

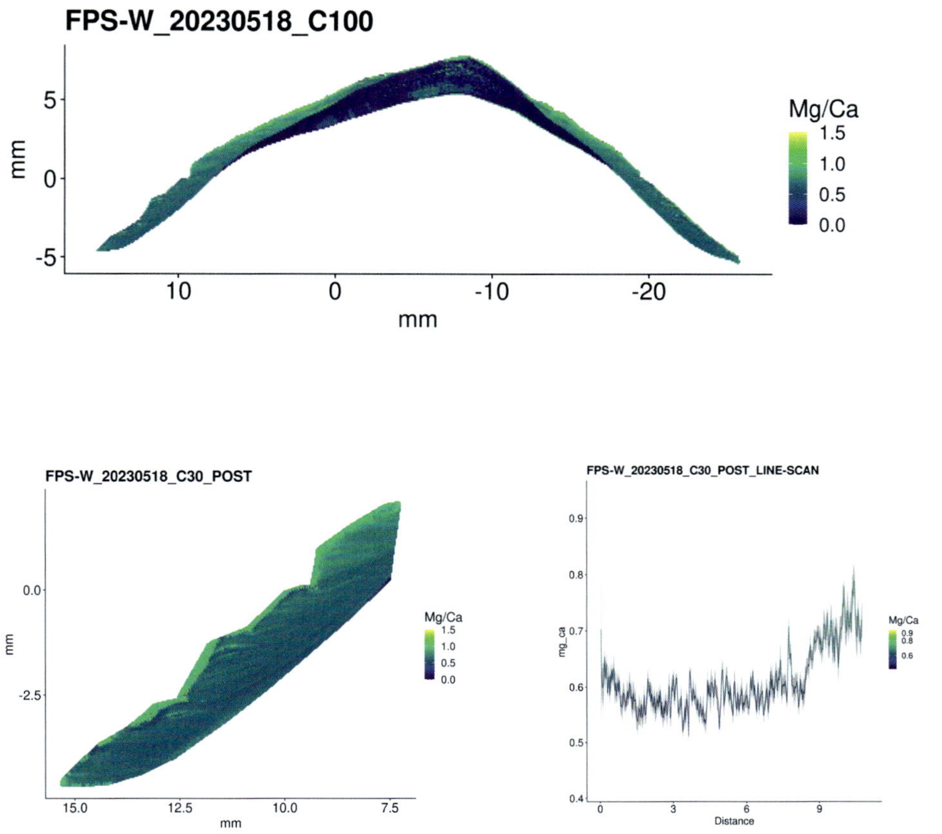

Fig. 47 A representative short-lived example from Faro Punta Sottile West with an extended winter increment: Sample FPS-W_C, 100 µm scan of the whole shell section, 30 µm scan of the posterior calcitic edge and the linear scan of the same edge.

4.1.1.6. Sousa, Libya

Specimens from Sousa (n = 12) were collected from a shore in the vicinity of the archaeological site of Haua Fteah, as part of the Cyrenaica Prehistory Project (CPP) by A. Prendergast. They bear three different code names: MP6 (n = 8), MO (n = 2) and MBRP (n = 2) and had been previously sampled for isotope analysis. Samples were collected from mid-February to mid-September of 2012 (Prendergast/Schöne 2017, tab. 1). This set of samples had been fixed with epoxy resin on glass slides and most of them were still partially preserved following the previous isotope study, as seen on the elemental maps. All of the samples produced seasonally-resolved elemental maps. However, as context about all of the twelve samples is missing, I will focus on samples MP64A, MO31A, MP67A and MP68A, which were part of subsequent studies (Prendergast/Schöne 2017; Hausmann et al. 2019a), all of which produced seasonally-resolved elemental data.

To begin with, sample MP64A recorded pronounced summer periods and was around 4 years old. The elemental linear scan compares well with the oxygen isotope profile of the shell, while the linear scan covers a large portion of the calcitic part of the shell. For this reason, the fourth (or the first) year of growth is only exhibited on the linear scan and not on the isotope profile where the latest three years are shown. Sample MP67A was collected in August and summer growth increments are present both in the linear and the elemental map scans. When comparing the linear scan of MP67A with the oxygen isotope profile, the edge high Mg/Ca values visible on the linear scan (in accordance with the summer-time collection) are not reflected on the isotope profile. Moving backwards in time, a small section of a third year is reflected on the oxygen isotope profile, something that is not reflected in the linear scan, probably because the linear scan was not drawn further up towards the apex. Sample MP68A produced the linear elemental scan that is

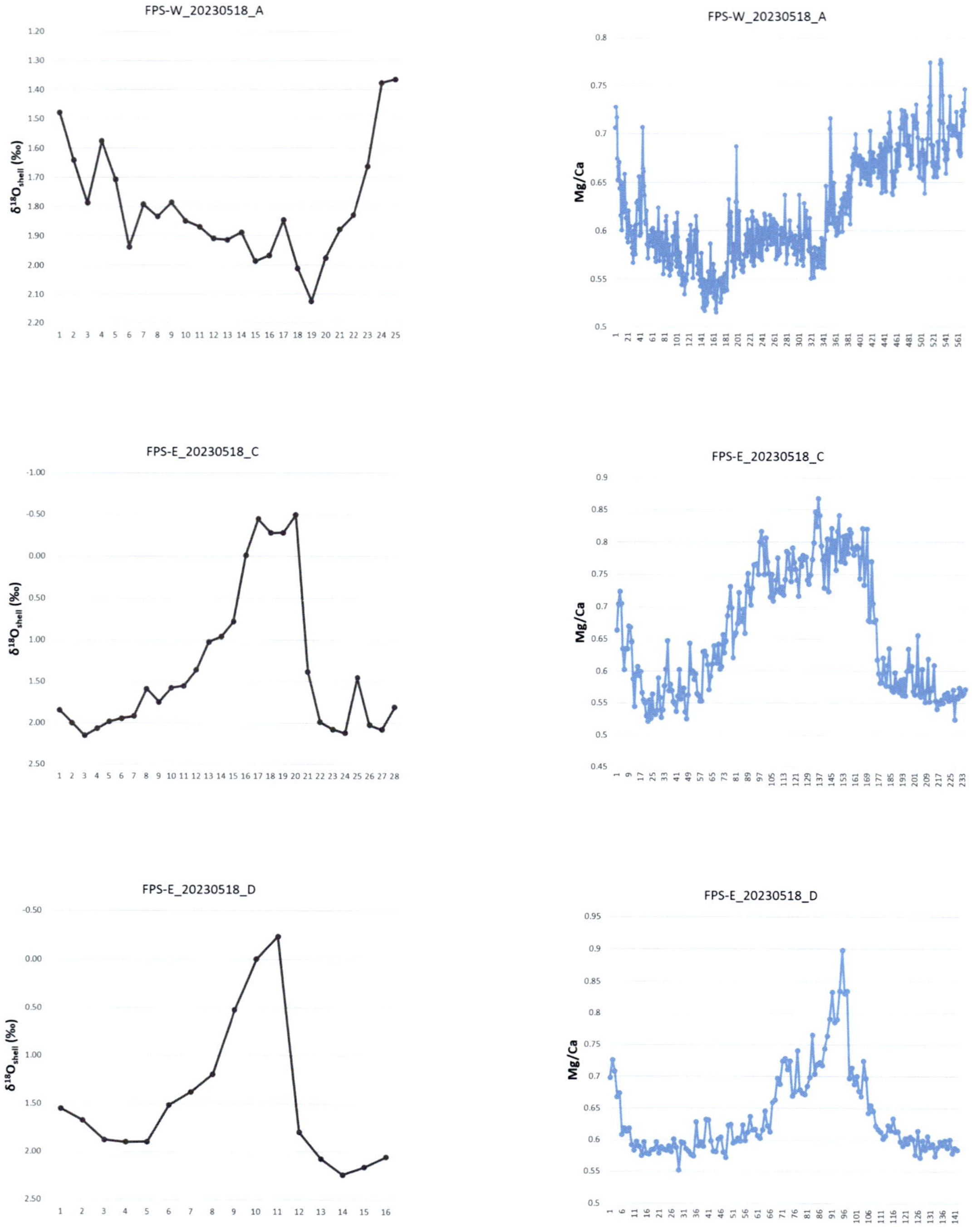

Fig. 48 Isotope micromilling results of samples from the Faro Punta Sottile location on the island of Favignana (FPS-W_A, FPS-E_C and D) compared to elemental line scans.

the least similar to that of the oxygen isotope profile. The sample was collected in September and the line scan reflects these high sea surface temperatures, however this is not reflected in the oxygen isotope profile, where edge values are evidently quite low (**fig. 49**).

4.1.1.7. Other Collection Sites

Following the specimens related to archaeological sites, I will present the rest of the elemental data from the north to the south of Greece. Twelve further locations are presented below that are not associated with archaeological shell specimens that have been

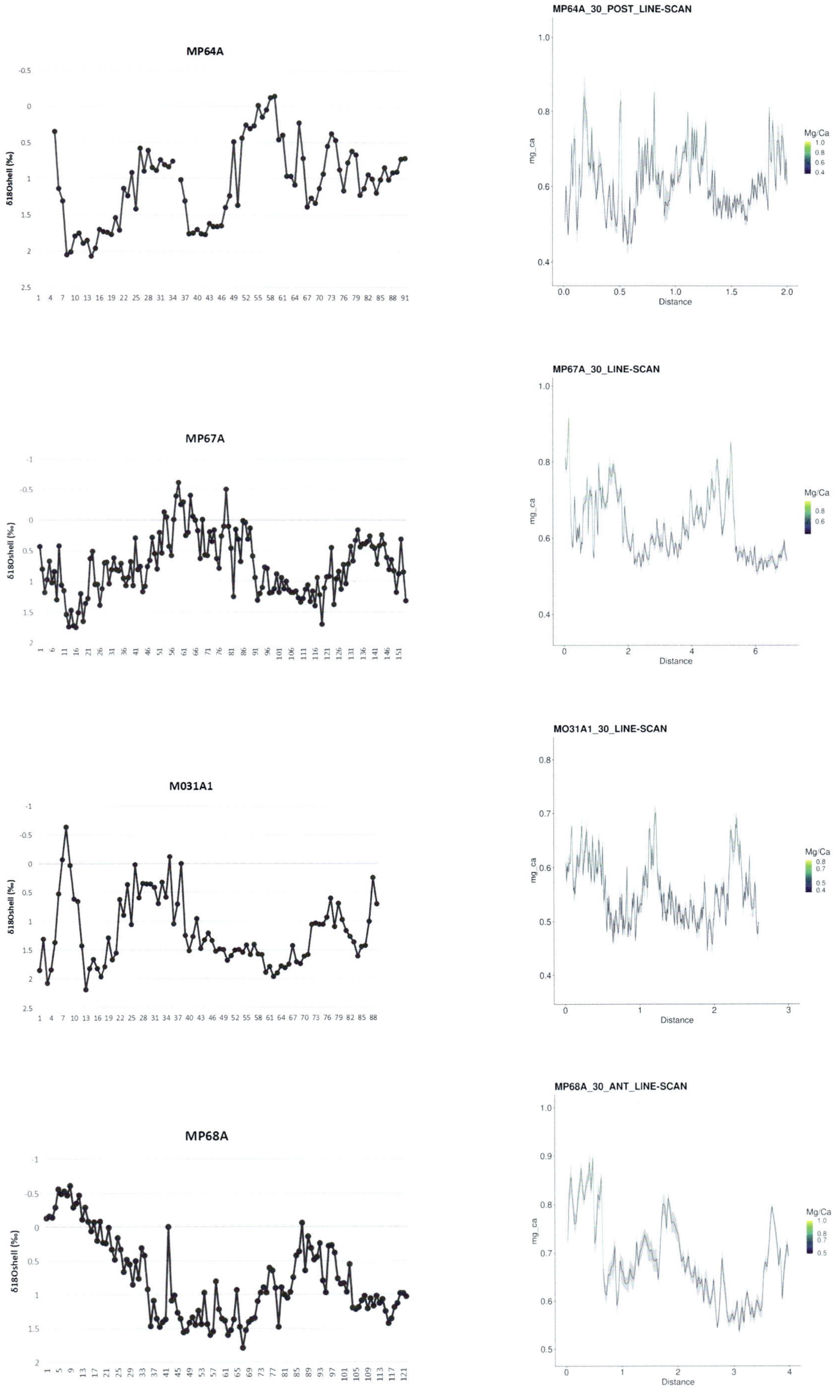

Fig. 49 Comparison of the oxygen isotope profiles of samples MP64A, MO31A, MP67A, MP68A with their respective elemental linear scans. – (After Prendergast/Schöne 2017, Supplementary data Appendix A).

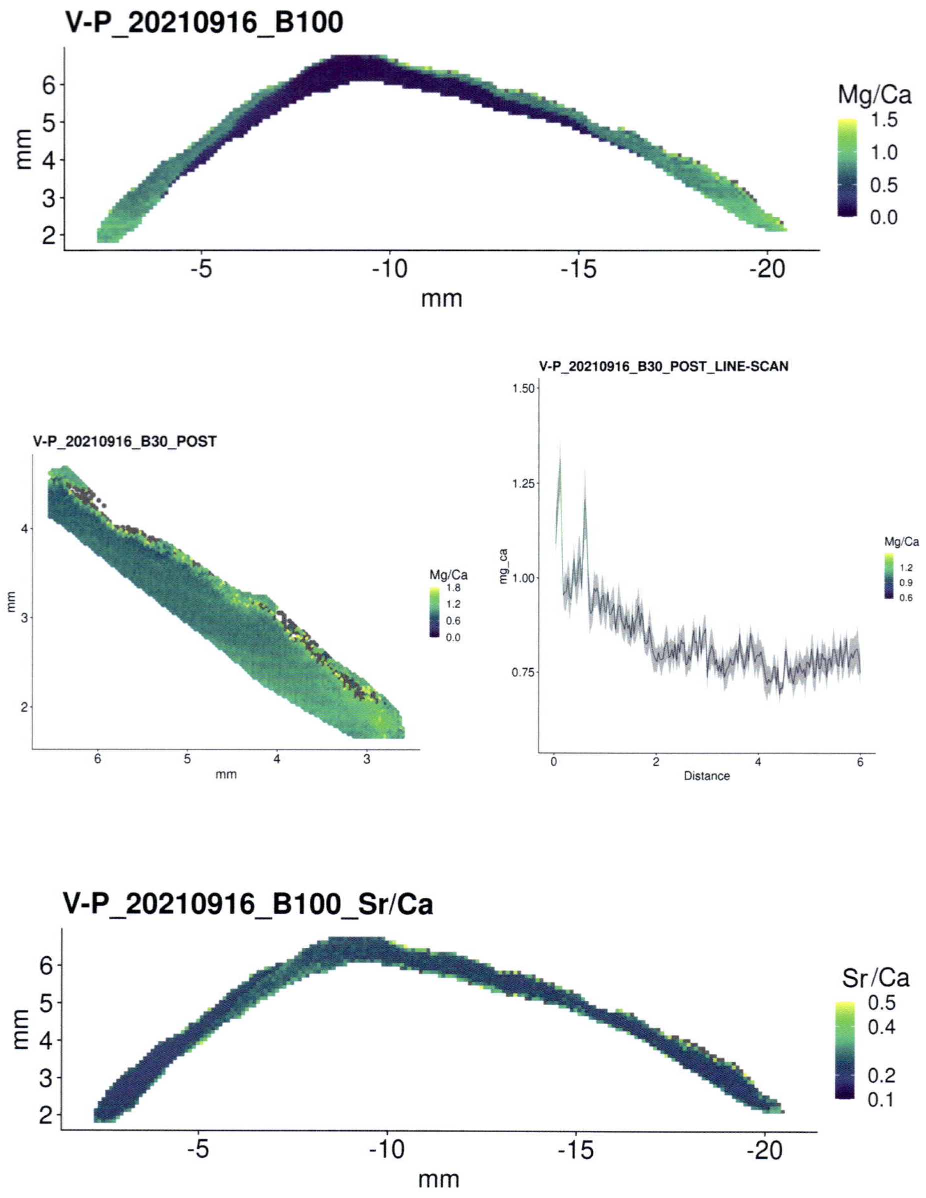

Fig. 50 A short-lived example from Plakes, Volos: Sample V-P_B, 100 µm scan of the whole shell section, 30 µm scan of the posterior calcitic edge, the linear scan of the same edge and a Sr/Ca 100 µm scan.

or will be analysed as part of this study. More specifically, I will present 39 shells from the locations of Pyrgos Apollonias near Kavala and Nea Kallikrateia in Chalkidiki, from Plakes in Volos, Koropi in the Pagasetic Gulf, Katiyorgis in mount Pelion, from Patitiri on the island of Alonnisos, from Alyko in Naxos and Ysterni in Paros, from Monemvasia in Laconia, as well as from Ligres, Glaros, Kedrodasos on the island of Crete.

4.1.1.8. Northern Greece

Starting from two locations in Northern Greece, I will present the shell samples from Pyrgos Apollonias in the vicinity of Kavala and Nea Kallikrateia in Chalkidiki. First, the shell specimens from Pyrgos Apollonias (n = 2), PARK_A and PARK_B were quite young specimens (less than 1 year old), even though they were not small in size (mean length: 31.1 mm, mean width: 22.7 mm). They both indicated the actual season of collection (summer/autumn), exhibiting high values with a cooling trend visible on the linear scan. They are both characterised by extensive summer and winter increments.

In the case of the limpets from Nea Kallikrateia in the region of Chalkidiki seasonal variability was very pronounced. All but one of the specimens (NKC-M_B) pointed towards a spring/summer collection with rising temperatures following extended winter increments. In addition, specimens were also in this

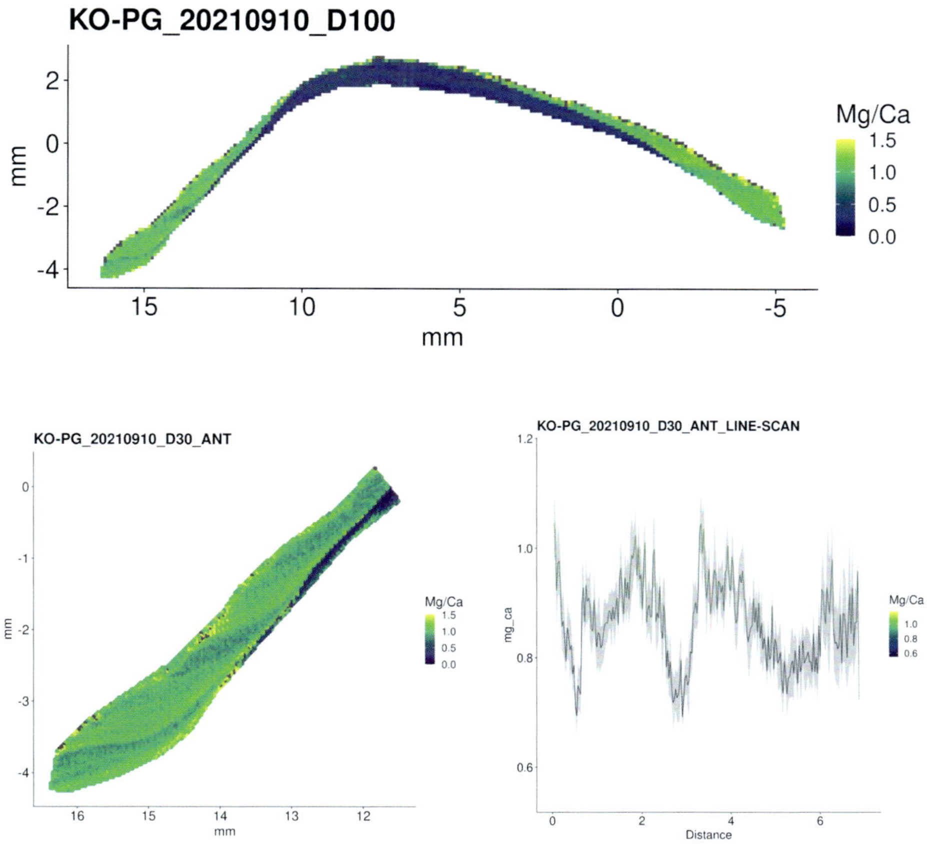

Fig. 51 Sample KO-PG_D from Koropi, Pagasetic Gulf with large summer increments: 100 µm scan of the whole shell section, 30 µm scan of the anterior and linear scan.

case quite young, with only one shell exhibiting just over a year's growth. Results of shells NKC-M_B and NKC-M_C were ambiguous and this could be due to their young age, which makes elemental data more difficult to interpret. This could also be due to the location, as the specific beach was quite shallow and making the mollusc organisms less prone to cold water influx or temperature changes.

4.1.1.9. Thessaly

Apart from specimens from the islands of Youra and Kyra Panagia described above, specimens were also collected from the location of Plakes (V-P) nearby the city of Volos, the beach of Koropi at the Pagasetic Gulf (KO-PG), the beach of Katiyorgis on the eastern side of mount Pelion (V-KB) and the location of Patitiri (PA), nearby the port of the island of Alonnisos. Shell specimens from Plakes (n = 3) collected in mid-September were short-lived, as all of them contained less than a year's growth. For this reason, the elemental maps did not indicate pronounced seasonality, especially in the case of samples V-P_A and V-P_C. Nonetheless, clear summer edge values indicative of the summer temperatures. A Sr/Ca elemental map was also produced for sample V-P_B, which did not reveal any seasonal cyclicity patterns. Slightly higher values were only noted in the aragonitic part of the apex (**fig. 50**).

Specimens from Koropi (n = 5) were collected in early September. This group of samples also revealed seasonality patterns in their elemental maps, even though most of them were short lived. Sample KO-PG_C was the only specimen that produced more ambiguous elemental maps that did not exhibit well-defined and distinct seasonal increments that could be interpreted. The rest of the samples pointed towards the correct season of collection displaying high summer edge increments. An interesting feature of this group of shells is their large summer increments, which tends to not be a common feature among the modern shell specimens. This may be due to the location being inside the closed and relatively shallow Pagasetic Gulf which is a large bay with relatively higher instrumental SSTs (copernicus.eu). An example of this is sample KO-PG_D (**fig. 51**), while the same is also visible on sample KO-PG_E.

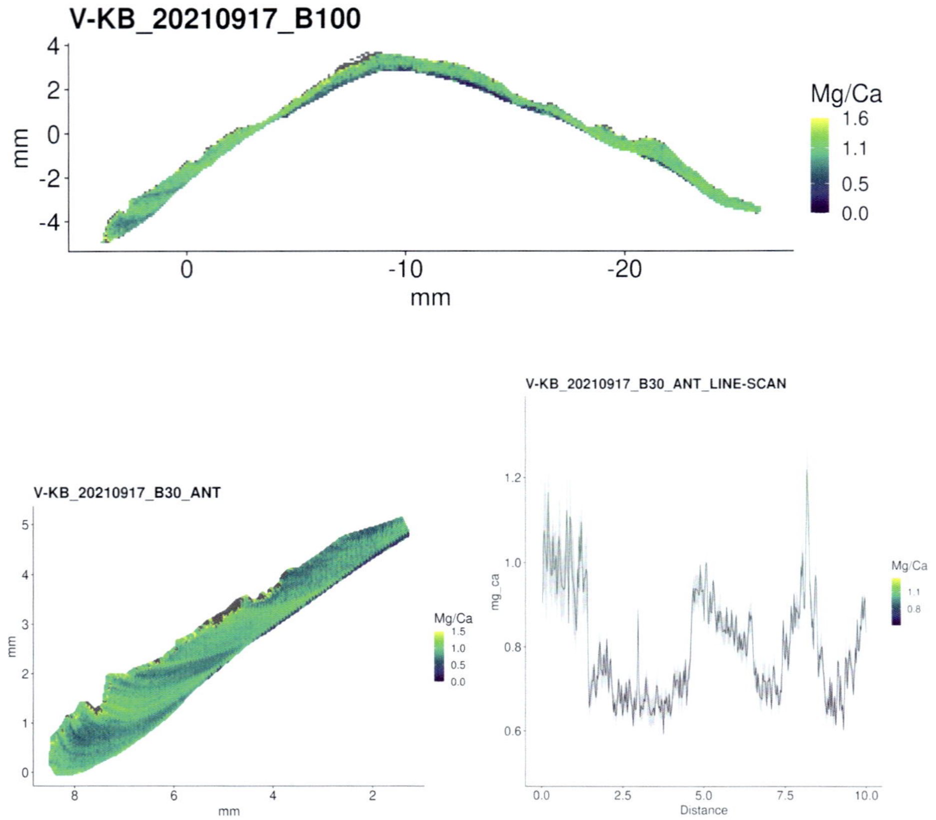

Fig. 52 A representative example from Katiyorgis where extended summer increments are recorded: Sample V-KB_B, 100 µm scan of the whole shell section, 30 µm scan of the anterior and linear scan.

Furthermore, the limpets from Katiyorgis on the eastern exterior side of the Pagasetic Gulf (n = 6) collected in mid-September produced well-defined seasonal cyclicity patterns (**fig. 52**). This set of samples was also characterised by relatively extensive summer growth increments (specimens V-KB_B, C, F). All of the samples pointed towards the actual season of collection (summer), with one specimen (V-KB_A) possibly presenting a cooling autumnal trend visible in the linear scan.

Elemental maps of the shell samples from Patitiri on the island of Alonnisos (n = 5) demonstrated poor seasonality patterns (**fig. 53**). Specimens were relatively small (mean length: 21.3 mm, mean width: 17.6 mm) and quite young to produce meaningful elemental results as four out of five specimens recorded less than a year's growth in their hard parts. A warming trend and high summer Mg/Ca ratios could be identified on the linear scans of all of the five specimens, however seasonal patterns were not trustworthy on the elemental maps as short-lived are difficult to interpret and any results cannot be considered as reliable.

4.1.1.10. Cyclades

Progressing further south, I will present two small groups of shells collected in August from two locations on two cycladic islands: Ysterni on the island of Paros (GYP) and Alyko on the island of Naxos (AN). In the case of the shells from Ysterni (n = 2), elemental results were ambiguous rendering their interpretation difficult. In the case of sample GYP_A, the high Mg/Ca ratios visible on the elemental map are not reflected on the line scan. It is a quite large specimen (length: 33.4 mm, width: 27.1 mm) and could record in its growth increments four years, however this estimation cannot be certain as seasonal patterns are not repetitive or consistent. The same was the case with sample GYP_B: A winter increment is visible, succeeded by several inconsistent summer growth lines, which probably represent one summer but are difficult to interpret.

Furthermore, the samples from Alyko (n = 3) also produced ambiguous results, as seasonal increments were unclear. Sample AN_C was the only one that reflected more consistent patterns of seasonal cyclicity (**fig. 54**).

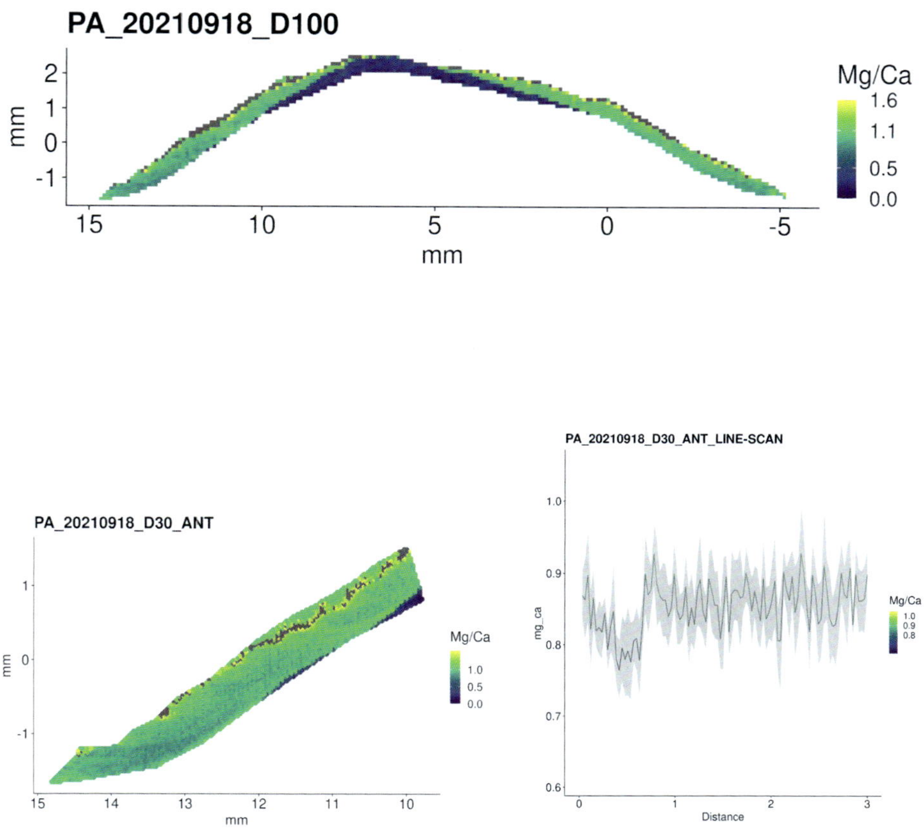

Fig. 53 A representative example from Patitiri: Sample PA_D demonstrating poorly-defined seasonal cyclicity patterns.

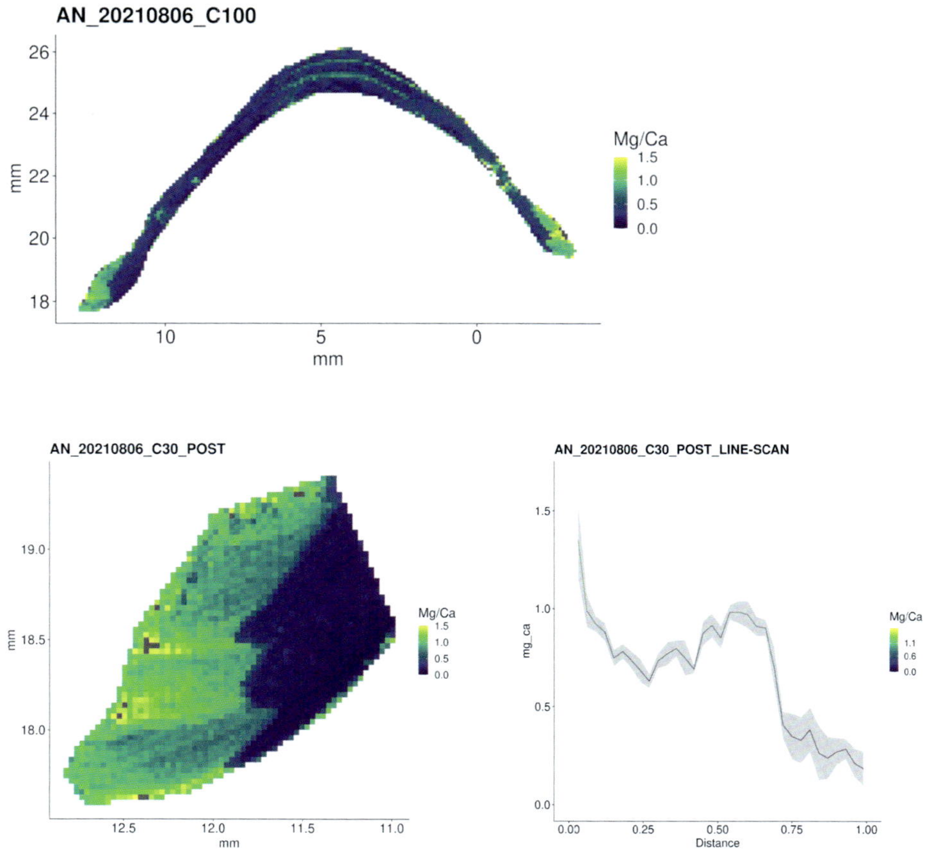

Fig. 54 An example from Alyko: Sample AN_C, 100 μm scan, 30 μm scan of the posterior, and linear scan of the posterior.

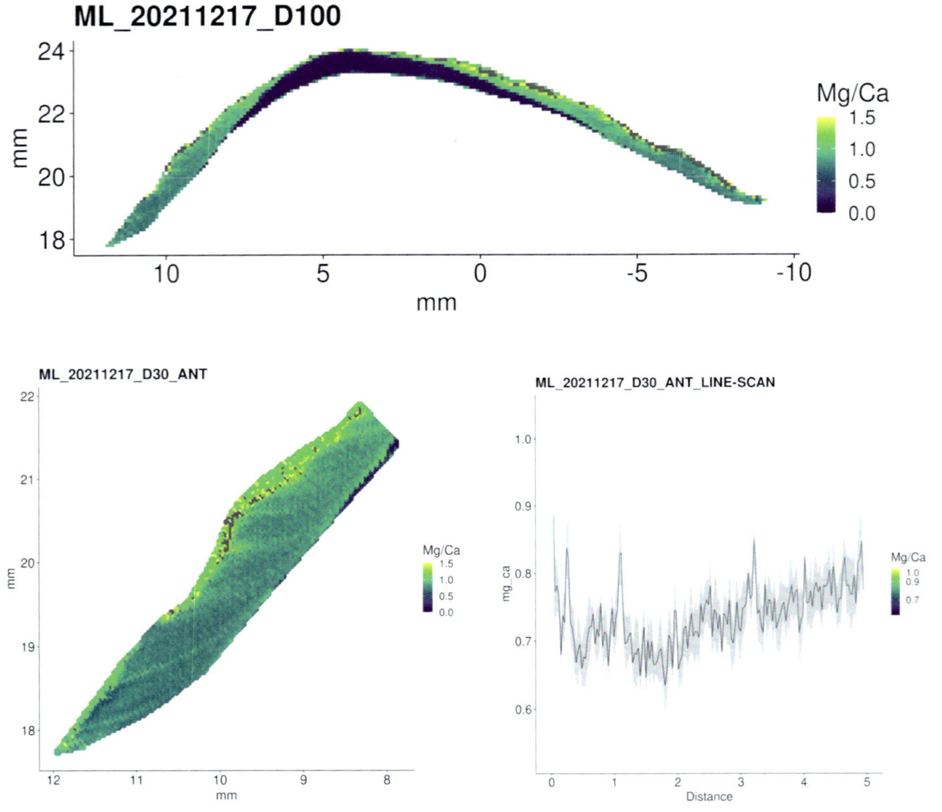

Fig. 55 An example of a young specimen from Monemvasia with homogeneous growth increments: Sample ML_D, 30 µm scan of the posterior and linear scan of the posterior.

4.1.1.11. Peloponnese

Apart from the samples from Franchthi, limpets were also collected from Monemvasia (n = 5) in the region of Laconia in mid-December. This was the only set of samples collected in winter, however as samples were quite short-lived (no specimens were more than 1 year old), only two specimens (ML_A and ML_B) indicated the actual season of collection. Patterns that could indicate seasonal SST variation were absent in the rest of the samples with elemental Mg/Ca maps being rather homogeneous. Specimens ML_C and ML_D (**fig. 55**) do show a few high Mg/Ca ratio lines, which nonetheless cannot be characterised as distinct summer lines with certainty.

4.1.1.12. Crete

The four specimens from the island of Crete that were analysed in this study come from three different locations: Kedrodasos in the region of Chania and Glaros and Ligres in the region of Rethymno, all collected during the summer. In the case of Kedrodasos, only one sample was analysed: K-CH_A. Even though the sample was short-lived (approximately 1 year old), it did reflect in its linear scan the high summer temperatures, while the elemental map did not present pronounced cyclicity patterns.

Regarding the specimen from Glaros, Rethymno, results were satisfactory. Specimen RE-G_A collected in early August, clearly reflected the summer season of collection both on the elemental 2D map and the linear scan (**fig. 56**). The interchange of the winter seasons is distinctly depicted on the elemental data rendering the season of collection estimation certain.

From the location of Ligres, in the region of Rethymno two samples were analysed. In both samples RE-L_A and B growth increments were puzzling and difficult to interpret. More specifically, sample RE-L did show high Mg/Ca ratios corresponding to the August collection, however seasonal growth increments were not consistent and well-defined in order to be securely evaluated (**fig. 57**). Various »summer« peaks are recorded in the linear scan, but the winter intervals are missing in order to interpret them as such. As for sample RE-L_B, potential summer increments seen on the elemental map were not well-defined and were thus difficult to identify.

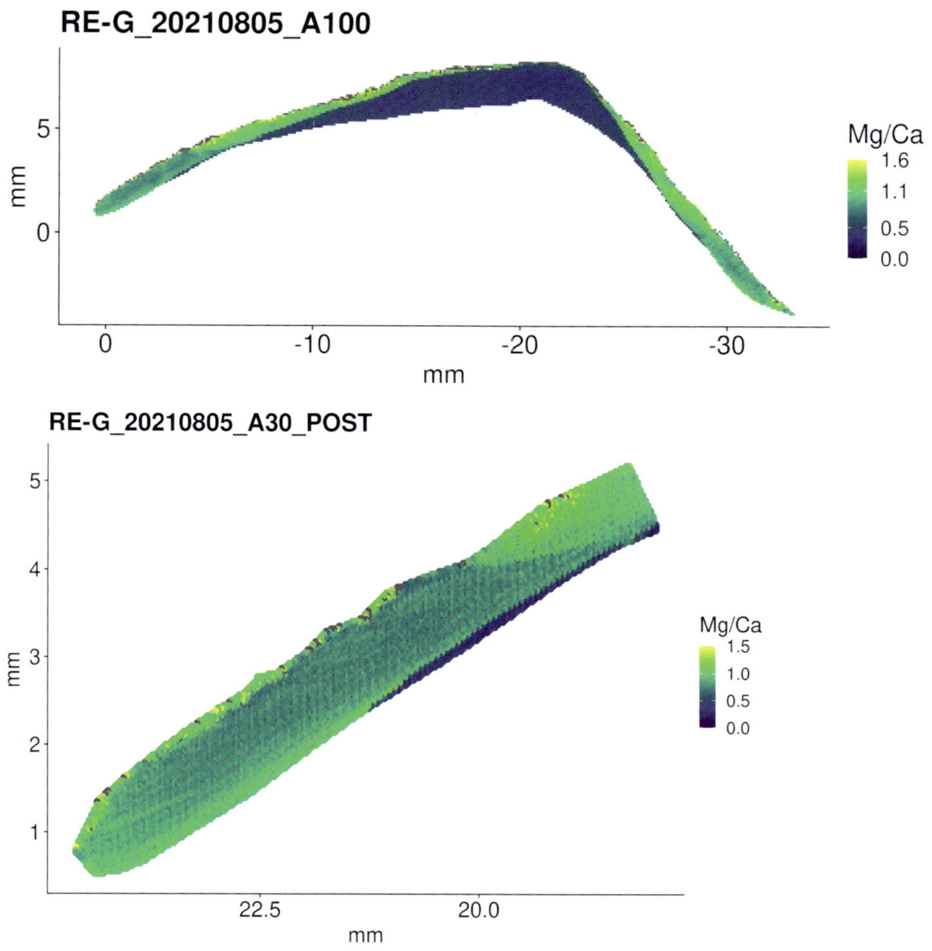

Fig. 56 Shell specimen from Glaros, Rethymno with distinct seasonal growth increments: Sample RE-G_A, 30 µm scan of the posterior and linear scan of the posterior.

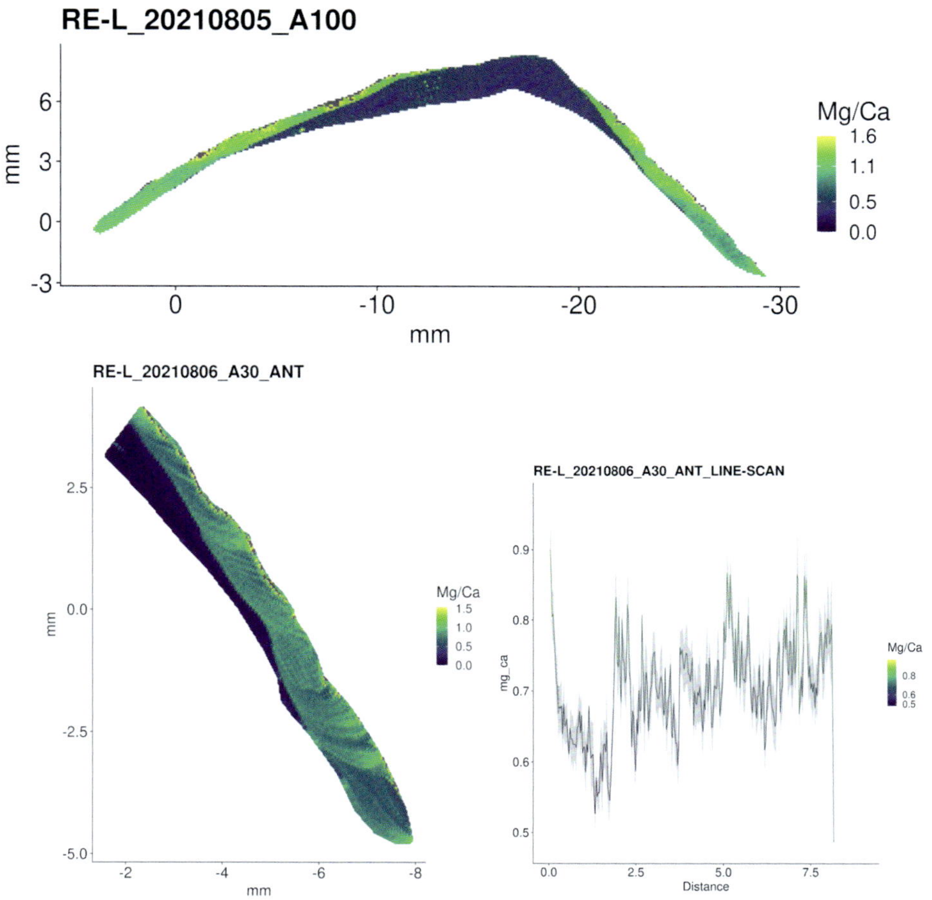

Fig. 57 A specimen from Ligres, Rethymno with ambiguous seasonal growth increments: Sample RE-L_A, 100 µm scan, 30 µm scan of the anterior side and linear scan of the anterior.

	Collection site	Coordinates	Number of samples	Mean age (years)	Average growth rate (mm per year)	Successful seasonal interpretation (%)
1	Franchthi (FR-P)	37.42280, 23.13084	24	1.6	15.7	70.8
2	Monemvasia (ML)	36.68648, 23.03960	5	Less than 1	24.6	40
3	Youra isl. (west) (Y-W)	39.37678, 24.14985	12	4	5.4	96
4	Youra isl. (south) (Y-S)	39.36861, 24.14602	8	3.2	7.9	81
5	Agios Petros isl. (KP-AP)	39.32127, 24.05445	15	4.4	5.9	100
6	Kyra-Panagia isl. (KP-KO)	39.31602, 24.06613	4	2.4	9.5	50
7	Patitiri, Alonnisos isl. (PA)	39.14285, 23.86363	5	1.5	14.2	70
8	Plakes, Volos (VP)	39.34755, 22.97251	3	Less than 1	22.4	100
9	Koropi, Pagasetic Gulf (KO-PG)	39.28927, 23.14363	5	2.3	11.8	80
10	Katiyorgis, Aegean Sea (V-KB)	39.17255, 23.34283	6	2.3	12.3	100
11	Ligres, Rethymno (RE-L)	35.13481, 24.53191	2	1	32.7	75
12	Glaros, Rethymno (RE-G)	35.41310, 24.82722	1	1	34.8	100
13	Kedrodasos, Chania (K-CH)	35.26844, 23.55886	1	1	26.3	100
14	Alyko, Naxos isl. (AN)	36.98486, 25.38561	3	Less than 1	18.2	100
15	Ysterni, Paros isl. (GYP)	37.08777, 25.26442	2	3	10.1	0
16	Nea Kallikrateia, Chalkidiki (NKC-M)	40.30134, 23.08626	4	Less than 1	22.4	75
17	Pyrgos Apollonias, Kavala (PARK)	40.74020, 24.14294	2	Less than 1	31.3	100
18	Tonnara del Secco (east) (TS-E)	38.16794, 12.77000	5	Less than 1	40.5	80
19	Tonnara del Secco (west) (TS-W)	38.16694, 12.76886	3	1.3	24.4	83
20	Faro Punta Sottile (east) (FPS-E)	37.93468, 12.27661	4	2.3	14.1	100
21	Faro Punta Sottile (west) (FPS-W)	37.93521, 12.27333	5	Less than 1	41.6	100
22	Sousa (MP6, M0, MBRP)	32.90801, 22.04439	12	-	-	-

Tab. 9 Mean age, average growth rate and successful seasonal interpretation percentages of modern samples by location.

4.1.2. Elemental Analysis Overview

To summarise the elemental data, it became evident from the previous sections that seasonal cyclicity is not reflected in a consistent way; the shape of the seasonal growth increments can vary in numerous ways as shown in the diverse modern specimen examples (**tab. 9**). The cyclicity of these seasonal variation patterns was not pronounced among specimens from the same location. That is, same-location specimens often produced similar results as was the case with the group from Agios Petros to give an example. Nonetheless, it appears that the more usual spatial pattern of growth and succession found was extensive winter increments and more short and pronounced summers.

4.1.2.1. Season of Collection

I was able to identify seasonally-resolved patterns in most (92 %) of the 131 analysed samples, as 11 of them did not exhibit interpretable patterns. Regarding the season of collection estimates, among the 122 samples with a known season of collection, results showed that in the majority (79 %) of the cases results agreed with the known season of collection (**fig. 58**). 96 samples out of 122 indicated that the seasonality estimate based on LIBS matched the actual season of capture, when evaluating both the elemental data and the (cooling or warming) trends of the linear scans. Another 3 % was characterised as actual/unclear, meaning it was potentially close to the actual one and another 4 % was characterised as preceding/actual, meaning that elemental data reflected a liminal phase where they could be interpreted as both. Furthermore, 7 % of the samples reflected the preceding season in the elemental data.

Furthermore, cyclicity and repetition constitute important aspects in order to draw results, however they are not indispensable. This means that even if distinct seasons were not always apparent or discernible, I was mostly able to identify the final season of growth that corresponds to the season of collection. In other words, even if there is not a clear sinusoi-

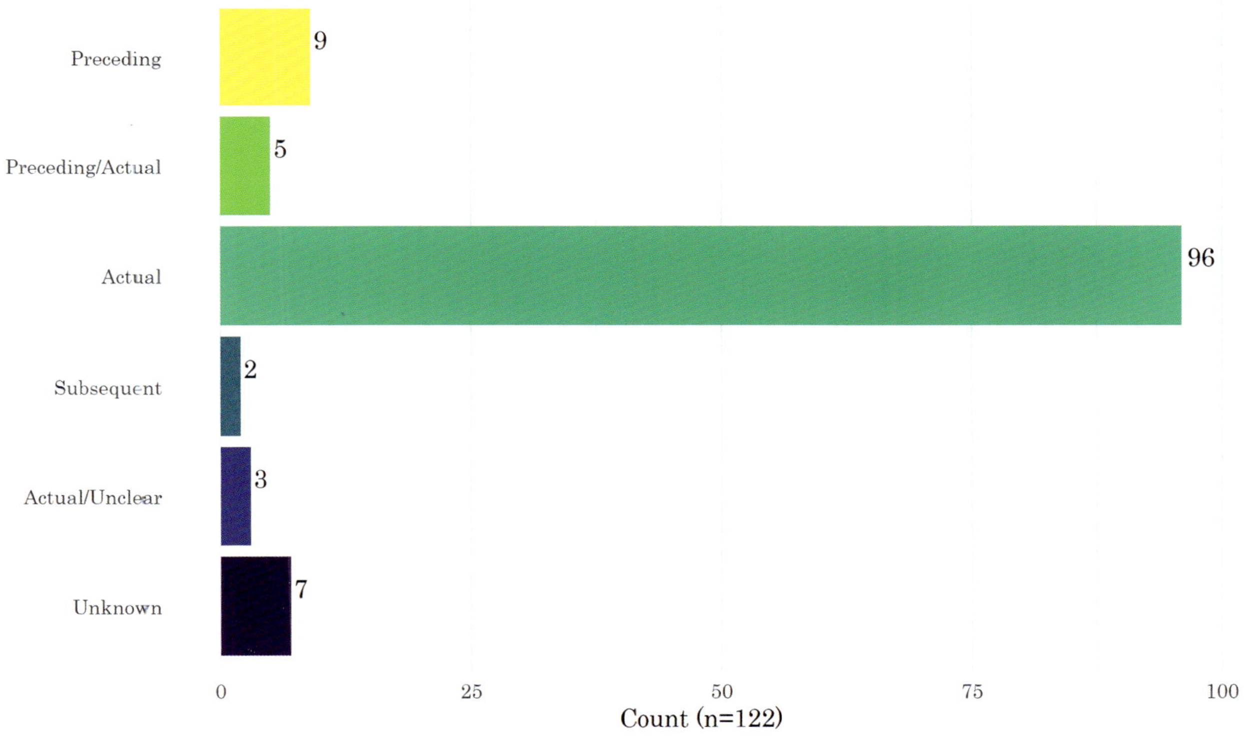

Fig. 58 Bar plot displaying the season that LIBS data reflected compared to the actual one by count of specimens.

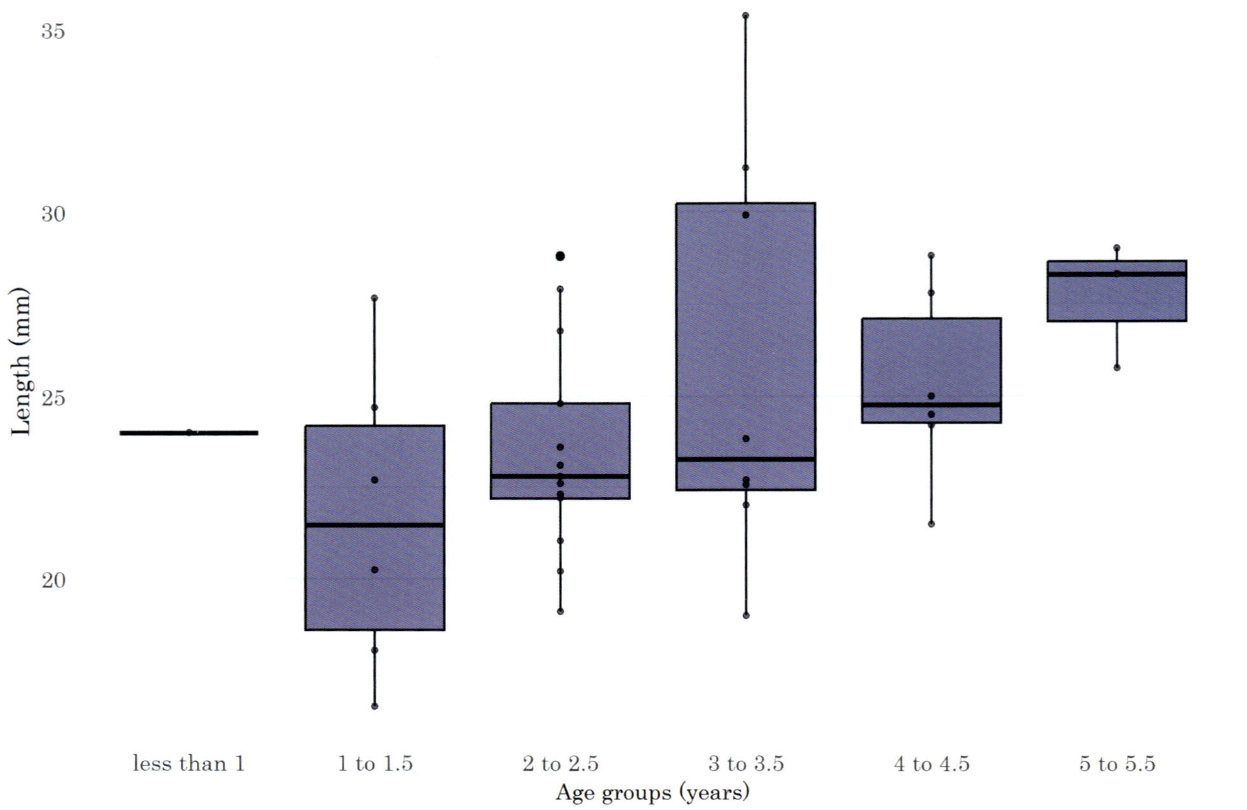

Fig. 59 Barplot comparing age with length (n = 102). This includes all modern samples that were analysed with LIBS and an age determination could be made.

dal curve in the linear scan, the elemental data of the shell edge can reflect the season of collection by comparing them with the preceding elemental data. Nonetheless, when the time-resolved data are repetitive, the interpretation of the data can be done in a more confident manner as seasonal extremes are easier to recognize.

4.1.2.2. Biological Age Determination

Furthermore, when comparing size with age among 102 modern shell specimens, it appears that there is a positive correlation between the two variables (**fig. 59**). As seen in the following plot, age increases along with size. The filtering of different sites of collection allowed for further observations on this matter. When examining the size-age relation of samples from less frequented areas such as Youra and Kyra Panagia, the positive correlation becomes more pronounced (**fig. 60**). Furthermore, when comparing the age and size of younger shell specimens such as those of Franchthi, this observation does not apply. The younger shell specimens which recorded less than a year of growth had more variable lengths (**fig. 61**).

Certainly, seasonal cyclicity depicted onto the shell increments per location is susceptible to the specific environmental conditions of each area, while high-resolution instrumental SST data are not available. Without doubt, younger specimens did not allow for precise season-of-collection estimations and these younger specimens were usually present in the more popular and highly frequented beaches where human impact and exploitation of the marine environment is more pronounced. This becomes especially prevalent when looking at specimens collected from more touristic and overexploited areas, such as Crete or the Cyclades (e.g. K-CH_A, AN_A, RE-L_B) tended to be younger in age and gave more poor results in terms of seasonal variability. Inversely, the shell specimens from Agios Petros, a rarely visited tiny islet, and Youra, an uninhabited island under protection as part of a maritime conservation zone. These two locations also had the longest-lived specimens among the 22 locations from where samples were collected (**fig. 62**).

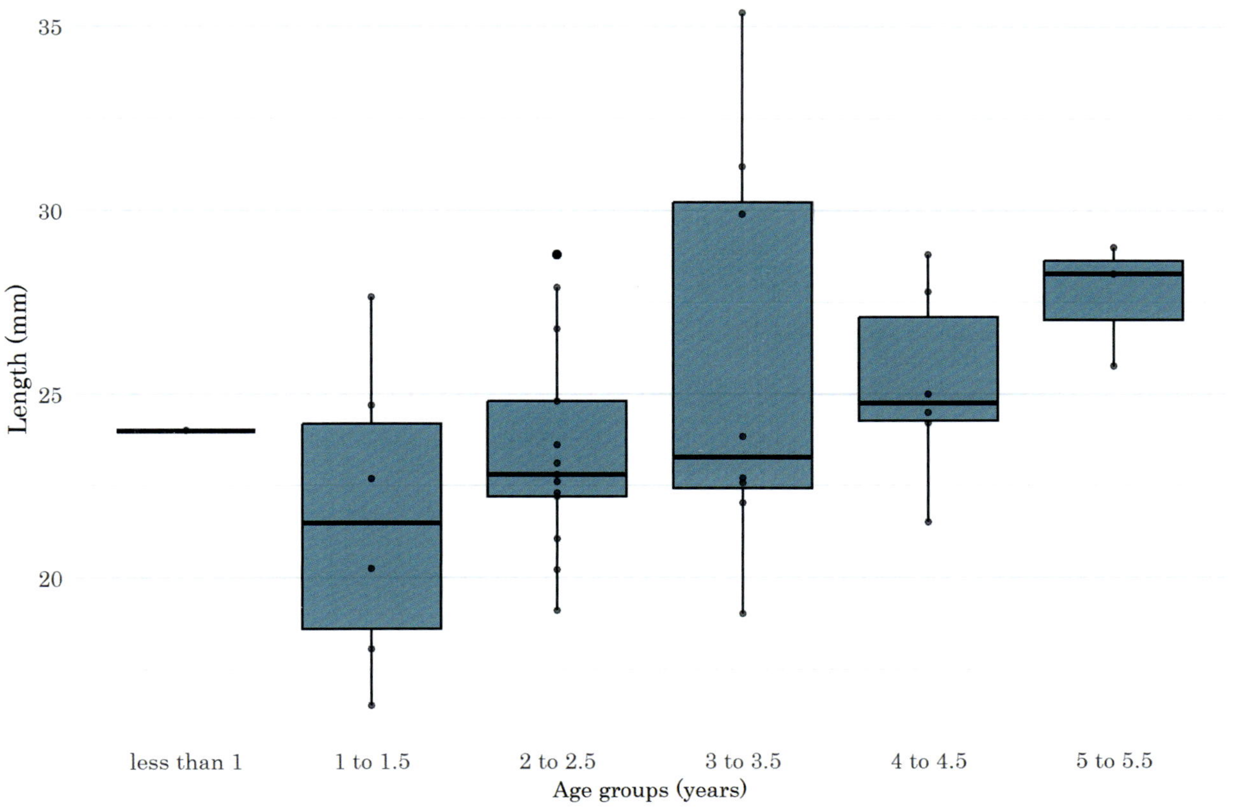

Fig. 60 Barplot comparing age with length of the Kyra Panagia (KP-AP and KP-KO) and Youra (Y-S and Y-W) specimens (n = 37).

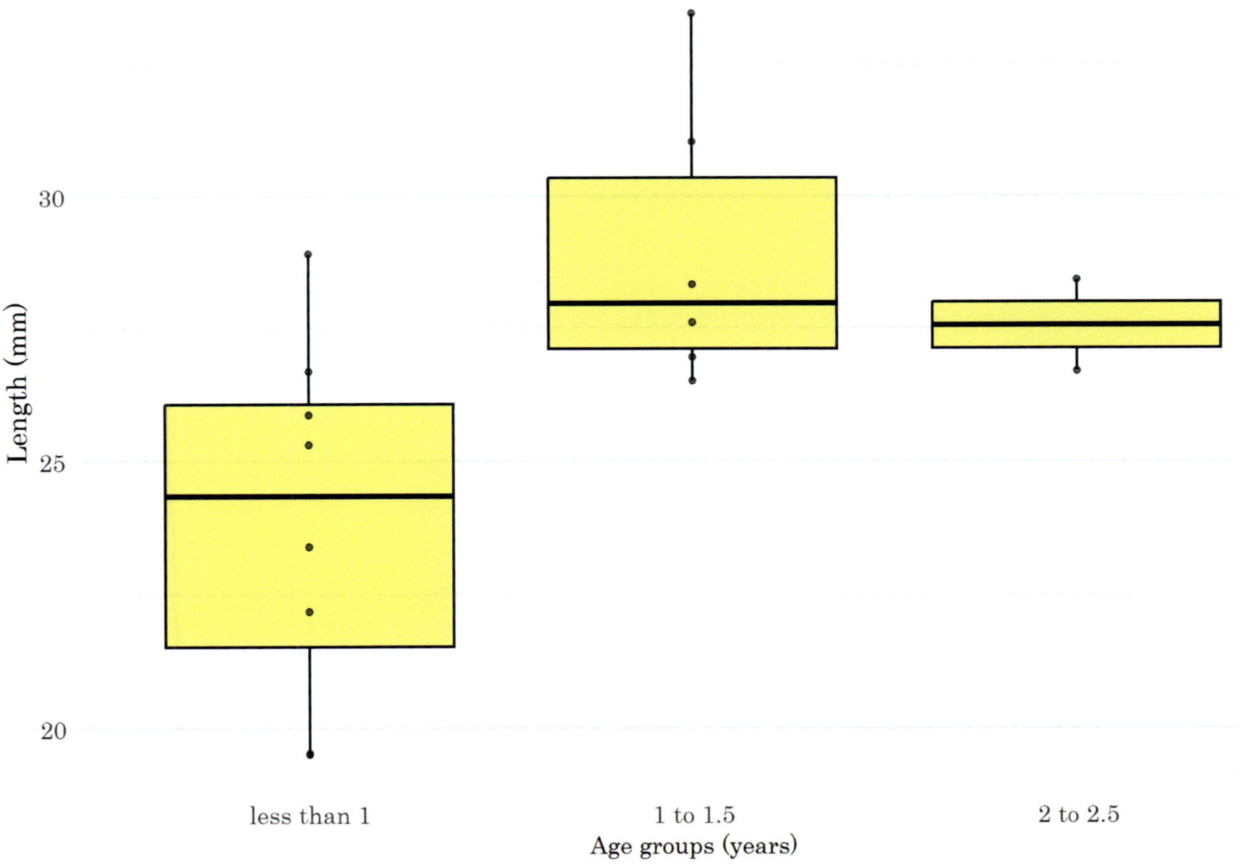

Fig. 61 Barplot ccmparing age with length of the Franchthi modern specimens (n = 16).

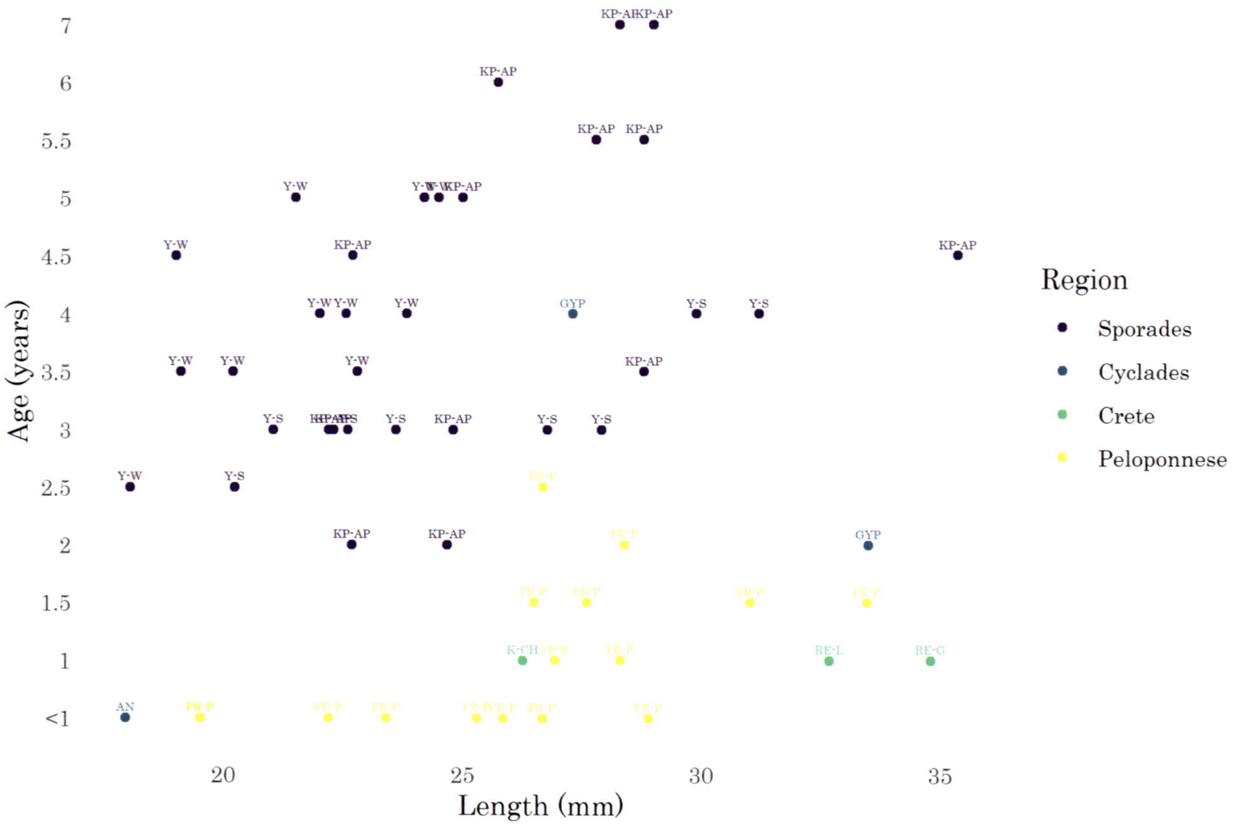

Fig. 62 Scatterplot comparing age with length from selected locations: Sporades (KP-AP, Y-W, Y-S), Cyclades (GYP, AN), Crete (RE-L, RE-G, K-CH) and the Peloponnese (FR-P). The distribution shows the fast growing shells of Franchthi compared to the more slow growing samples from the Sporades complex.

4.1.3. Oxygen Isotope Analysis

Oxygen isotope analysis was undertaken on a number of shell samples as well as on water samples that were collected at the same time as the shells. Both materials produced satisfactory results. Furthermore, the oxygen isotope analysis of the shell specimens demonstrated seasonal cyclicity in $\delta^{18}O$ values and results produced quasi-sinusoidal patterns.

4.1.3.1. Shell Samples

As mentioned in the introduction, 20 modern shell specimens were sampled for isotope analysis. 10 modern shells from Franchthi Paralia (FR-P_A to J), 4 from the small island of Youra in Central Greece (Y-S_A and C, Y-W_B and G), 3 from Faro Punta Sottile (FPS-W_A, FPS-E_C and D) and 3 from the Tonnara del Secco (TS-E_A and B, TS-W_B) location in Sicily (**tabs A.5–A.6**). Edge and sequential carbonate samples were extracted from the shell sections while the sampling methods varied according to the available methods at hand (see Chapter 3).

Overall, the main difference between elemental mapping and oxygen isotope analysis is certainly the sampling resolution. Elemental linear scans provide higher resolution data as the sampling resolution was at 30 µm and a larger number of points was measured. With LIBS the human competence factor involved in sampling is minimal due to the automated sampling stages that enable a consistent sampling resolution regardless of the sample surface. With isotope continuous milling sampling even though increments of similar size were removed each time, the exact location and size of each powder sample was not as precise and consistent, something that in the case of LIBS is an automated procedure. Seasonal cyclicity patterns are certainly visible in the isotope results, even though time-averaged in certain cases, while elemental linear scans present the data in a higher resolution and consistent manner.

Finally, as it became evident from the samples from Youra, micromilling sampling cannot be as successful and precise as elemental linear scans when only a small surface of a few millimetres is targeted for sampling. Nonetheless, what is important is that all of the samples produced quasi-sinusoidal patterns and gave evidence of cyclical variations in $\delta^{18}O$ values, even if many of the shells were quite young.

The coefficient of determination (R^2) was calculated between the oxygen isotope values and the Mg/Ca linear scan ratio variables (**figs A.27–A.43**). For the creation of the linear correlation graphs using Mg/Ca and $\delta^{18}O$, I used the dynamic time-warping method, as suggested by Hausmann et al., in order to align the records and account for the differences in the growth rates, the degrees of time-averaging, and the sampling procedures and resolutions (2019a, 8). The R^2 results of all of the modern samples showed a high degree of correlation with values of a minimum of 0.71 (Y-S_C) and a maximum of 0.98 (TS-E_B, TS-W_B) (**tab. 10**). The 17 samples analysed sequentially (either via micromilling or drilling) had a mean value of R^2 = 0.89. Among the different groups of specimens, those from Tonnara del Secco had on average the highest degree of correlation (average of 0.98) and those from Youra the lowest (average of 0.84). Even the samples that were sequentially drilled (FR-P_A, B and D) produced good correlation results (0.85, 0.95, 0.83 respectively) (**fig. 63**).

The comparatively low values of the Youra specimens could be due to the fact that two out of four of them (Y-S_C, Y-W_G), were characterised by the small calcitic layers, a feature discussed in section 4.1.1.2, which inevitably increased time-averaging during sampling. The more scattered data of the Youra group is also further illustrated in the form of outliers. On the other hand, the Tonnara del Secco group of samples had the highest R^2 values that equalled 0.97, 0.98 and 0.98 (for samples TS-E_A and B and TS-W_B respectively). This set of data is better clustered together in comparison to specimens from other regions, possibly because of the large calcitic layers of the specimens as well as the relatively broad seasonal increments (as described in section 4.1.1.2) which allowed for a more accurate sampling for oxygen isotopes. This subsequently led to better oxygen isotope results with these values matching closely those of the elemental mapping. A similar case is that of the Favignana shells, the mean R^2 of which equalled 0.91. The elemental overall image of this group of shells is similar to that of Tonnara del Secco: Large seasonal increments which allowed for more precise and matching oxygen isotope measurements. Lastly, the Franchthi group had a mean R^2 of 0.87, which shows that the two variables are quite well associated with one another. This result is not as high as in other groups of data, either because seasonal increments are as well-resolved and distinguishable as in other cases (e. g. FR-P_G) either due to the more coarse results obtained via sequential drilling.

	Sample name	R^2 value
1	FR-P_20210515_A	0.85
2	FR-P_20210515_B	0.95
3	FR-P_20210515_C	0.91
4	FR-P_20210515_D	0.83
5	FR-P_20210515_G	0.84
6	FR-P_20210515_H	0.96
7	FR-P_20210515_I	0.78
8	Y-S_20210917_A	0.93
9	Y-S_20210917_C	0.71
10	Y-W_20210917_B	0.89
11	Y-W_20210917_G	0.82
12	FPS-W_20230518_A	0.94
13	FPS-E_20230518_C	0.95
14	FPS-E_20230518_D	0.84
15	TS-E_20230516_A	0.97
16	TS-E_20230516_B	0.98
17	TS-W_20230516_B	0.98

Tab. 10 R^2 correlation values between Mg/Ca ratio and $\delta^{18}O_{shell}$ values of the modern shell specimens.

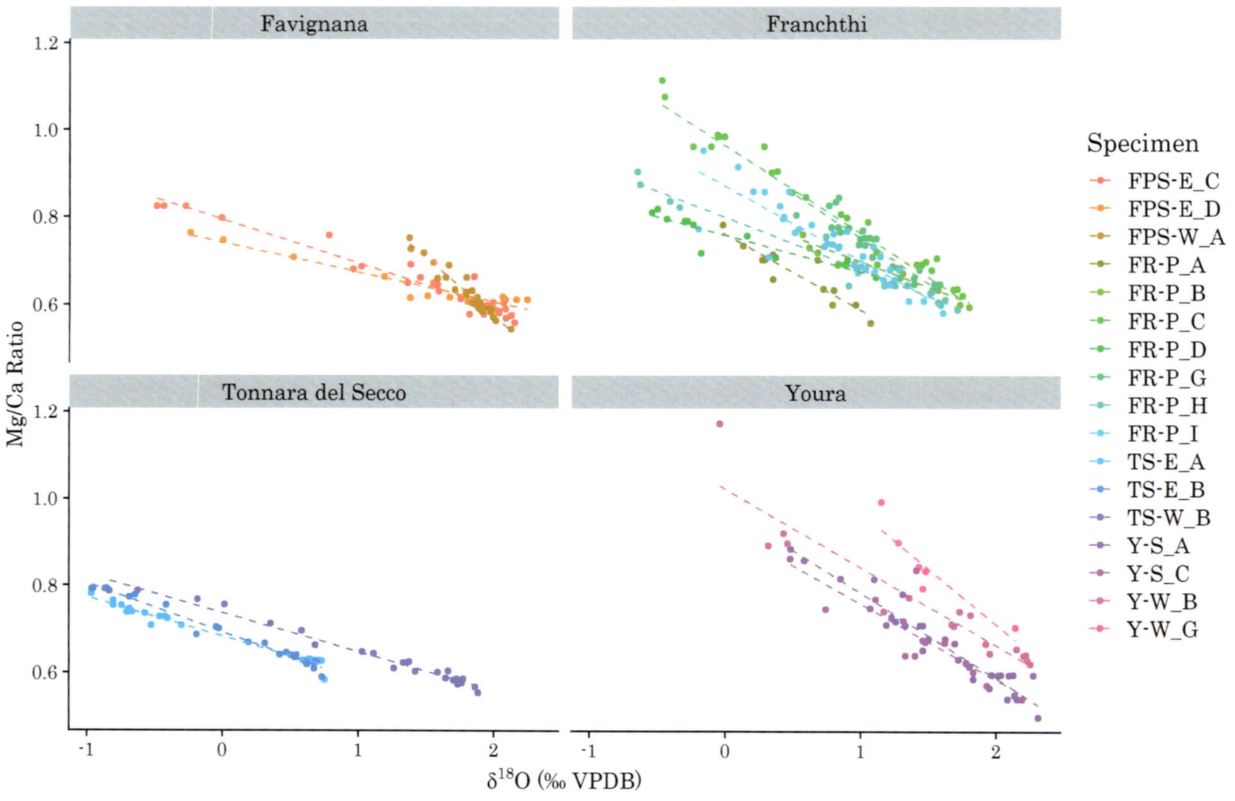

Fig. 63 Collated correlation graphs grouped by region comparing the elemental data and oxygen isotope data of all 17 modern specimens that were sampled sequentially.

4.1.3.2. Water Samples and Temperature Approximations

As presented in section 4.1, six different water samples were collected from Greek and Italian shores and run for oxygen isotope analysis (**tab. 11**). H. Vonhof from the Max Planck Institute for Chemistry in Mainz performed measurements on a Picarro (L2140-i) spectroscopy instrument, according to the method by de Graaf et al. 2021. The data produced was post-processed for scale conversion to VSMOW based on a 2-point calibration using internal standards.

In order to compare the calculated isotope derived SST data with the instrumental data, no offset was applied to the water samples. The standard −0.27 ‰ adjustment – typically used to align the VSMOW scale for water with the VPDB scale for carbonates – was deemed unnecessary, as it is not required by the Hays and Grossman (1991) equation (Grossman 2012).

The isotope values that resulted from the analysis needed calibration in order to be applied to the equation of Hays and Grossman (1991). The $\delta^{18}O_{shell}$ edge values from the location of Franchthi ranged from 0.52 to 1.11 ‰ and had a mean value of 0.78 ‰. The derived SST data from the calcium carbonate $\delta^{18}O$ values without an offset were too high compared to the actual instrumental temperature. Without calibration, the average temperature found was 21.9 °C (**tab. 12**). I calculated the SST values for the specific

Location	$\delta^{18}O_{water}$ (‰ VSMOW)
Franchthi Paralia	2.38
Youra isl. (south)	1.27
Agios Petros isl.	1.58
Tonnara del Secco (east)	0.37
Tonnara del Secco (west)	1.09
Faro Punta Sottile (west)	1.12

Tab. 11 $\delta^{18}O_{water}$ average value (‰ VSMOW) from locations in Greece and Italy derived from two separate measurements.

	Sample name	$\delta^{18}O_{water}$ (‰ VSMOW)	Instru-mental SST* (°C)	$\delta^{18}O_{shell}$ edge (‰ VPDB) (no offset)	Calculated SST (°C) (no off-set)**	Instrumen-tal and cal-culated SST (no offset) difference	Applied offset (‰)	Calculated SST (°C) (with offset)	Instrumen-tal and cal-culated SST (with offset) difference
1	FR-P_20210515_A	2.38	19.01	0.80	22.93	-3.92	+0.85	18.99	0.02
2	FR-P_20210515_B	2.38	19.01	0.83	22.78	-3.77	+0.85	18.84	0.17
3	FR-P_20210515_C	2.38	19.01	0.82	22.84	-3.83	+0.85	18.90	0.11
4	FR-P_20210515_D	2.38	19.01	1.11	21.44	-2.43	+0.85	17.56	1.45
5	FR-P_20210515_E	2.38	19.01	0.52	24.25	-5.24	+0.85	20.25	-1.24
6	FR-P_20210515_F	2.38	19.01	0.78	22.99	-3.98	+0.85	19.04	-0.03
7	FR-P_20210515_G	2.38	19.01	0.81	22.85	-3.84	+0.85	18.91	0.10
8	FR-P_20210515_H	2.38	19.01	0.69	23.43	-4.42	+0.85	19.46	-0.45
9	FR-P_20210515_I	2.38	19.01	0.84	22.71	-3.70	+0.85	18.78	0.23
10	FR-P_20210515_J	2.38	19.01	0.64	23.65	-4.64	+0.85	19.68	-0.69
11	Y-S_20210917_A	1.27	26.87	1.19	16.05	10.82	-2.35	27.00	0.13
12	Y-S_20210917_C	1.27	26.87	1.32	15.48	11.39	-2.35	26.35	-0.51
13	Y-W_20210917_B	1.27	26.87	-0.04	21.62	5.25	-2.35	33.27	6.40
14	Y-W_20210917_G	1.27	26.87	1.15	16.23	10.64	-2.35	27.20	0.33
15	FPS-W_20230518_A	1.12	20.66	1.47	14.19	6.46	-1.43	20.52	-0.15
16	FPS-E_20230518_C	1.12	20.66	1.84	12.61	8.05	-1.43	20.04	-0.62
17	FPS-E_20230518_D	1.12	20.66	1.54	13.86	6.80	-1.43	21.40	0.74
18	TS-E_20230516_A	0.37	18.59	0.63	22.93	-4.34	-0.46	16.57	-2.02
19	TS-E_20230516_B	0.37	18.59	-0.18	18.18	0.41	-0.46	20.28	1.69
20	TS-W_20230516_B	1.09	18.59	1.03	15.98	2.61	-0.46	19.69	0.63

Tab. 12 $\delta^{18}O_{water}$, instrumental SSTs, $\delta^{18}O_{shell}$ and differences between $\delta^{18}O_{shell}$-calculated SST and instrumental SST (after Prendergast/Schöne 2017, tab. 2). SST was calculated from measured $\delta^{18}O_{shell}$ edge values and $\delta^{18}O_{water}$ using the equation by Hays and Grossman (1991). Calculated SSTs are shown both with and without the calculated for each location offset applied to $\delta^{18}O_{shell}$.

* Month of collection SST averages were retrieved from the Copernicus Marine Service (copernicus.eu).

** Calculated using the equation of Hays and Grossman (1991) (see section 3.1.2).

area of Franchthi using the edge $\delta^{18}O$ values of the sampled shells and compared them with the instrumental SST values.

Based on the measurements of the modern Franchthi shells, a positive offset of +0.85 ‰ to the $\delta^{18}O_{shell}$ values was calculated empirically using the Hays and Grossman (1991) equation. To calculate this offset, I applied different offsets on the $\delta^{18}O_{shell}$ values and calculated the difference between the instrumental and the calculated SST for each shell. I then chose the offset that gave the lowest value for differences. After applying this calibration, the $\delta^{18}O_{shell}$-derived SSTs from the ten Franchthi shells provided results that approximated the instrumental SST records. As shown in the box plot (**fig. 64**), the data are distributed closely around the known SST of the specific time period (19 °C in the case of Franchthi). The calculated mean SST after calibration was 19.0 °C, with a minimum outlier temperature of 17.5 °C (FR-P_D) and a maximum of 20.2 °C (FR-P_D).

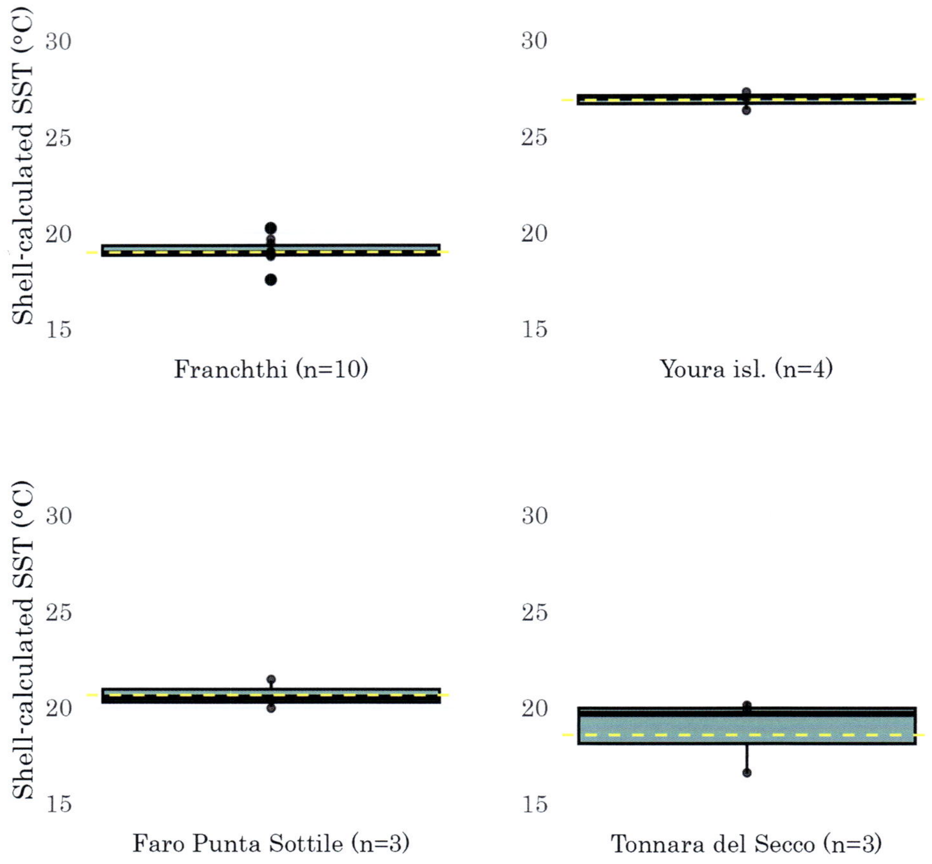

Fig. 64 Comparison of boxplots of $\delta^{18}O_{shell}$-derived SST calibrated values. Yellow dashed lines represent the instrumental SST values at the time of collection.

The positive 0.85 ‰ offset found in this case falls within the ranges of offsets found by several previous studies (+0.36 and +1.01 ‰; Shackleton 1973; Fenger et al. 2007; Ferguson et al. 2011; Gutiérrez-Zugasti et al. 2017). However, it is not very close to the negative offset of −0.72 ‰ found in a study by Prendergast and Schöne (2017, 37) including several shells from different central and eastern Mediterranean locations. When this offset was applied to the samples under discussion here, they did not provide comparable results.

Furthermore, the positive 0.85 ‰ offset I found for Franchthi was not applicable to the other groups of modern shells from Youra, Tonnara del Secco and Faro Punta Sottile and different offsets were calculated for each location (**tab. 13**). It is worth mentioning that the samples from these three locations were not as numerous as those from Franchthi and produced more diverse outcomes that were not deemed adequate to draw definite conclusions regarding their reliability.

Location	Minimum $\delta^{18}O_{shell}$ (‰)	Maximum $\delta^{18}O_{shell}$ (‰)	Mean $\delta^{18}O_{shell}$ (‰)	Calculated offset (‰)
Franchthi	0.52	1.11	0.78	+0.85
Youra isl.	-0.04	1.31	0.90	-2.35
Faro Punta Sottile	1.47	1.84	1.62	-1.43
Tonnara del Secco	-0.18	1.03	0.49	-0.46

Tab. 13 Summary of $\delta^{18}O_{shell}$ value ranges (‰) prior to calibration and calculated offset for each location.

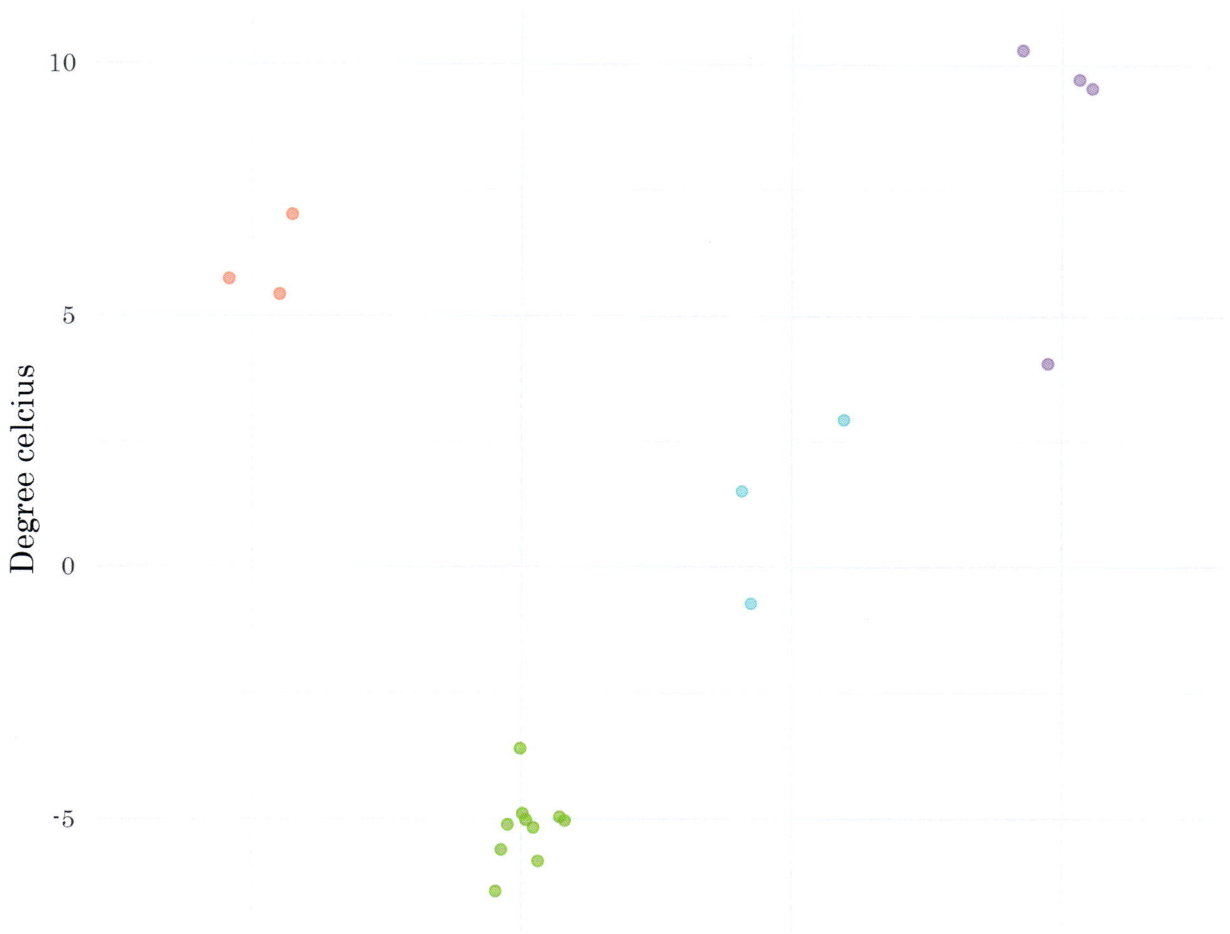

Fig. 65 Jitter plot where points represent the differences (in °C) between the instrumental and the calculated SST per location when applying the universal negative 0.25 ‰ offset.

More specifically, $\delta^{18}O_{shell}$ edge values from Youra ranged from −0.04 to +1.31 ‰ and had a mean value of 0.90 ‰. The $\delta^{18}O_{shell}$-derived SST data values without an offset were generally too low compared to the actual instrumental temperature with an average temperature of 17.34 °C (**tab. 12**). In the case of the four samples from Youra, an offset of −2.35 ‰ was found to approximate in the best way the satellite data, which in the specific area was quite high (approximately 27 °C). A similar approach was followed for the six modern shell samples from Sicily. The $\delta^{18}O_{shell}$ edge values of the three samples from Faro Punta Sottile ranged from 1.47 to 1.84 ‰ and had a mean value of 1.62 ‰. The calculated mean SST after applying an offset of −1.43 ‰ was 20.64 °C, with a minimum temperature of 19.6 °C (sample FPS_E_C) and a maximum of 21.3 °C (sample FPS-W_A). In the case of Tonnara del Secco, $\delta^{18}O_{shell}$ edge values ranged between −0.18 and 1.03 ‰ and a mean value of 0.49 ‰. An offset of −0.46 ‰ was found to best approximate the instrumental data. After calibration, $\delta^{18}O_{shell}$-derived SSTs had a minimum of 20.6 °C and a minimum of 16.9 °C and a mean temperature of 18.6 °C.

4.1.3.3. Outcomes of Oxygen Isotope Analysis

A universal offset from all 20 shells from the four locations in Greece and Italy was also calculated to be −0.25 ‰. However, when applying this universal offset to approximate the instrumental data, results are less precise. The ranges of the differences between the instrumental and the calculated SSTs become quite large ranging from −6.64 to +9.75 °C (**tab. A.7**). Inevitably, site-specific offsets allow for more secure temperature approximations. As shown in **figures 65** and **66**, the aforementioned differences when applying site-specific offsets produce more

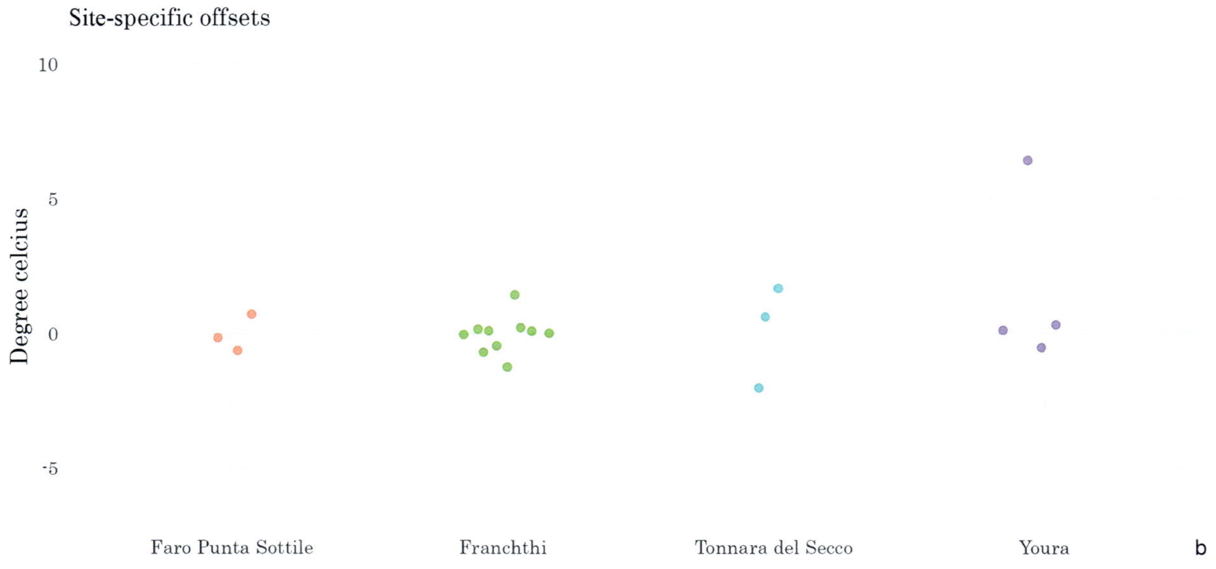

Fig. 66 Two jitter plots where points represent the differences (in °C) between the instrumental and the calculated SST per location: **a** Results with no offset. – **b** Results when applying individual site-specific offset.

accurate results with values clustering closer to 0 in all of the cases.

Certainly, more samples are needed to restrain to a certain extent the degree of variability of the $\delta^{18}O$ values. Nonetheless, temperature loggers that would record the water temperature over the course of a year at a specific location of interest, could provide more accurate temperature data and allow for more secure offset calculations. Unfortunately, placing temperature loggers at Franchthi was not possible at this time.

4.1.4. Summary

To summarise, 131 modern shell specimens with a known season of collection from 22 locations in Greece, Italy and Libya were successfully processed using elemental mapping and oxygen isotope analysis. The 400 elemental maps and 132 line scans that were produced with the use of LIBS as part of this modern study, revealed seasonally-resolved patterns with lower Mg/Ca ratio areas representing winter periods and higher Mg/Ca ratio areas representing summer periods. Among the 122 samples with a known season of collection, in 79 % of the cases elemental results indicated that the seasonality estimate based on LIBS matched the actual season of capture. Another 7 % of the analysed modern samples indicated a season that was close to the expected result. Elemental analysis also allows for the estimation of the age of the molluscs and the comparison of their dimensions to their biological age. These two variables are positively correlated with the size of the specimens increasing along with their age. This tendency is more obvious when comparing specimens that are over 1 year old.

Regarding the oxygen isotope results of the 20 modern shell specimens that were analysed, all of the samples produced quasi-sinusoidal patterns and gave evidence of cyclical variations in $\delta^{18}O$ values. The sampling technique used, drilling or micromilling, was shown to make a difference, as micromilling results produced results of higher resolution. Nonetheless, elemental line scans yielded higher resolution results than both isotope techniques, while the quality of the results is less dependent on the experience of the person that performs the sampling. Furthermore, the comparison of the sequential $\delta^{18}O$ results with the elemental data had a mean coefficient of determination (R^2) value of 0.89, showing that these two variables in all of the cases are positively correlated. This good correlation result strengthens the credibility of the two methods employed and allows for further confidence in the evaluation of the archaeological results. Six water samples collected from Greece and Italy were also analysed in order to proceed further with temperature approximations. When applying the temperature approximations on the results of the isotope analysis of the 20 modern specimens (see **tab. 12**), a standard deviation of 1.47 °C was found between the estimated and actual sea surface temperature.

In addition, based on the edge isotope measurements of the modern shells, offsets to the $\delta^{18}O_{shell}$ values were calculated empirically using the Hays and Grossman (1991) equation by comparing the edge values with the instrumental satellite data. Thus, an offset was calculated for each of the four locations from which shell specimens were sampled as previously reported offsets were not applicable in these samples. When the calculated SSTs from the different locations are compared, it becomes evident that the shell-derived temperatures from Franchthi are more consistent presenting a smaller distribution in terms of calculated temperature or of $\delta^{18}O$ edge values. In the case of Youra, Faro Punta Sottile and Tonnara del Secco, results on their respective offsets cannot be conclusive as isotope samples are not numerous enough. A universal offset that covered all four locations, evidently yielded less accurate shell-derived SSTs compared to the site-specific offsets.

4.2. Archaeological Samples

The 180 archaeological samples analysed with the use of LIBS and oxygen isotope analysis produced various and multi-faceted results, while AAR analysis yielded outcomes that were not meaningful for the establishment of a new chronology. Elemental analysis entailed a total of 617 scans, of which 157 were 100 µm scans, 284 were 30 µm scans and 176 linear scans (**tab. A.8**). As described in Chapter 3, elemental analysis also involved the marking with the use of the laser of the annual minima and maxima areas (see star notations in the **Appendix I**). These markings were the guides for isotope carbonate powder extraction, which led to the sampling of a total of 381 calcium carbonate powder samples that underwent isotope analysis.

In the following section, I will first describe the outcome of the elemental mapping analysis, including the season of collection estimates and the determination of the molluscs' biological age. This is followed by the $\delta^{18}O$ data derived from the targeted analysis based on the elemental maps, to gain information about annual temperature ranges and variability. Finally, the results of the AAR analysis will be presented.

4.2.1. Main Results

In this section, I will describe in detail the elemental mapping and isotope outcomes. Using the Mg/Ca ratio, the instances of summer and winter are revealed and characterised as such when visible and well-defined. Counting these instances provides the biological age of the mollusc, while looking at the most recent season contributes to the determination of the season of collection, which can also be described as the season of death, capture or consumption. The isotope results are reported here for each specimen and are further evaluated in section 4.2.2 that follows. All of the results can be found in the **Appendix I**[10] in alphabetical order. Here they will be described by trench and excavation unit as well as in chronological order starting from the oldest.

4.2.1.1. Trench FAS

FAS 207

Excavation unit FAS 207 is the oldest of the units analysed in this study and dates to the Final Palaeolithic period. The excavation unit winter increment isotope average is 2.62 ‰ and the summer average is 0.83 ‰ (**fig. 67**).

Two shells belong to this unit: Specimen FAS_207_A has well-defined seasonal increments, two of which are summers and three of which are winters. Two winters are visible on the posterior side while three on the anterior. For this reason the specimen appears to be 2.5 years old. The specimen had been collected between spring and summer as relative temperature appears to be rising in the scan of the posterior side. Two oxygen isotope measurements were made in this shell, one on the last summer (1.17 ‰) and one on the penultimate winter (3.22 ‰) increment. This is the coldest winter increments measured in this study, while both the winter and summer values of this shell are quite low, but are to be expected at this point in time, the end of the Pleistocene.

Specimen FAS_207_B does not have as distinct seasonal increments as the previous specimen. Nonetheless, it appears to reflect two winter periods and one summer period. Therefore, it had an age of 1.5 years. Both from the elemental map as well as from the line scan it appears that it was collected during winter. The trend of the line scan is stable. Two carbonate samples were collected from this shell, one from the last winter (2.02 ‰) and one from the summer increment (0.50 ‰). These values are, also in this case, quite low and indicative of this time period.

10 The star denotations on the elemental maps in **Appendix I** indicate the position of the extraction of the oxygen isotope sample, while the numbering of the oxygen isotope samples always starts from the sample closer to the edge which corresponds to the isotope sample number one in the graphs, and so on.

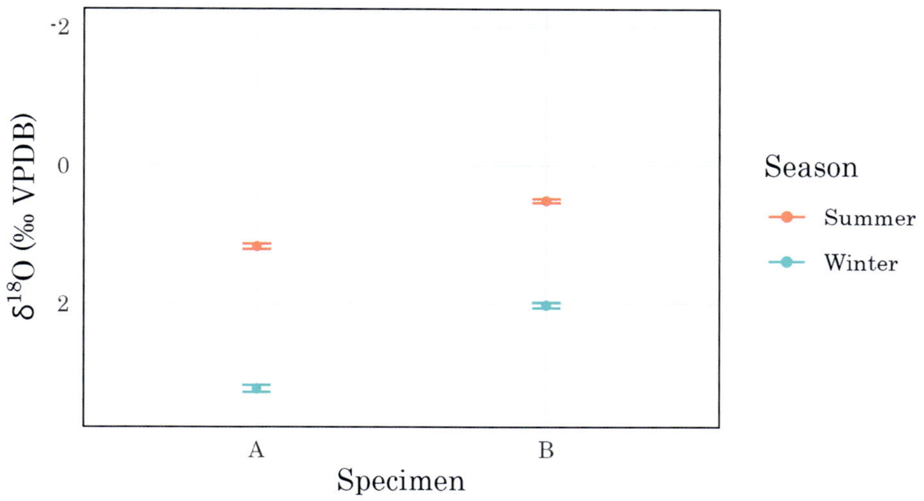

Fig. 67 Seasonal δ¹⁸O values for unit FAS 207.

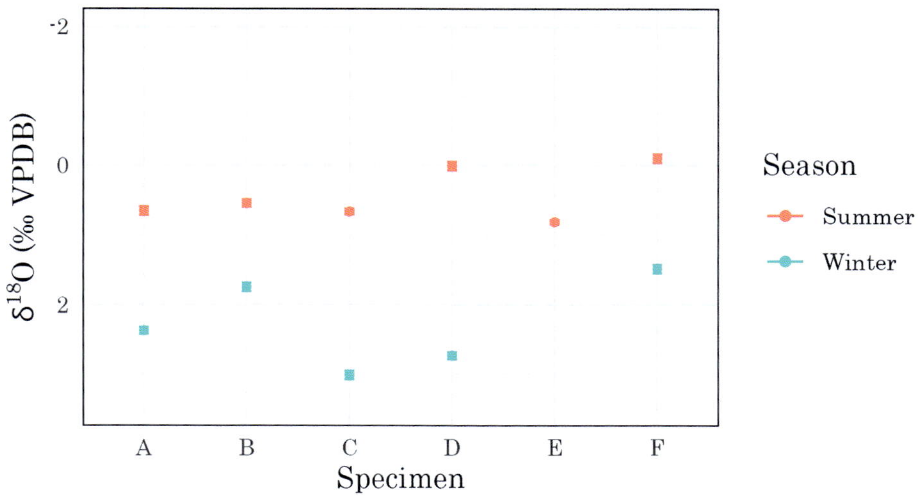

Fig. 68 Seasonal δ¹⁸O values for unit FAS 206.

FAS 206

Excavation unit FAS 206 dates to the Final Palaeolithic period and contains six specimens. The excavation unit winter increment isotope average is 2.18 ‰ and the summer average is 0.46 ‰ (**fig. 68**).

Specimen FAS_206_A had broad seasonal increments and recorded two winter periods and a summer one. Thus, the mollusc was 1.5 years old. The mollusc was collected during winter as the edge increment presents low Mg/Ca ratios. The trend of the line scan is stable. Three samples were collected for oxygen isotope analysis, one from each winter increment (last winter: 2.44, penultimate winter: 2.31 ‰) and one from the summer increment (0.65 ‰). These $\delta^{18}O$ winter values are high and indicative of the Final Palaeolithic period.

Specimen FAS_206_B did not produce clear results as seasonal increments are ambiguous. This is probably because of its young age, which was less than a year. Even though the elemental map shows low Mg/Ca values at the last growth of increment, no seasonal cyclicity is present and for this reason the shell was not attributed to a certain season of collection. The line scan is reflecting a cooling trend. Two samples were collected for oxygen isotope analysis: The winter one had a value of 1.75 ‰, while the summer sample 0.54 ‰.

Specimen FAS_206_C presented clear and broad seasonal increments especially on the posterior edge while the anterior edge was not as clear as there was some noise in the laser signal. The elemental data revealed two summer periods and a winter one indi-

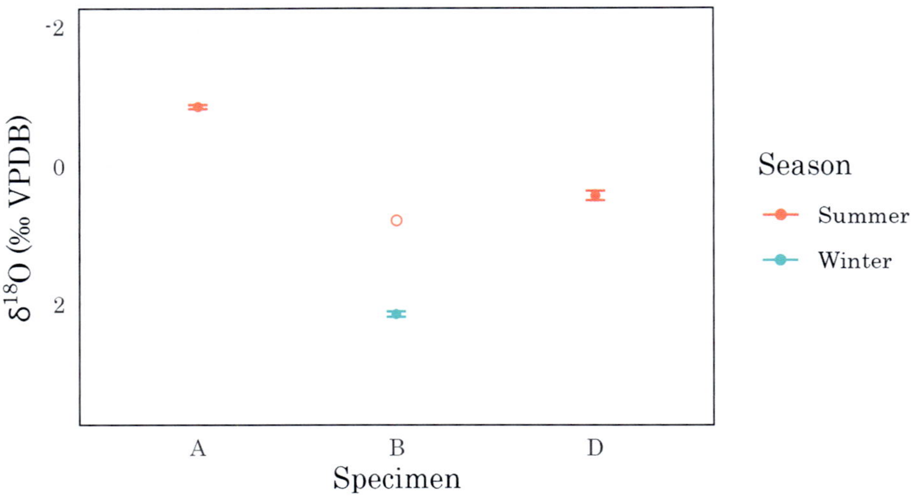

Fig. 69 Seasonal $\delta^{18}O$ values for unit FAS 203.

cating that the specimen was 1.5 years old. From the cooling trend recorded in the quasi-sinusoidal line scan, I concluded that the shell was collected during autumn. Three oxygen isotope samples were retrieved from this specimen: Two summer ones (last summer: 0.79 ‰, penultimate summer: 0.53 ‰) and a winter one (3.02 ‰). Interestingly, the second summer is slightly warmer than the first one.

Specimen FAS_206_D had broad seasonal increments and the elemental maps reflected a winter and a summer period that is followed by an edge increment that can be attributed to an autumn/winter season. The mollusc was approximately 1 year old. The line scan is not as clear as the elemental map as it presents some noise. Nonetheless, a cooling trend is visible. The abrupt peak at point 0 of the line scan is noise of the edge and not a summer signal. Furthermore, two samples were collected for oxygen isotope analysis: A summer increment sample (0.01 ‰) and a winter one (2.74 ‰).

Specimen FAS_206_E has some noise on the borders of the posterior edge, which may be due to the uneven surface of the shell. The same is the case for the apex of the shell. Even though the data is not particularly clear, one may observe that there are two winters recorded in the shell and two thin summer increments. This makes the shell approximately 2 years old and was collected during the summer. The two summer peaks are also seen on the line scan, however the last peak is characterised by noise. The line scan indicates an upward warming trend. One oxygen isotope sample was extracted from the shell from the last summer growth increment. The summer value (0.81 ‰) may appear somehow high and could be due to the fact that the summer increments are very faint in this specimen. Nonetheless, it is not dissimilar to other relatively »cold« summers of this excavation unit (e. g. the summer increment of specimen A: 0.65 or specimen C: 0.53 ‰).

The elemental data of specimen FAS_206_F were ambiguous and growth increments could not be attributed to a specific season. This is probably because the specimen was quite young. No season of collection was attributed to this specimen and no age determination was made. Apart from the unclear increments of the elemental maps, the line scan also presented a lot of noise, and no seasonal cyclicity could be observed. Three oxygen isotope samples were extracted from this shell: One towards the edge, from an area that seemed to have higher Mg/Ca ratio and was characterised as a potential summer (−0.10 ‰) followed by two winter samples that gave very similar results (1.30 and 1.68 ‰), with the second one being higher.

FAS 203

Excavation unit FAS 203 also dates to the Final Palaeolithic period and contains four specimens. The excavation unit summer increment isotope average is −0.21 ‰ (**fig. 69**).

Specimen FAS_203_A is characterised by a thick apex and a large aragonitic area. Regarding the limited seasonal variation, one may discern a winter increment followed by summer increment, which is then followed by an autumn or a winter season. However, as the edge is not preserved in the sam-

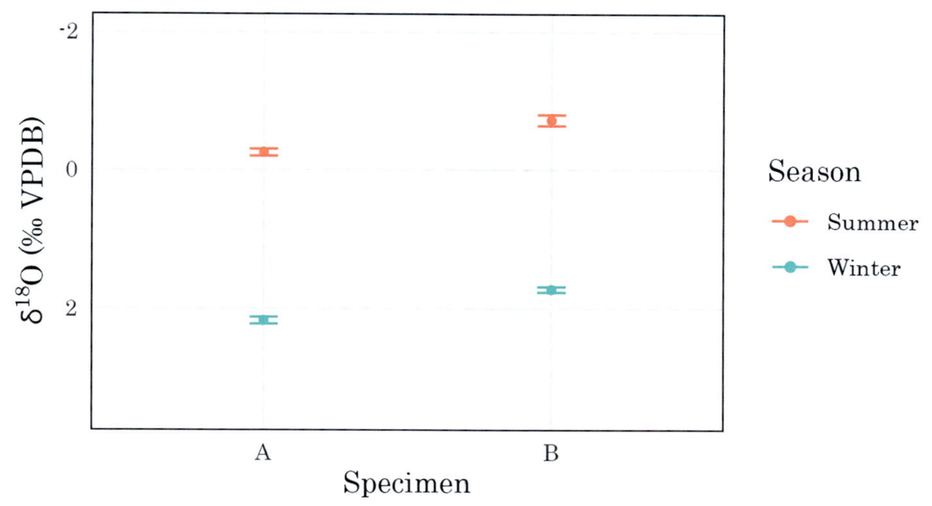

Fig. 70 Seasonal δ18O values for unit FAS 195.

ple, no season of collection was determined nor an age determination was made. An oxygen isotope sample was extracted from the summer increment (−0.85 ‰).

Specimen FAS_203_B is similar to sample FAS_203_A as the calcitic area of the shell is relatively homogeneous. There are slightly higher Mg/Ca ratios towards the apex, while there is an area of lower Mg/Ca ratio towards the edge. Only the apex and one side of the sample has been preserved. As the last growth increment has not been preserved no season of collection was attributed to the specimen nor was there an age determination made. Two oxygen isotope samples were extracted: One towards the edge that represents a winter with a value of 2.15 ‰, and another one towards the apex that was thought to represent another seasonal maximum (summer), however the value was quite high and was marked as an uncertain one, as this could represent an autumn rather than a summer.

Specimen FAS_203_C is quite fragmented and only a small amount of the calcitic layers is preserved. From the small calcitic layer of the posterior seasonal increments are not well-defined as is visible on the elemental map, while the line scan is also unclear. None of the edges were preserved and for this reason no season of collection was identified and no age determination was made. Due to the specimen's fragmented state, no oxygen isotope samples were retrieved.

Specimen FAS_203_D is a fragment where a summer and an autumn/winter increment are present. The mollusc was probably collected during autumn/winter, while the oxygen isotope sample extracted from the summer increment indicated a relatively cold summer (0.43 ‰), which nonetheless approximates the summers of this chronological period (e. g. specimen FAS_206_C). The mollusc's age could not be determined as the shell is a fragment and is not adequately preserved.

FAS 195

Advancing across time, excavation unit FAS 195 dates to the Lower Mesolithic period and contains three specimens. The excavation unit winter increment isotope average is 1.94 ‰ and the summer average is −0.49 ‰ (**fig. 70**).

Specimen FAS_195_A exhibited seasonal variation patterns, however the two extremities do not show similar results, as the anterior edge is not fully preserved. Based on the posterior edge, a winter period is followed by a warmer one towards the edge, that could represent a spring or a summer season of collection. This makes the specimen approximately 1 year old. The line scan clearly reflects the winter and the following rising of the temperature towards the warmer period. Two oxygen isotope samples were collected from the specimen: The warmer period summer increment had a value of −0.25 ‰, while the winter the value of 2.17 ‰. The winter value is similar to one of the warmer winters of the previous chronological period (the winter value of specimen 203 B, 2.15 ‰), while the summer value is slightly warmer than the previous warmer summers of the Final Palaeolithic (specimen FAS_206_D recorded a summer of 0.01 ‰, FAS_206_F −0.10 ‰).

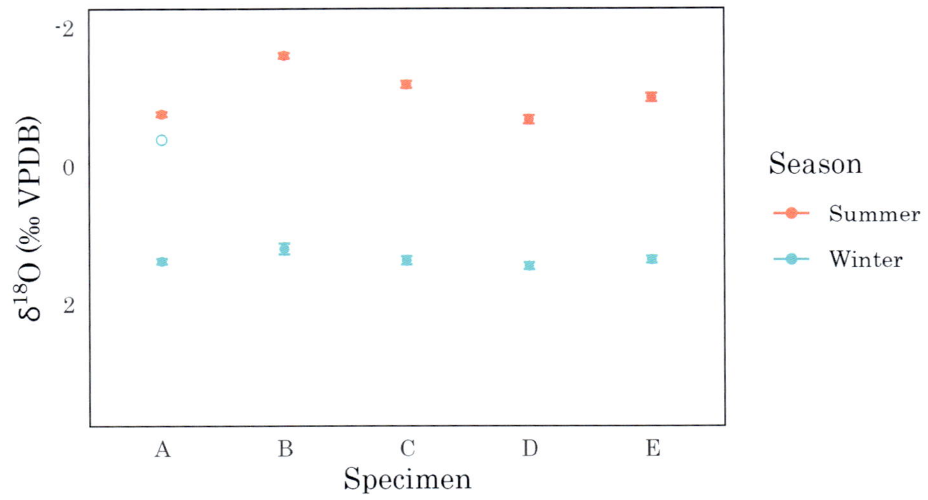

Fig. 71 Seasonal $\delta^{18}O$ values for unit FAS 189.

Specimen FAS_195_B does not preserve any edge increment. At the 30 µm scan of one side, a warmer period is present followed by a colder one. This is also evident in the line scan where one may notice a cooling trend. However, given that the edges are absent, no season of collection was determined, nor was an age of the sample deducted. Nonetheless, two oxygen isotope samples were extracted: The winter increment had the value of 1.72 ‰, while the summer increment had a value of −0.72 ‰. Interestingly, the summer value is lower compared to the previous summers of the specimens presented until this point, as well as the winter increment.

Mainly the apex area is preserved from specimen FAS_195_C. For this reason, no significant information could be extracted from the specimen. The line scan run in the very limited calcitic area, only provides a very partial picture of seasonal cyclicity.

FAS 189

Specimen FAS_189_A belongs to the Lower Mesolithic period. The specimen does not preserve any edge increments. As the posterior side does not preserve enough calcitic material, a 30 µm scan was run on the anterior side. On the anterior side calcitic area, a colder winter period is present followed by a warmer summer period. The line scan depicted this warming temperature trend. As the shell does not preserve an edge, no season of collection was determined nor an age determination was made. Two oxygen isotope samples were extracted: The summer increment had a value of −0.84 ‰ (close to the respective value of specimen FAS_195_B of the previous unit), while the winter increment had a relatively low value of 1.39 ‰, representing a warmer winter than the previous specimens (**fig. 71**).

FAS 186

Specimen FAS_186_A also belongs to the Lower Mesolithic period. The specimen exhibits clear patterns of seasonal cyclicity, which are better visible on the posterior edge. A summer period is present followed by a winter period, which is then followed by a summer period. After the latest in time summer period, a cooling trend appears both on the linear scan, as well as on the elemental map of the posterior. This indicates that the shell was collected during autumn and that the specimen is at least 1.5 years old. Three oxygen isotope samples were extracted from this specimen: The last summer increment had a value of −0.60 ‰, the winter one 0.67 ‰ and the penultimate summer 0.09 ‰ (**fig. 72**). Wintertime appears to be relatively short and a time when the mollusc did not grow significantly. Both from the Mg/Ca ratio (lowest value approximately 0.50) as well as from the isotope results, the winter recorded in this specimen appears to be relatively warm.

FAS 182

Specimen FAS_182_A belongs to the Lower Mesolithic period. The specimen gave well-defined seasonal cyclicity results. An extended winter increment is followed by a summer increment as seen on the elemental maps. This is clearly visible also in the linear scan of the anterior, which shows a stable trend. All of the above indicate that the specimen

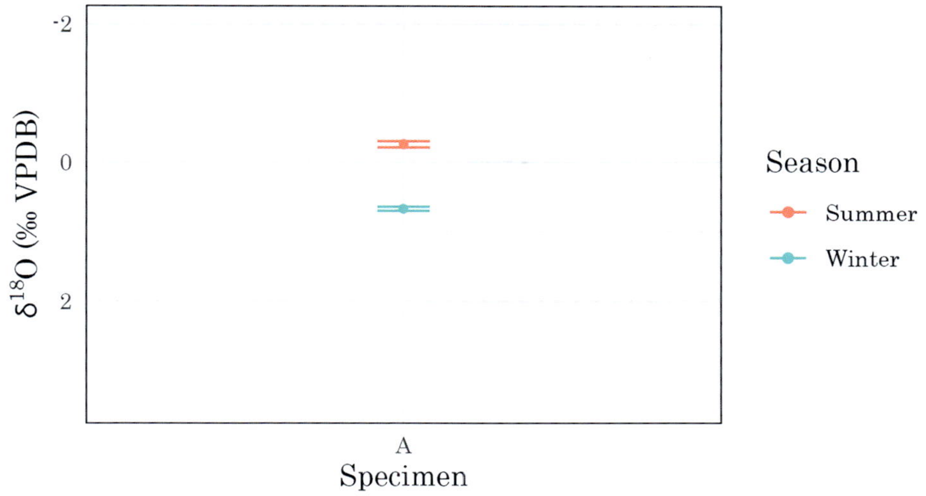

Fig. 72 Seasonal $\delta^{18}O$ values for unit FAS 186.

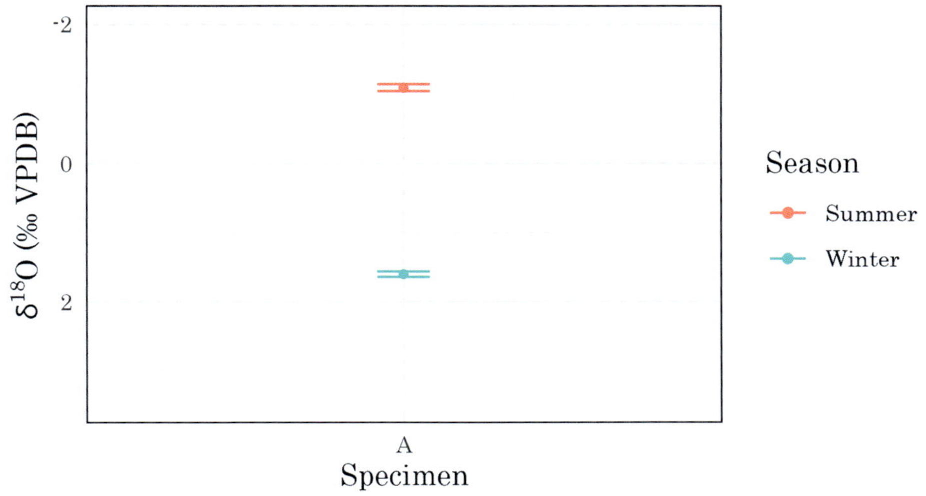

Fig. 73 Seasonal $\delta^{18}O$ values for unit FAS 182.

was collected during the summer and that the mollusc was at least 1 year old when it was collected. In this case, two oxygen isotope samples were extracted from the specimen: The summer increment close to the edge had a value of −1.09 ‰, and the winter increment a value of 1.60 ‰ (**fig. 73**). The summer value is the lowest in the stratigraphy up to this point.

FAS 174

Excavation unit FAS 174 is dated to the Lower Mesolithic period and contains two specimens. Low Mg/Ca ratio is present in the elemental maps, however the specimen does not exhibit significant patterns of seasonal cyclicity. As the edges are not preserved, no season of collection was identified in this case, while no age determination could be made. Furthermore, the line scan of the posterior does not provide any cyclical patterns that could indicate a season of collection. Due to the above, the specimen was not sampled for isotope analysis. Thus, the specific specimen did not provide any meaningful outcomes.

Specimen FAS_174_B shows patterns of seasonal cyclicity, while elemental data are more clear on the posterior rather than the anterior. A winter period is present followed by a warmer period that can be interpreted as a liminal spring/summer period. This is further supported by the warming trend that is well visible in the linear scan of the posterior. The specimen is approximately 1 year old. Two oxygen isotope samples were extracted: One from the warmer increment area that yielded a value of 0.29 ‰, and one from the colder area that gave a value of 2.23 ‰ (**fig. 74**). The winter value is similar to the cold winters of unit FAS 195, while the summer value is not

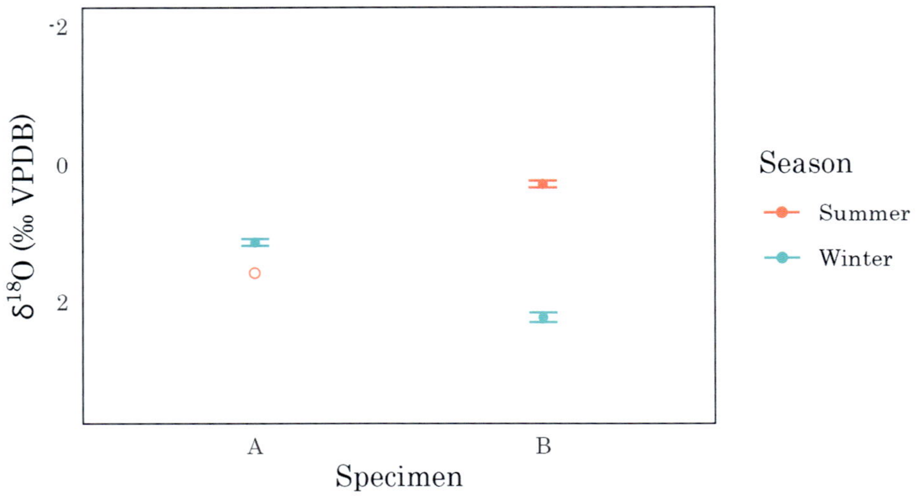

Fig. 74 Seasonal $\delta^{18}O$ values for unit FAS 174.

as low as possibly expected, which is in accordance with the liminal period between spring and summer.

FAS 156

Moving forward in time, excavation unit FAS 156 dates to the Upper Mesolithic period and contains twelve specimens. The excavation unit winter increment isotope average is 1.82 ‰ and the summer average is −0.52 ‰ (**fig. 75**).

Specimen FAS_156_A is quite fragmented and no seasonal cyclicity patterns could be discerned in the elemental data. Low Mg/Ca ratios are visible both in the elemental maps as well as the line scan, however as the edge is not preserved neither a season of collection nor the age could be determined. The specimen was subsequently not sampled for oxygen isotopes.

Specimen FAS_156_B also constitutes a fragmented sample. Elemental Mg/Ca results are illegible, while the line scan is also incomprehensible and noisy. Seasonal increments are not discernible, cyclical or repetitive rendering the data unclear. No conclusions were drawn from the specific specimen.

Specimen FAS_156_C presented a clear picture of seasonal cyclicity. An extended winter increment is present followed by a small area of higher Mg/Ca ratio. Possibly, there is also a warmer period prior to the winter one. The sample appears to have been collected during the summer and the line scan shows a warming trend. The sample must have been approximately 1.5 years old. One oxygen isotope sample was extracted from the winter increment and had a value of 1.96 ‰. This winter value is slightly warmer from the preceding excavation unit (e. g. FAS_174_B: 2.23 ‰) and quite similar to other specimens of the previous period (e. g. FAS_195_B: 1.72 ‰)

Specimen FAS_156_D presented articulate seasonal variation patterns. The oldest period recorded in the sample is possibly a summer period (near the apex) followed by a winter period, which is then followed by a very thin summer increment followed by a winter period, which subsequently shows signs of warming. The sample is approximately 2.5 years old and was collected during spring. The oldest winter period increment is quite extended while the summer period is characterised by a very thin increment, indicating that the mollusc stopped growing during that time. Two oxygen isotope samples were extracted from the specimen: One from the same summer increment which had a value of 0.79 ‰ and one from the penultimate long winter increment that yielded a value of 2.35 ‰. The winter increment indicates a winter as cold as others of the previous Upper Mesolithic period (e. g. FAS_174_B: 2.23 ‰).

Specimen FAS_156_E reflected clear and distinct seasonal variability patterns. There is a potentially summer increment close to the apex which is followed by a broad winter increment, followed by a summer increment, which is then followed by a declining Mg/Ca ratio area. The drop in the Mg/Ca ratio is visible on the posterior extremity of the shell as well as in the line scan, which consists of a quasi-sinusoidal curve. Thus, the shell was collected during autumn and was at least 1 or 1.5 years old. Two oxygen isotope samples were retrieved from this specimen: One from the last summer of growth that had a

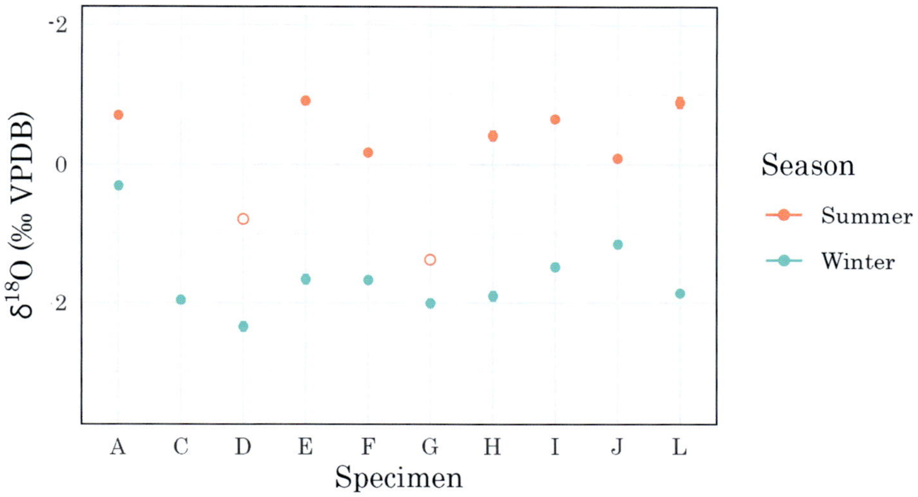

Fig. 75 Seasonal $\delta^{18}O$ values for unit FAS 156.

value of −0.91 ‰, and one from the broad winter increment that had a value of 1.66 ‰. Both values are quite similar to that of the Lower Mesolithic specimen FAS_182_A: summer: −1.09 ‰, winter 1.60 ‰.

In the case of specimen FAS_156_F, the apex was not preserved. The remaining shell material preserved quite restricted calcitic parts, with the posterior presenting a lot of noise in the 30 µm scan and the anterior presenting a winter period, followed by a potentially summer period. As the shell sample is fragmented and small, no conclusion regarding the season of collection was drawn nor was an age determination made. The line scan depicts a warming trend. An oxygen isotope sample was extracted from the potential summer increment and had a value of −0.17 ‰, as well as from the winter increment that had a value of 1.67 ‰.

Specimen FAS_156_G recorded three winter periods and three thin summer periods (accompanied by growth stops), which are well visible on the anterior edge. Three summer peaks are present in the line scan, whereas the last peak is not an actual peak, but noise in the laser signal. The declining cooling trend of the Mg/Ca ratio after the last summer, indicates an autumn season of collection. The mollusc was 3 years old at the time that it was collected. Five carbonate powder samples were extracted for oxygen isotope analysis: Three from each winter increment that had a value of 2.14, 1.86, 2.02 ‰ respectively and two from the summer increments (excluding the last summer) with values of 1.74 and 1.01 ‰. The winter increment values gave consistently high values that match well the winters of this period, whereas the summer increments and especially the first summer value (1.74 ‰) yielded also quite high values, reflecting cold summers. This may mean that the summers recorded in this specimen were unusually cold or it may mean that these results are a product of time-averaging as the increments are too thin to be accurately sampled.

Specimen FAS_156_H does not preserve nor a large amount of calcitic parts nor an edge. For this reason, no season of collection was identified in this case and no age determination was made. A higher Mg/Ca ratio area is present in the elemental map towards the apex and lower towards the extremities. Furthermore, the line scan also depicts this seasonal variation and follows a cooling trend. Two oxygen isotope samples were retrieved from this specimen: One from the broader winter increment towards the edge that had a value of 1.90 ‰ and one from the warmer seasonal increment area that had a value of −0.41 ‰. This winter value is very close to the mean winter value of this chronological period (1.83 ‰). Lastly, even though the edges are not preserved in certain specimens, they can still provide useful isotope results, as in this case.

Specimen FAS_156_I gave very similar results to that of specimen FAS_156_H: Higher Mg/Ca ratio towards the apex and lower towards the edge. In addition, isotope results were similar to some degree although slightly warmer: The winter increment had a value of 1.48 ‰ and the summer increment a value of −0.65 ‰. The specimen was collected during winter and the line scan illustrates a cooling effect, while it was less than 1 year old.

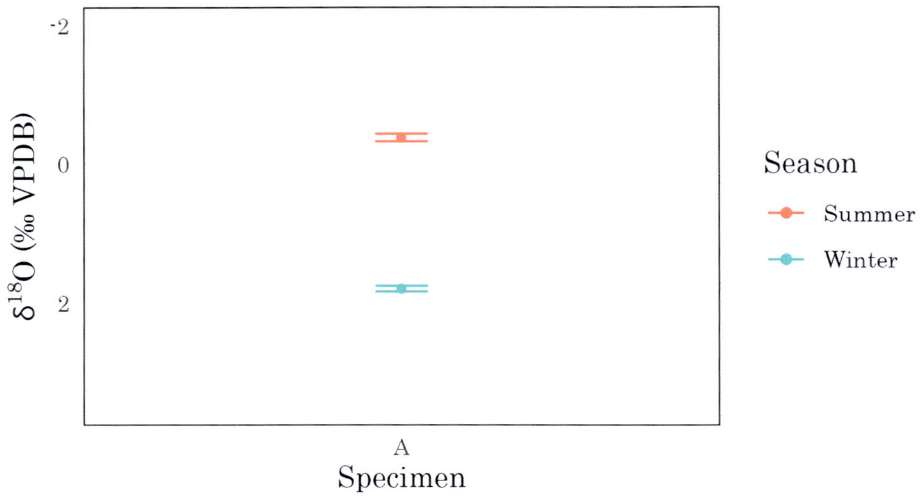

Fig. 76 Seasonal $\delta^{18}O$ values for unit FAS 155.

Specimen FAS_156_J recorded a summer and a winter period. These two periods are also recorded in the linear scan which exhibits a quasi-sinusoidal curve. The edges are not preserved in the specimen and thus no season of collection was attributed to the shell nor was an age determination made. Two oxygen isotopes samples were collected: One from the latest winter period (1.15 ‰) and one from the preceding summer period (−0.09 ‰). The winter period is relatively warmer than the period average (1.83 ‰).

Specimen FAS_156_K recorded two winters and two summers in the posterior edge, whereas the anterior edge is not fully preserved. The specimen must have been 2 years old when it was collected. The latest winter is not as pronounced as the penultimate winter, indicating that the shell did not grow during that time. The specimen was collected during the summertime as the latest increment of growth has a high Mg/Ca ratio. No oxygen isotope samples were extracted from this sample.

Specimen FAS_156_L is only partially preserved, while there are no edges which would allow the determination of the season of collection. The sequence of the seasonal increments is not particularly clear in this specimen: On the posterior side, a noisy higher relative temperature signal is present followed by a colder seasonal increment. The line scan of the posterior has noise in the higher Mg/Ca ratio area. Two oxygen isotope samples were extracted: One from the winter increment (1.86 ‰) and one from the summer increment (−0.89 ‰). They both gave typical results for this chronological period.

FAS 155

Excavation unit FAS 155 dates to the Upper Mesolithic period and contained only specimen FAS_155_A. The specimen recorded two summers and one winter. A cooling trend is present in the line scan, and based on the anterior side of the shell where a winter increment of growth is the last recorded, I concluded that the shell was collected during the winter. The specimen was at least 1.5 years old. Three oxygen isotope samples were collected: The latest summer increment had the value of −0.84 ‰ and the penultimate summer had a value of 0.09 ‰. The winter increment had a value of 1.79 ‰, which is quite close to the period average (1.83 ‰) (**fig. 76**).

FAS 151

This excavation unit dates to the Final Mesolithic period and contained one specimen. FAS_151_A recorded approximately three or four seasonal increments. There is a potential winter increment close to the apex, followed by a summer increment, then a winter increment and the last increment of growth is a spring/summer. Thus, the specimen was at least 1.5 years old, if not older. On the linear scan, which captures the three seasonal increments, there is a lot of noise in the area of the penultimate summer increment, while a warming trend is reflected close to the edge of the specimen, which corresponds to a spring/summer period. Three oxygen isotope samples were collected: The two summer increments yielded −0.50 ‰ (the latest) and −0.87 ‰ (summer average 0.69 ‰) and the winter one 1.98 ‰ (**fig. 77**). The last summer increment has a slightly higher val-

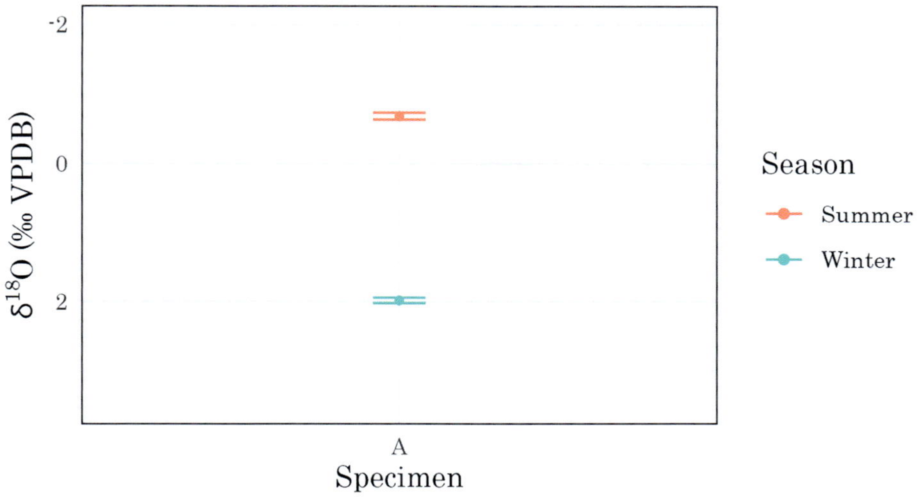

Fig. 77 Seasonal δ18O values for unit FAS 151.

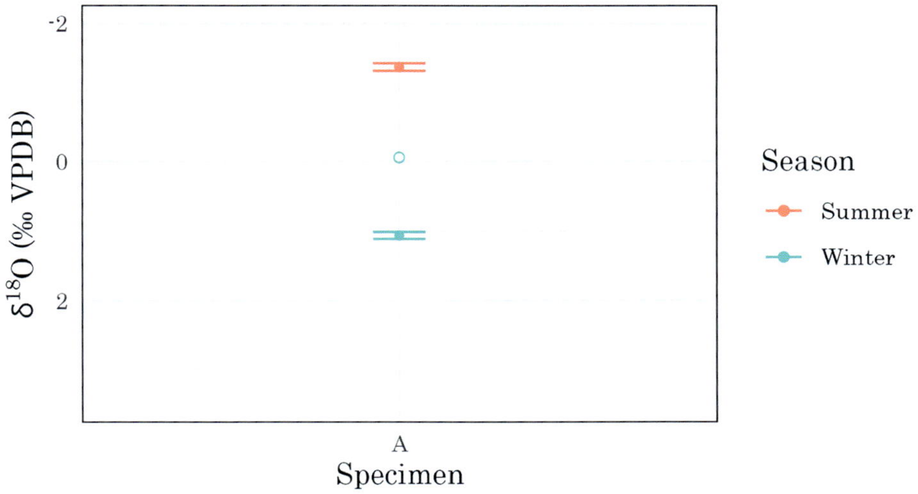

Fig. 78 Seasonal δ18O values for unit FAS 148.

ue than the penultimate, which could be due to the fact that the shell was collected during late spring or early summer. The winter increment is colder than the period average (1.63 ‰).

FAS 148

Excavation unit FAS 148 dates to the Final Mesolithic/Initial Neolithic period and contained one specimen. FAS_148_A reflects two summers and one winter. The increments are well visible on the anterior side of the shell, whereas the posterior has a crack also seen in the respective line scan. The line scan of the anterior has a lot of noise in the area of the penultimate summer and the line scan was repeated on the posterior which clearly reflects the rising summer temperatures. The oxygen isotope sample of the last summer had a value of −1.60 ‰, whereas the penultimate summer had a value of −1.14 ‰ (summer average 1.37 ‰) (**fig. 78**). The winter oxygen isotope sample had a value of 1.06 ‰. The specimen was most probably collected during summer and was 1.5 years old. The winter increment is warmer than the period average (1.63 ‰), as well as warmer than the winter of the previous excavation unit (1.98 ‰).

FAS 145

Excavation unit FAS 145 dates to the Initial Neolithic period. There are two summer increments and two winter increments reflected in specimen FAS_145_A. The shell was at least 2 years old. In addition, the shell was probably collected during the summer, given that there is a summer increment visible on the edge of the anterior side of the shell, which is not

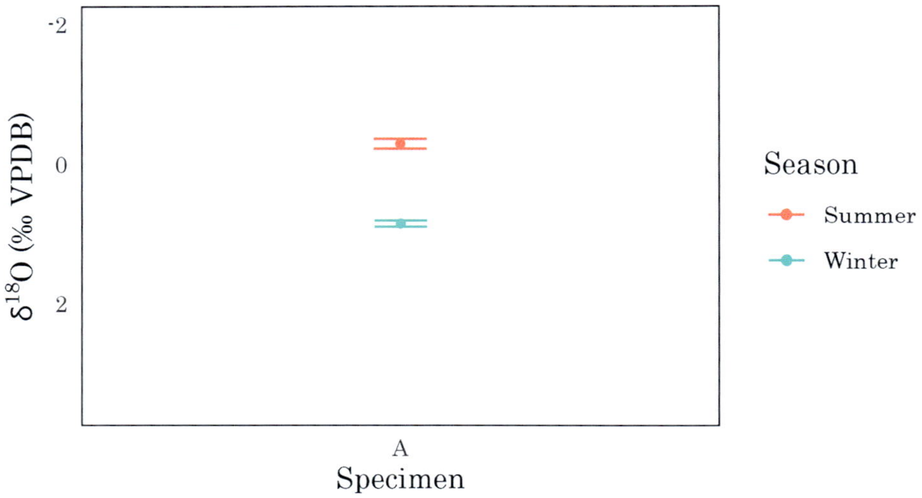

Fig. 79 Seasonal δ^{18}O values for unit FAS 145.

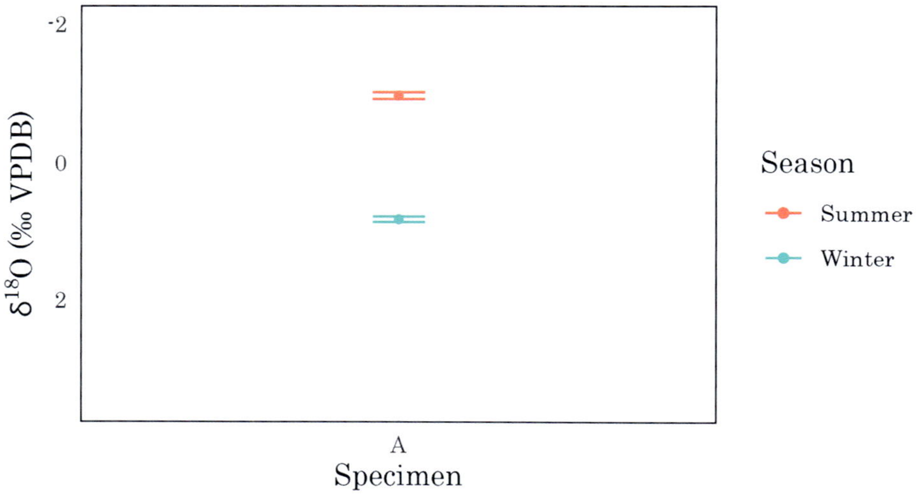

Fig. 80 Seasonal δ^{18}O values for unit FAS 142.

fully preserved. Three oxygen isotope samples were retrieved from the posterior of the shell (**fig. 79**). The last winter increment had a value of 1.18 ‰, while the penultimate winter increment had a value of 0.51 ‰ (winter average 0.85 ‰). The penultimate summer had a value of −0.29 ‰. Intriguingly, the winters of the Initial Neolithic period are on average warmer (0.64 ‰) than the preceding and succeeding winter averages.

FAS 142

Excavation unit FAS 142 also dates to the Initial Neolithic period. Specimen FAS_142_A reflects two summers and one winter, which is not very pronounced. The shell was collected during the summer, as the last increment of growth has a high Mg/Ca ratio, which is also visible in the line scan of the posterior. The summer increments had an oxygen isotope value of −1.03 ‰ and −0.92 ‰, while the winter increment had a value of 0.82 ‰, which indicates a colder winter than the period average (0.64 ‰) (**fig. 80**).

FAS 139

Excavation unit FAS 139 dates to the Initial Neolithic period. Specimen FAS_139_A reflects a winter followed by a summer increment. The shell was probably collected during a summer/autumn period as both the elemental map and the line scan indicate a cooling trend following the high Mg/Ca values. Its age could not be confidently determined. The summer increment is quite thin and difficult to sample for oxygen isotopes. Nonetheless, it yielded a value of −0.22 ‰, while the winter increment yielded a value of 0.70 ‰ (**fig. 81**).

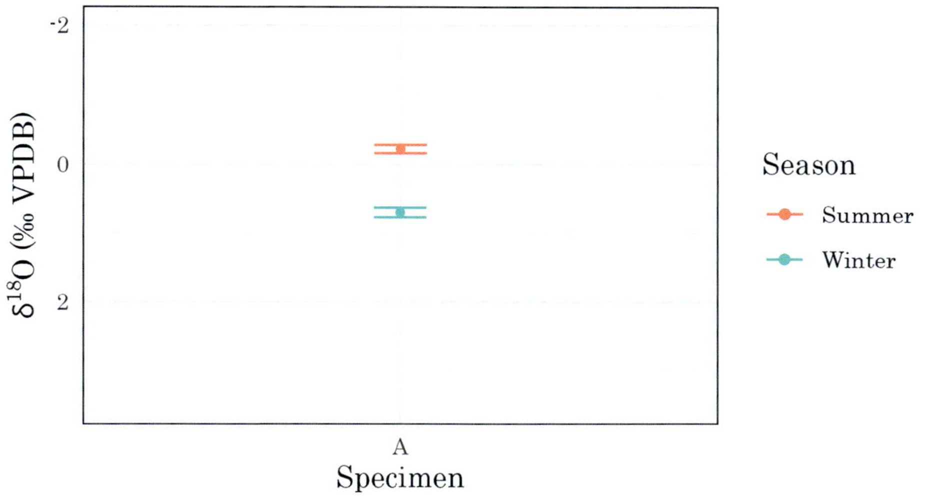

Fig. 81 Seasonal $\delta^{18}O$ values for unit FAS 139.

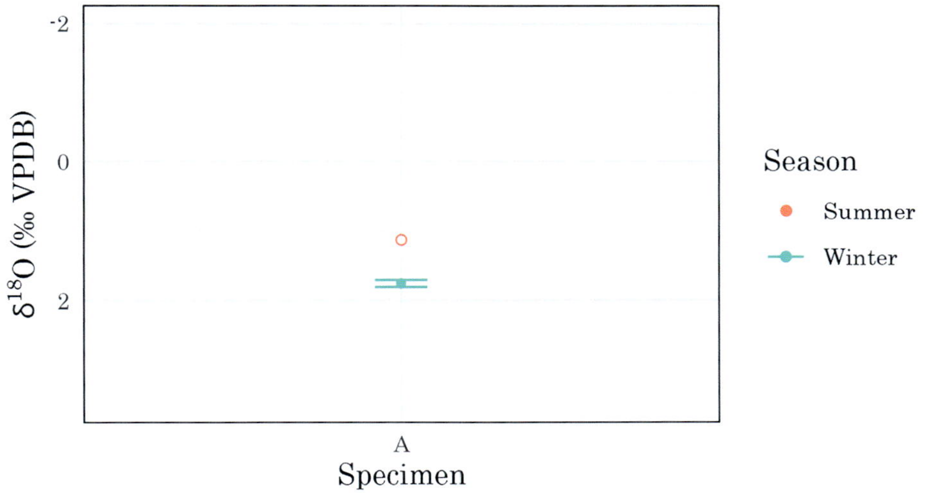

Fig. 82 Seasonal $\delta^{18}O$ values for unit FAS 128.

FAS 128

Excavation unit FAS 128 dates to the Early-Middle Neolithic period. Specimen FAS_128_A recorded two winter periods and one thin summer increment. A warming trend is visible in the elemental data which indicates that the specimen was collected during spring. Furthermore, the specimen was 1.5 years old. The last winter increment had a value of 1.76 ‰ and the thin summer increment a value of 1.13 ‰, which is quite high and maybe not accurate due to time-averaging while sampling (**fig. 82**). The winter increment is colder than the period average (1.43 ‰).

FAS 124

Excavation unit FAS 124 dates to the Middle Neolithic period and contains two specimens. The excavation unit winter increment isotope average is 1.92 ‰ (**fig. 83**).

Specimen FAS_124_A recorded well-defined and broad seasonal increments. More specifically, it recorded three winter periods and three summer periods. The shell was probably collected during summer and was 3 years old. A warming trend is visible in the line scan of the posterior indicating that it was collected during early summer. Three oxygen isotope samples were extracted from the posterior edge of the shell: The penultimate summer increment had a value of 0.14 ‰ and the third summer increment (the oldest) 0.32 ‰. These summers are relatively cold. The winter increment between these two values gave a result of 1.39 ‰.

Specimen FAS_124_B recorded two summer periods and a winter period. The last increment of

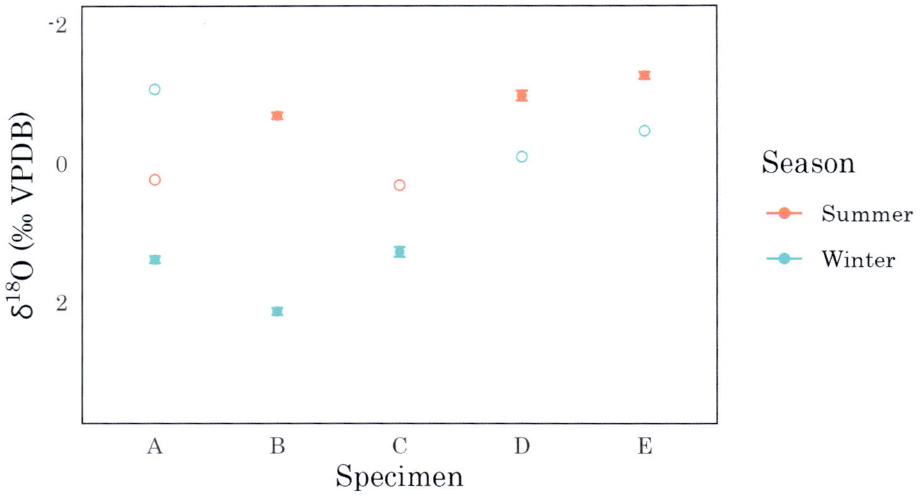

Fig. 83 Seasonal $\delta^{18}O$ values for unit FAS 124.

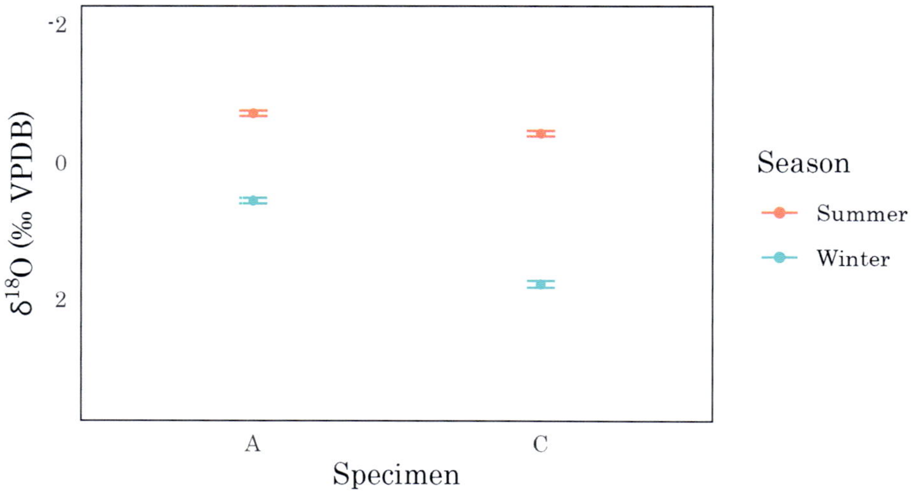

Fig. 84 Seasonal $\delta^{18}O$ values for unit FAS 122.

growth indicates a summer/autumn season of collection, which is visible in the line scan of the posterior. The specimen was 1.5 years old when it was collected. The summer increment had an oxygen isotope value of −0.76 ‰, while the winter oxygen isotope sample had the value of 2.44 ‰. This winter is particularly cold, and as well as colder than the period average (1.27 ‰). A wide range of seasonal extremes are present in this shell.

FAS 122

Excavation unit FAS 122 dates to the Middle Neolithic period and contains three specimens. The excavation unit winter increment isotope average is 1.17 ‰ and the summer average is −0.57 ‰ (**fig. 84**).

Specimen FAS_122_A recorded a winter period which is then followed by a warmer period towards the edge. This is also visible in the line scan of the posterior. Nonetheless, the edge is not preserved and thus no season of collection could be determined, nor an age determination made. Two oxygen isotope samples were extracted from the sample: The $\delta^{18}O$ sample from the potential summer increment which had a relatively low value for a summer (−0.72 ‰), which could indicate that this warmer increment is not actually a summer. On the other hand, this summer value is not dissimilar to other summer increments of this period (e. g. FAS_124_B, −0.76 ‰). The winter increment was relatively low (0.56 ‰).

Specimen FAS_122_B did not show interpretable elemental patterns. The elemental maps show a homogeneous low Mg/Ca ratio across the shell. The edge is not preserved, thus no results could be drawn from the specimen.

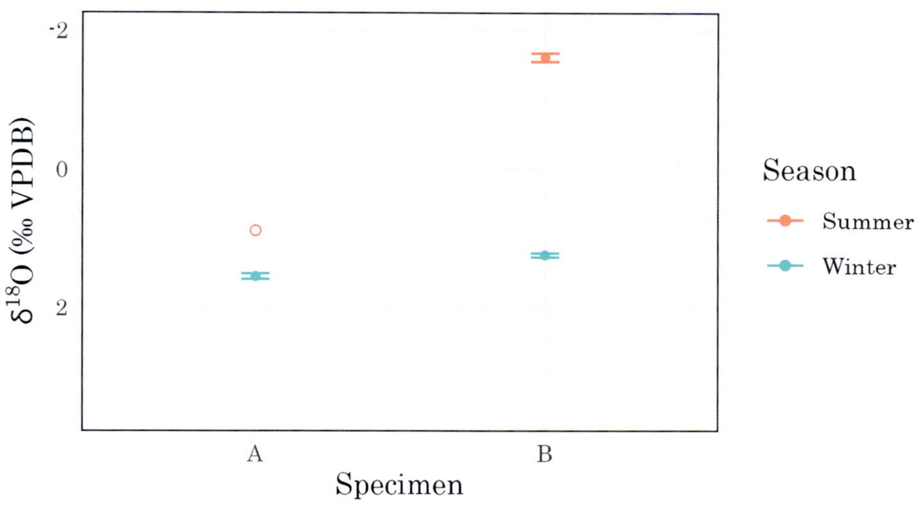

Fig. 85 Seasonal δ18O values for unit FAS 91.

Specimen FAS_122_C reflects a summer and a winter period, which is probably followed by another summer towards the edge. However, the specimen is not fully preserved and for this reason I could not conclude on a certain season of collection or an age estimation. Two oxygen isotope samples were extracted: The winter increment had a value of 1.78 ‰, while the oldest summer increment towards the apex had a value of −0.43 ‰. The winter increment shows a colder relative temperature compared to the period average (1.27 ‰).

FAS 91

Excavation unit FAS 91 dates to the Late Neolithic period and contains two specimens. The excavation unit winter increment isotope average is 1.40 ‰ (**fig. 85**).

Specimen FAS_91_A recorded three winter and two summer periods. The summer increments are particularly thin, rendering them difficult to sample for oxygen isotopes. The last increment of growth represents a winter season, while the extremely high Mg/Ca ratio towards the edge constitutes noise due to the focusing of the laser, rather than a summer season. In addition, the specimen was approximately 2.5 years old. Two oxygen isotope samples were extracted from the specimen: The summer increment had a value of 0.88 ‰, while the winter increment had a value of 1.55 ‰. The summer increment gave a relatively high value which may not be particularly accurate due to the increment's thinness, while the winter value is quite close to the average value of this chronological period (1.51 ‰).

Specimen FAS_91_B does not represent similar seasonal cyclicity patterns with those of specimen FAS_91_A. In other words, the two seasonal increments of FAS_91_B are both quite broad. The elemental data show a summer period followed by a winter one, which is also the last. Thus, the specimen was collected during the winter and was 1 year old. Two oxygen isotope samples were extracted: One from the winter increment (1.24 ‰), warmer than the period average, and one from the summer increment, which gave a quite low δ18O value (−1.63 ‰).

FAS 83

Excavation unit FAS 83 dates to the Late Neolithic period and contains three specimens. The excavation unit winter increment isotope average is 1.59 ‰ (**fig. 86**).

Specimen FAS_83_A recorded two winter periods and an unusually broad summer period. The line scan of the anterior illustrates a quasi-sinusoidal curve, which clearly indicates a winter season of collection. The specimen was 1.5 years old when it was collected. Three oxygen isotope samples were retrieved from this shell: The winter increments had values of 1.62 ‰ and 1.08 ‰, whereas the summer increment had a value of −1.39 ‰.

Specimen FAS_83_B recorded four winters and five summers. Interestingly, the different seasonal increments do not have regular distributions in the elemental maps. At least three summer peaks are present in the linear scan, and a fourth peak is also present towards the edge of the sample. The high Mg/Ca ratio is also present in the elemental

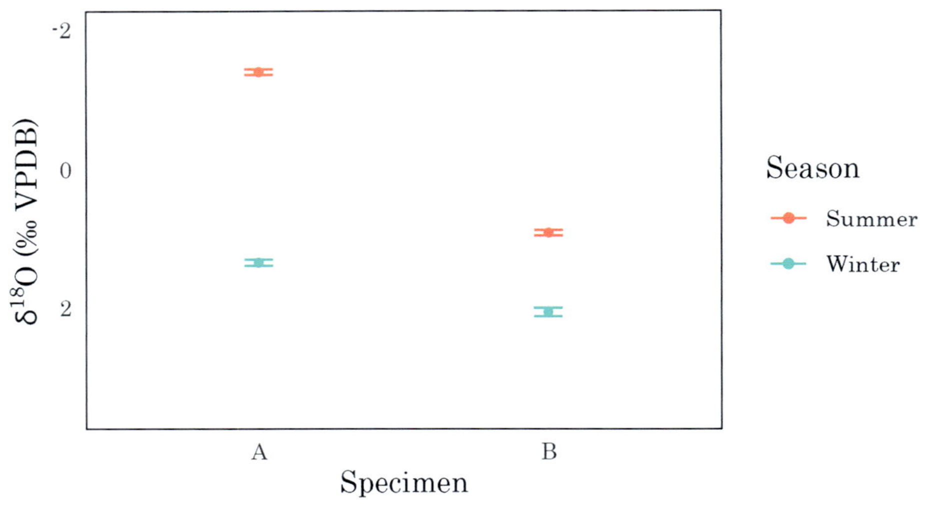

Fig. 86 Seasonal $\delta^{18}O$ values for unit FAS 83.

map. The above indicates that the specimen was at least 4.5 years old, and was collected during summer. Two oxygen isotope samples were collected from this specimen: A summer increment yielded a value of 0.92 ‰, which is relatively higher than expected, and a winter increment yielded a $\delta^{18}O$ value of 2.07 ‰, which is colder than the period average (1.51 ‰).

Little information could be extracted from specimen FAS_83_C as it did not preserve a significant amount of calcitic material. The elemental data both from the 2D maps as well as from the line scan were unclear and could not be interpreted with confidence.

FAS 74

The last excavation unit of this trench FAS 74 contained three shells and dates to the Final Neolithic period. Specimen FAS_74_A is among the oldest specimens that was analysed. It reflects five winters and five summers and was thus 5 years old. The summer increments are quite thin indicating that the specimen could have stopped growing during the summer. The last increment of growth has a high Mg/Ca ratio indicating that the shell was collected during the summer, while the line scan shows a warming trend. Two oxygen isotope samples were retrieved from this specimen: The winter increment had the value of 1.47 ‰, and the summer increment a value of 0.09 ‰ (**fig. 87**). The value is relatively high and could be due to time-averaging effects.

Specimen FAS_74_B provided unclear patterns of seasonal cyclicity. It appears that there are two winters and one summer recorded in the shell, which was probably collected during winter. The age of the specimen was unclear and could be close to 1 year old. Given that the elemental data were not clear enough and the seasonal extremes not obvious, no oxygen isotope samples were collected from this specimen.

Specimen FAS_74_C presented rather homogeneous seasonal increments. This specimen is characterised by low Mg/Ca ratios in its entire extent. This is probably because the shell was less than 1 year old. The mollusc was collected during winter. Due to the homogeneity of its calcitic increments, no oxygen isotope sampling took place.

4.2.1.2. Trench FAN

FAN 175

Excavation unit FAN 175 dates to the Upper Mesolithic period and contains three specimens. Specimen FAN_175_A presented higher Mg/Ca ratios towards the apex and lower in the rest of its calcitic parts towards the edge. However, the edge is not preserved and for this reason no season of collection could be identified nor an age determination made. The specimen was not sampled for isotope analysis.

Specimen FAN_175_B is a fragment and did not preserve any calcitic material.

Specimen FAN_175_C reflected a colder period towards the apex and then a warmer one. However, the edge is not preserved and thus no further conclusions could be drawn.

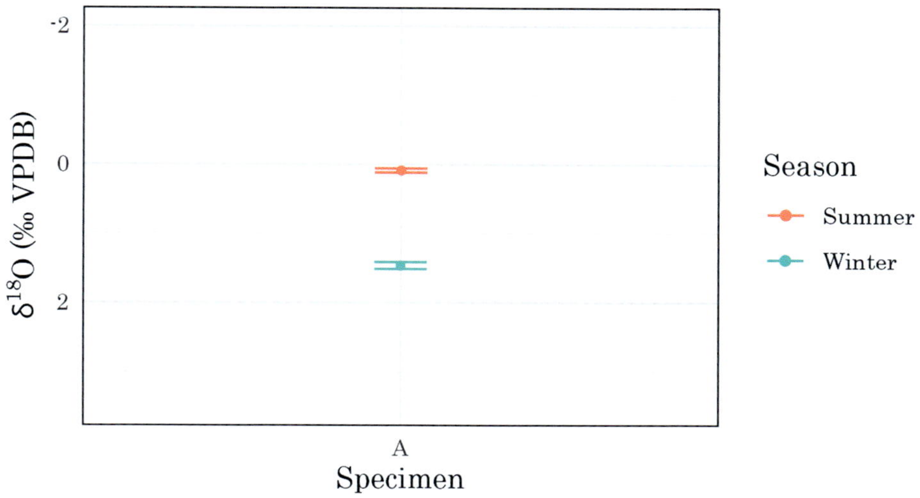

Fig. 87 Seasonal $\delta^{18}O$ values for unit FAS 74.

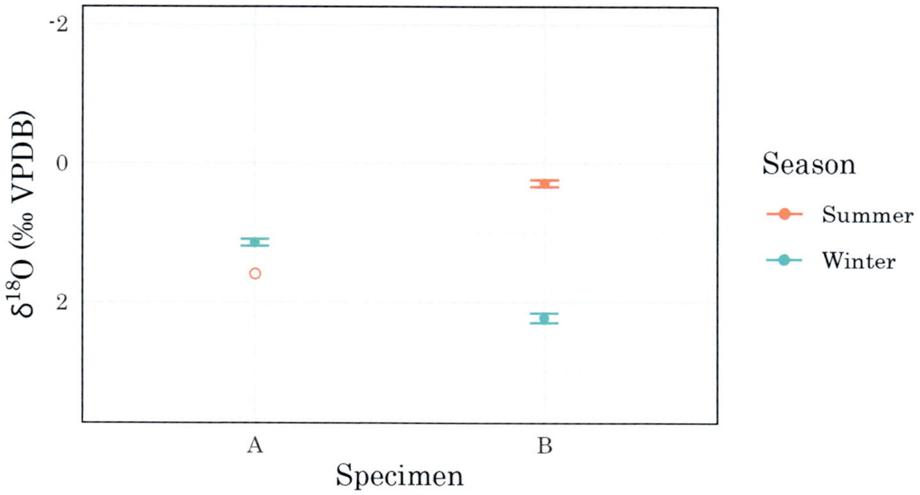

Fig. 88 Seasonal $\delta^{18}O$ values for unit FAN 174.

FAN 174

Excavation unit FAN 174 dates to the Final Mesolithic period. Specimen FAN_174_A recorded two winter periods and a very thin summer increment in between. The specimen was collected during winter and was approximately 2 years old. Two oxygen isotope samples were retrieved from this specimen: The summer increment had a very high value of 1.59 ‰, which is probably because it could not be accurately sampled, while the winter increment had a value of 1.14 ‰, which is warmer than the period average (1.63 ‰) (**fig. 88**).

FAN 172

Excavation unit FAN 172 dates to the Final Mesolithic period and contains three specimens. The excavation unit winter increment isotope average is 1.70 ‰ (**fig. 89**).

Specimen FAN_172_A is characterised by a very thick apex, which is visible in the 100 µm scan. The sample reflects three winter and three summer periods. The last increment of growth has higher Mg/Ca ratios following a relatively short winter increment. This warming trend indicates that the shell was collected during spring. The mollusc was approximately 3 years old when it died. Two oxygen isotope samples were extracted: The winter increment had a value of 1.73 ‰, while the summer increment had a value of 0.04 ‰. The winter value indicates a colder winter compared to the average value of the Final Mesolithic (1.63 ‰).

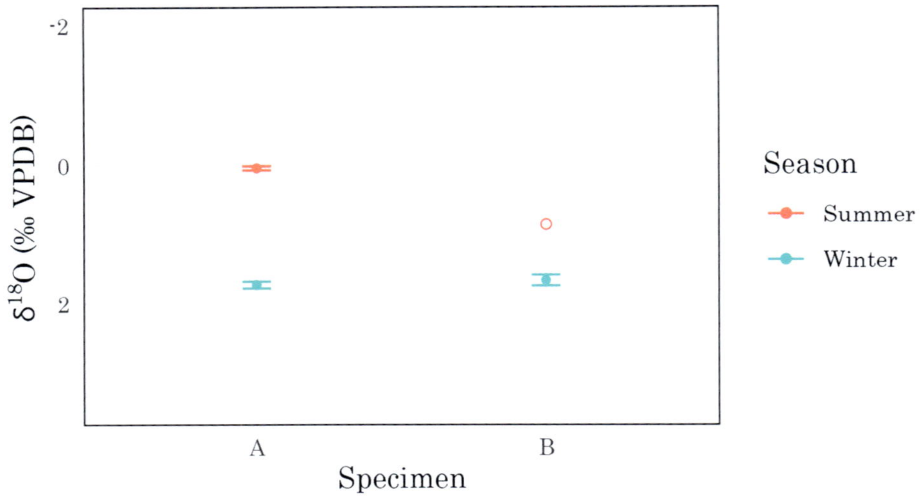

Fig. 89 Seasonal $\delta^{18}O$ values for unit FAN 172.

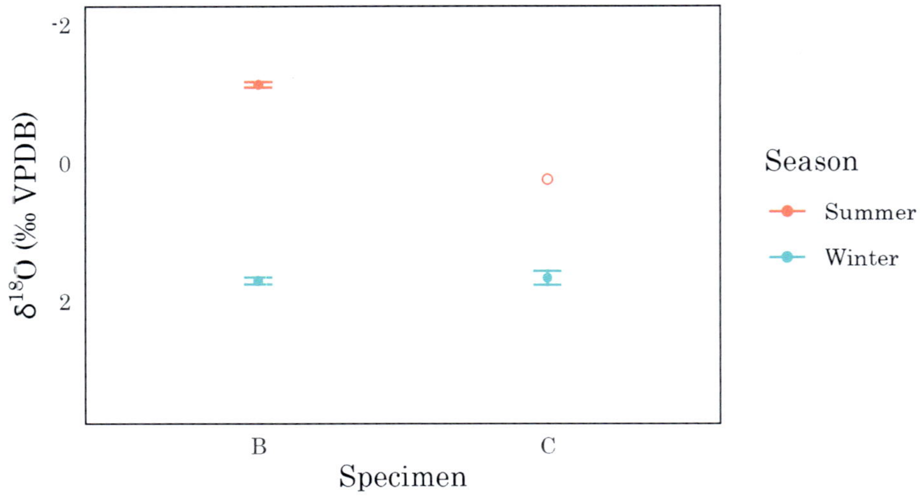

Fig. 90 Seasonal $\delta^{18}O$ values for unit FAN 163.

Specimen FAN_172_B recorded three winter periods and two summer periods and was 2.5 years old. Based on the anterior side of the shell, the specimen was collected during winter. Two oxygen isotope samples were extracted: The summer increment had a value of 0.86 ‰, while the winter increment had a value of 1.67 ‰.

Specimen FAN_172_C recorded potentially two summer increments and an extended winter increment. The broad winter increment is rather homogeneous and well visible on both the posterior and the anterior. The mollusc was 1.5 years old and was collected during late spring or early summer as documented by the rising temperatures visible in the line scan as well as the elemental maps. No oxygen isotope samples were extracted from this specimen.

FAN 163

Excavation unit FAN 163 dates to the Final Mesolithic/Initial Neolithic period and contains four specimens. The excavation unit winter increment isotope average is 1.69 ‰ (**fig. 90**).

Specimen FAN_163_A recorded the seasonal increments of three summer periods and two or three winter periods. This renders the mollusc approximately 3 years old. It was probably collected during late summer or early autumn as is visible from the last increment of growth. No oxygen isotope samples were collected from this specimen.

Specimen FAN_163_B has three winter and three summer increments as shown on the anterior edge scan. The shell was 3 years old and was collected during summer. One of the relatively broad summer

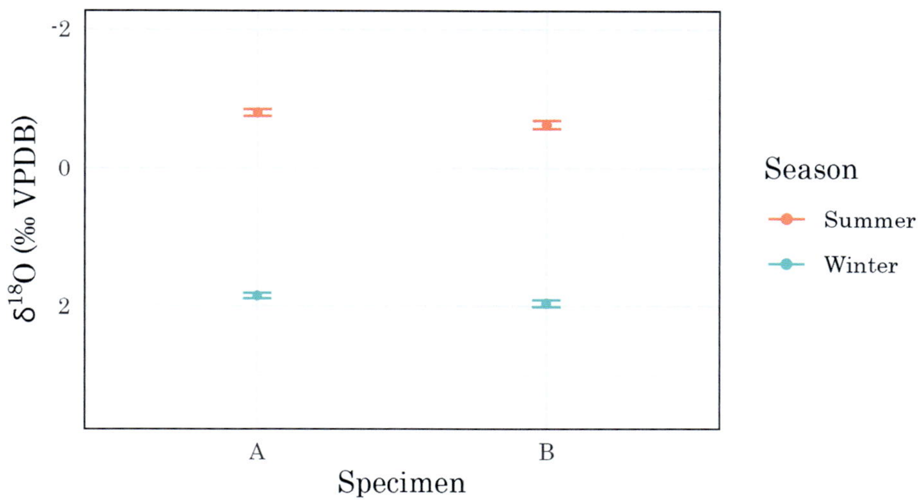

Fig. 91 Seasonal δ¹⁸O values for unit FAN 162.

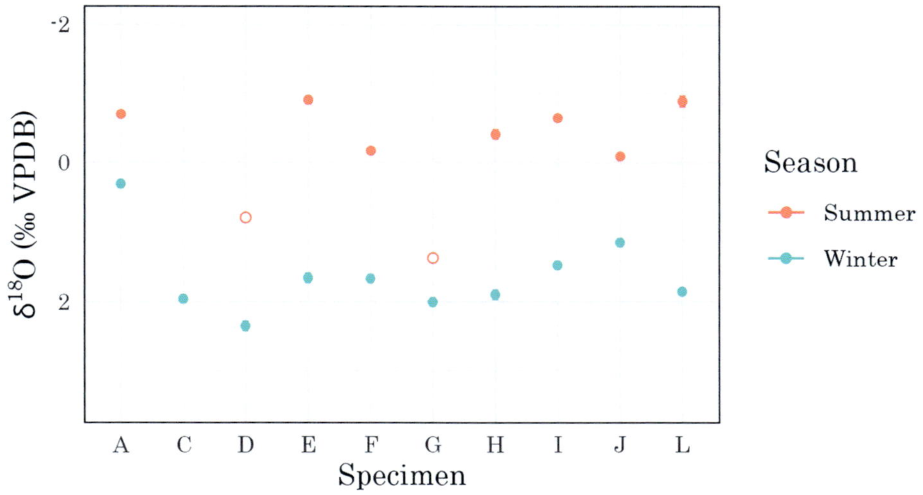

Fig. 92 Seasonal δ¹⁸O values for unit FAN 156.

increments had a value of −1.13 ‰, while the winter increment that was sampled had a value of 1.71 ‰. The winter value indicates a colder winter than the period average (1.65 ‰).

FAN_163_C also recorded three winter and three summer periods in its increments. Nonetheless, the edge of the sample is not preserved and for this reason no season of collection was identified nor an age determination was made. Two oxygen isotopes samples were extracted from this shell: The winter increment yielded a value of 1.67 ‰, while the summer increment a value of 0.25 ‰. The winter value is very close to the period average (1.65 ‰).

Specimen FAN_163_D did not preserve sufficient calcitic layers, nor any of its edges. The 30 µm scan of the apex shows some signs of seasonal cyclicity, where colder increments are followed by warmer seasonal increments. No further conclusions could be drawn from this specimen.

FAN 162

Excavation unit FAN 162 dates to the Final Mesolithic/Initial Neolithic period and contains three specimens. The excavation unit winter increment isotope average is 1.90 ‰ and the summer average is −0.69 ‰ (**fig. 91**).

Specimen FAN_162_A recorded approximately two or three summers and two winters. The specimen was at least 2 years old and was collected during summer. Two oxygen isotopes samples were extracted from this shell: The winter increment had a value of 1.85 ‰, while the summer increment had a

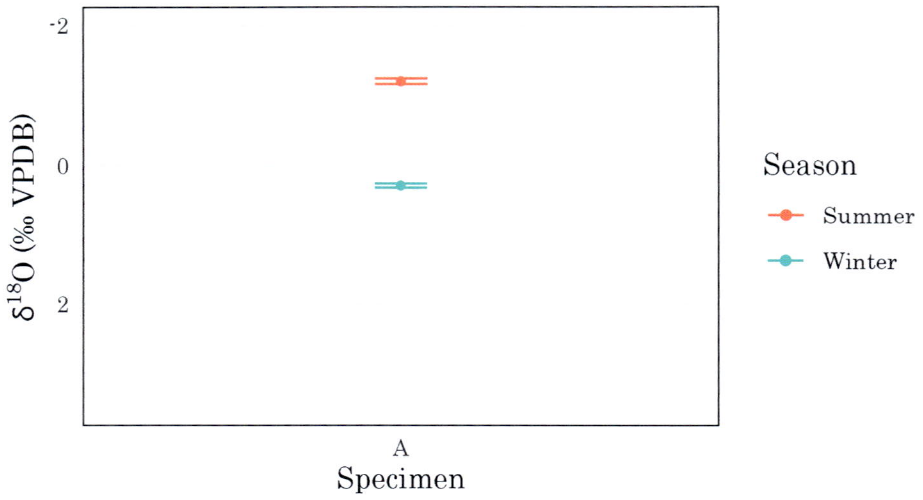

Fig. 93 Seasonal $\delta^{18}O$ values for unit FAN 152.

value of −0.80 ‰. The winter value is colder than the period average (1.65 ‰).

FAN_162_B reflects two summer and one winter period. Thus, the mollusc was 1.5 years old when it was collected. Based on the elemental map of the anterior, the specimen was collected during late summer or early autumn. The two summer increments yielded values of −1.00 ‰ and −0.26 ‰, while the winter had a value of 1.96 ‰ which is colder than the period average.

Specimen FAN_162_C presented low Mg/Ca ratios in its calcitic parts. However, the edge is not preserved and for this reason no season of collection could be identified nor an age determination made. For this reason, the specimen was not sampled for isotopes.

FAN 156

Moving forward in time, excavation unit FAN 156 belongs to the Initial Neolithic period. Specimen FAN_156_A reflected seasonal cyclicity to some extent, with two colder periods and a warmer one present in its increments. As none of the edges was preserved no season of collection could be identified nor an age determination made. Isotope sampling results revealed a relatively warm winter (0.31 ‰), which is warmer than the period average (0.64 ‰) as well as a more typical summer (−0.70 ‰) (**fig. 92**).

FAN 152

FAN_152_A, which also dates to the Initial Neolithic period, is a somewhat peculiar specimen in which seasonal increments are not clearly defined. The anterior edge is not well-preserved, while the posterior indicates the presence of two or three summers, and two winters. Thus, the shell was approximately 2.5 years old and was collected during winter. Two oxygen isotope samples were extracted from this specimen: The winter increment yielded a value of 0.29 ‰, which is also in this case relatively warm compared to the previous periods as well as the period average, while the summer had a value of −1.20 ‰ (**fig. 93**).

FAN 148

Excavation unit FAN 148 belongs to the Early Neolithic period. Specimen FAN_148_A is a fragment and preserved one of its calcitic edges. A warmer period is present towards the apex, followed by a colder winter period, which is then followed by a small area that shows signs of warming again. The specimen was probably collected during the end of spring or early summer. As the shell is a fragment, no age determination could be made. One oxygen isotope sample was extracted from the winter increment which yielded a quite warm value of −0.06 ‰ (**fig. 94**).

FAN 144

Excavation unit FAN 144 belongs to the Early Neolithic period. Specimen FAN_144_A yielded rather homogeneous elemental data of low Mg/Ca values, and a smaller area of higher values towards the apex. The specimen was less than 1 year old and was collected during winter. Two oxygen isotope samples were extracted: The winter sample had a value of 1.71 ‰ and the summer a value of −1.92 ‰ (**fig. 95**).

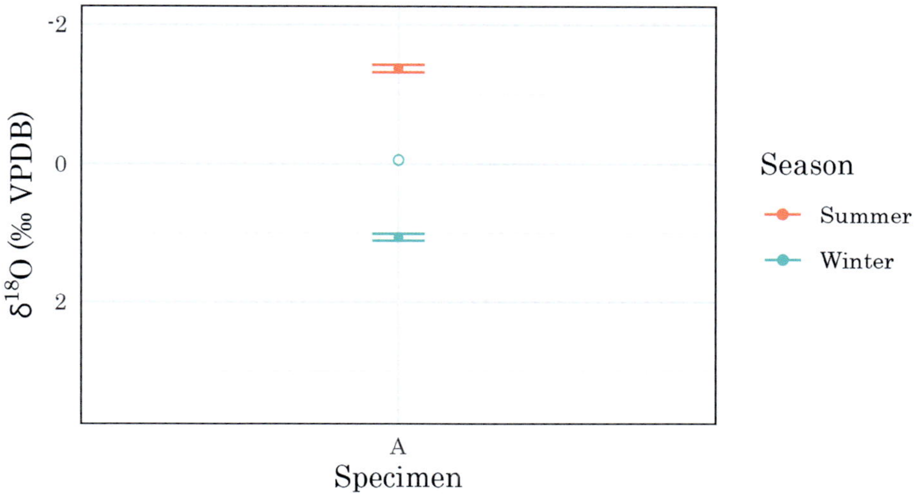

Fig. 94 Seasonal $\delta^{18}O$ values for unit FAN 148.

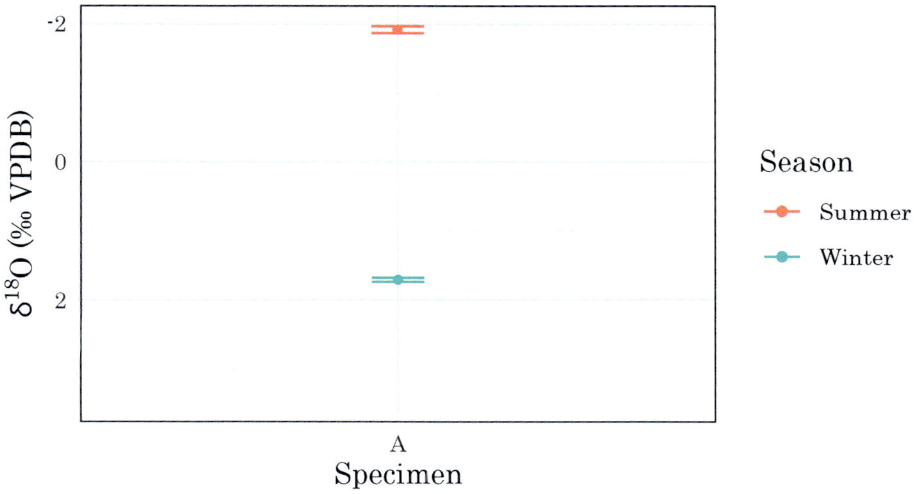

Fig. 95 Seasonal $\delta^{18}O$ values for unit FAN 144.

The winter measurement is colder than this period's average (1.18 ‰), while the specific summer increment yielded the warmest summer value of all the samples analysed in this study.

FAN 134

Excavation unit FAN 134 dates to the Middle Neolithic period and contains six specimens. The excavation unit winter increment isotope average is 1.33 ‰ and the summer average is −0.78 ‰ (**fig. 96**).

Specimen FAN_134_A exhibits well-defined and broad seasonal increments. It reflects in its increments two winter and two summer periods as documented in the 30 µm scan of the posterior. As the shell is a fragment no age determination could be made, while the shell was collected during summer. Two oxygen isotope samples were extracted from the shell: The winter sample had a value of 1.40 ‰, and the summer sample −1.04 ‰.

Specimen FAN_134_B recorded a winter and one or two summers. The mollusc was approximately 1 year old and was collected during autumn as is visible on the line scan which shows a clear cooling trend. Two oxygen isotope samples were extracted from the shell: The summer increment yielded a value of −1.00 ‰, while the winter increment a value of 0.33 ‰. This winter increment is quite warm in comparison to the winter average of the Middle Neolithic (1.27 ‰).

Specimen FAN_134_C shows clear patterns of seasonal alternating increments. More specifically, the specimen shows the presence of two extensive summers and one more faint winter. This is also well-depicted on the line scan of the anterior side.

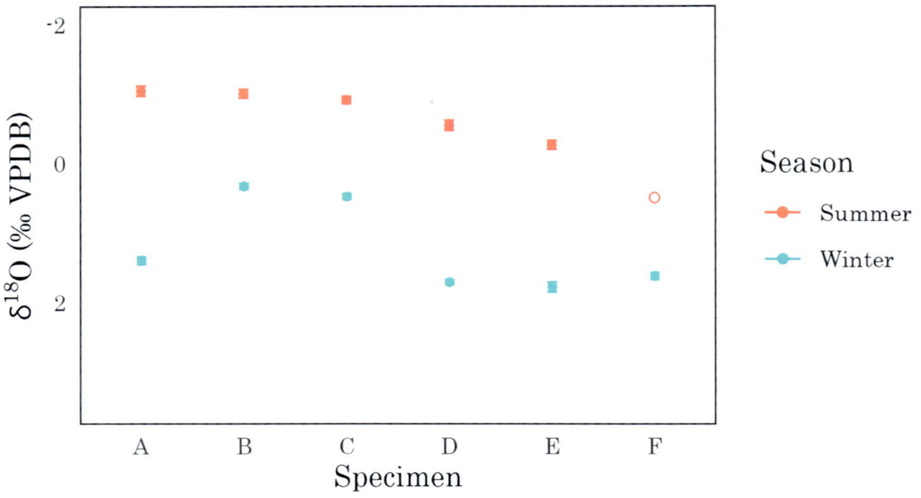

Fig. 96 Seasonal δ18O values for unit FAN 134.

The mollusc was collected during the summer and was 1.5 years old. Three oxygen isotope samples were extracted from the shell: The summer samples had a value of −1.09 ‰ and −0.72 ‰, and the winter sample 0.48 ‰. The Mg/Ca ratio of the winter increment is not particularly low (approximately 0.55), and this »warmer« winter is also reflected in the relatively low oxygen isotope value.

Sample FAN_134_D is a fragment while it presents some noise along its edges, rendering the scan not as clear. The specimen reflects one or two winters and one summer. The warming trend visible in the line indicates that the shell was collected during spring. As the shell is a fragment no age determination could be made. Sampling for isotope analysis yielded a winter increment value of 1.71 ‰ and a summer value of −0.54 ‰. The summer sample is relatively high, while it comes from an area of the shell that is not particularly clear in the 2D elemental map.

Specimen FAN_134_E reflects an extended winter period followed by a warmer summer period. The shell was collected during late summer or early autumn and was less than 1 year old. Two oxygen isotope samples were extracted from the shell: The summer sample had a value of −0.26 ‰, and the winter sample a value of 1.78 ‰. The summer value is relatively high and could potentially represent a spring season rather than a summer, as it was not extracted from the edge of the shell, while the winter value is colder than the period average (1.27 ‰).

Specimen FAN_134_F has recorded in its increments two winter periods and one very thin summer period. The line scan presents an irregularity as a number of smaller peaks are visible but are barely visible in the elemental map. This makes it difficult to characterise them as summer peaks. Thus, the specimen was probably 1.5 years old and was collected during the summer. Three oxygen isotope samples were extracted: The winter samples had a value of 1.77 ‰ and 1.47 ‰ and the summer a value of 0.50 ‰. The thin summer increment was sampled twice in an attempt to best capture the summer period and the lowest value of the two was kept. Nonetheless, the result remains relatively high and could be due to time-averaging.

FAN 132

Excavation unit FAN 132 dates to the Middle Neolithic period and contains five specimens. The excavation unit winter increment isotope average is 0.83 ‰ and the summer average is −1.16 ‰ (**fig. 97**).

Starting from specimen FAN_132_A, indicates the presence of two warmer and one colder period. This is better visible in the line scan, whereas the elemental map does not show pronounced seasonal cyclicity. Even though the elemental map seems rather homogeneous, it appears that the shell was collected during late spring or late summer as it shows a warming trend and was 1.5 years old. Three oxygen isotope samples were extracted from this specimen: The two summer values had values of −1.03 ‰ (last summer) and −1.22 ‰ (penultimate summer), while the winter increment yielded a value of 0.76 ‰, which is very similar to the period average (1.27 ‰).

Specimen FAN_132_B recorded in its increments two winters and two summers. The winter incre-

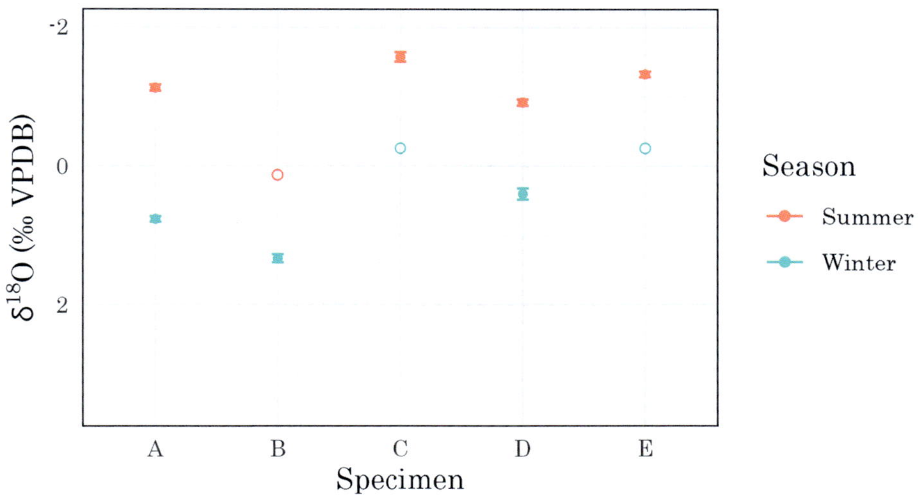

Fig. 97 Seasonal $\delta^{18}O$ values for unit FAN 132.

ments are quite broad, whereas the summer increments are quite thin. Seasonal cyclicity is well-illustrated in the line scan of the posterior side, which indicates that the specimen was 2 years old when it was collected. Furthermore, the shell was collected during winter. Two oxygen isotope samples were collected: The summer sample had a value of 0.13 ‰, while the winter sample 1.33 ‰. The thin summer increment yielded a relatively high result which could be due to time-averaging.

Elemental data of shell FAN_132_C reflect a winter and a summer period. The specimen was probably collected during the summer and was 1 year old. Seasonal increments are broad and not particularly well-defined, while the line scan is difficult to interpret and shows signs of noise towards the edge. This could be due to its young age. Two oxygen isotope samples were collected: The summer sample had a value of −1.57 ‰, while the winter sample had a value of −0.26 ‰. This last value is particularly elevated for a winter increment.

Specimen FAN_132_D recorded two winters and two summers based on the anterior edge of the shell. The last increment of growth shows a summer collection, and the shell was 2 years old when it was collected. Seasonal increments in this sample are clear and distinct. One particular feature of this shell is the abrupt change from the penultimate summer towards the last winter, which is visible in the line scan. Three oxygen isotope samples were collected: The summer samples had values of −1.30 ‰ and −0.53 ‰, whereas the winter increment had a value of 0.40 ‰, which is warmer than the period average (1.27 ‰).

Specimen FAN_132_E has relatively homogeneous increments which are difficult to differentiate. Towards the apex the elemental data have a low Mg/Ca ratio, while in the larger extent of the shell the ratio is higher. The specimen was collected during the summer, and must have been less than 1 year old. Two oxygen isotope samples were collected: The summer sample had a value of −1.32 ‰, while the potential winter had a rather low value of −0.26 ‰.

FAN 131

Excavation unit FAN 131 dates to the Middle Neolithic period and contains four specimens. The excavation unit summer increment isotope average is −0.87 ‰ (**fig. 98**).

Specimen FAN_131_A recorded one or two summers and a winter. The specimen was collected during the summer and was approximately 1 year old. Three oxygen isotope samples were extracted from this shell: The summer increments had values of −1.73 ‰ and −0.71 ‰, whereas the winter increment had a value of 0.78 ‰. The oxygen isotope value of the last summer was quite high.

Specimen FAN_131_B recorded two winters and two or three thin summers. It is difficult to identify the increments towards the edge of the shell which are not particularly well-defined, as there is a thin winter-looking increment between two summer increments. However, the winter-looking increment does not have a low enough Mg/Ca ratio (see line scan). Nonetheless, the mollusc was collected during the summer and was at least 2 years old. The apex was also scanned at a 30 µm resolution, however no

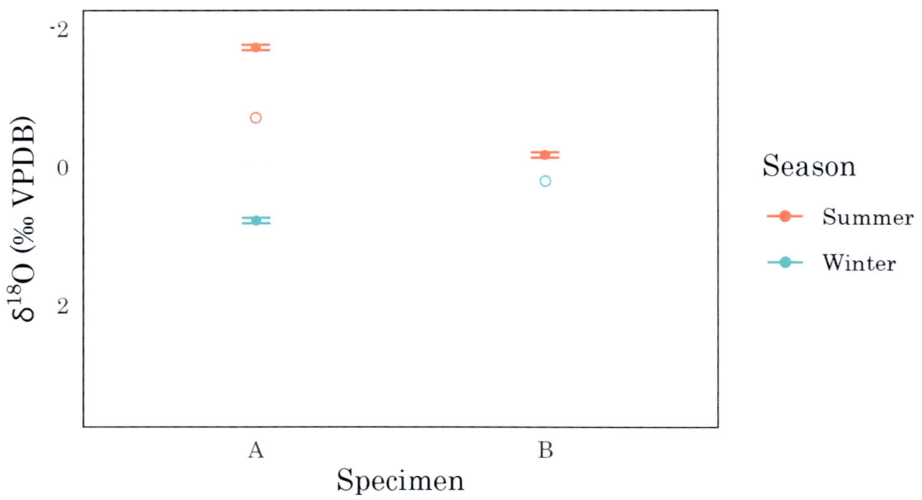

Fig. 98 Seasonal $\delta^{18}O$ values for unit FAN 131.

significant information could be extracted. Two oxygen isotope samples were extracted from this shell: The winter sample yielded a very warm winter as it had a value of 0.21 ‰, which could be a result of time-averaging because of its position, while the summer had a value of −0.17 ‰.

Specimen FAN_131_C does not preserve enough calcitic material or an edge. For this reason, no further conclusions were drawn.

The same is the case for specimen FAN_131_D. Low Mg/Ca ratio is recorded in the calcitic layers of the shell, however no meaningful results could be drawn.

FAN 129

Excavation unit FAN 129 dates to the Middle Neolithic period and contains two specimens. Specimen FAN_129_A presented increments with a small degree of cyclicity. Lower Mg/Ca ratio is present towards the apex of the shell, which gradually increases towards the edge. The mollusc was collected during the summer and was less than 1 year old. Two oxygen isotope samples were extracted: The summer increment produced a value of −1.03 ‰, and the winter a value of −0.50 ‰ (**fig. 99**). The winter value is quite low and this could be due to the specimen's young age, which did not allow the accurate targeting of seasonal extremes.

Specimen FAN_129_B presented very clear patterns of seasonal cyclicity. The posterior of the shell reflects two broad summer and one broad winter increments. Thus, the shell was 1.5 years old when it was collected. A cooling trend at the sample's edge is present both in the elemental map and the line scan which indicate that the shell was collected in autumn. Three oxygen isotope samples were extracted: The summer increments yielded two very consistent results of −0.71 ‰ and −0.65 ‰, while the winter increment yielded a value of 0.74 ‰. An interesting feature of this shell is the relative abrupt increase of the Mg/Ca ratio of the winter increment towards the last summer increment.

FAN 127

Excavation unit FAN 127 dates to the Middle Neolithic period and contains seven specimens. The excavation unit summer increment isotope average is −1.19 ‰ (**fig. 100**).

Specimen FAN_127_A presented unclear seasonal cyclicity patterns which is due to its young age (less than 1 year old). Relatively lower Mg/Ca ratio is visible towards the apex and higher towards the edge. Based on the line scan as well as the elemental map, the shell was collected in late summer or early autumn. Two oxygen isotope samples were extracted from this shell: The summer sample had a value of −1.32 ‰, while the winter −0.38 ‰. This »winter« value is particularly low and could potentially reflect a spring value. Alternatively, it could be due to the specimen's young age that did not allow the accurate targeting of seasonal extremes. This unclear homogeneous overall picture is representative of short-lived specimens.

Specimen FAN_127_B also presented ambiguous seasonal cyclicity patterns due to its young age. The shell is characterised by low Mg/Ca ratio in its

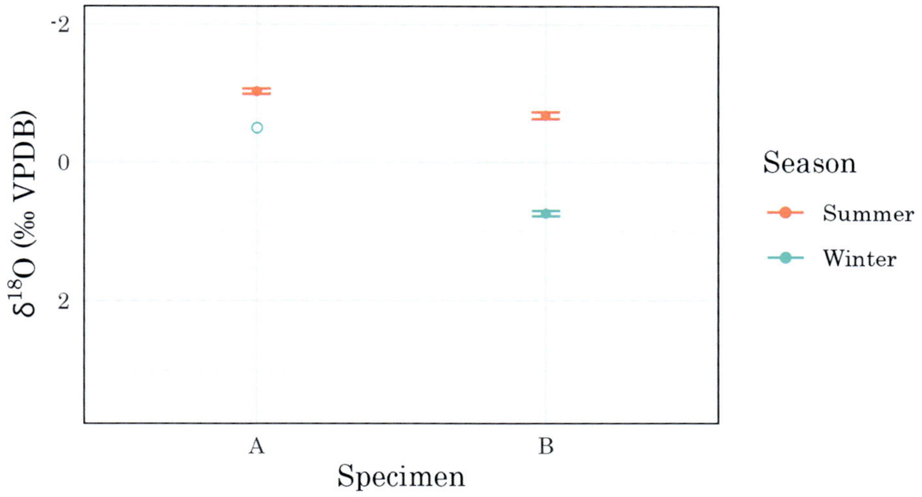

Fig. 99 Seasonal δ18O values for unit FAN 129.

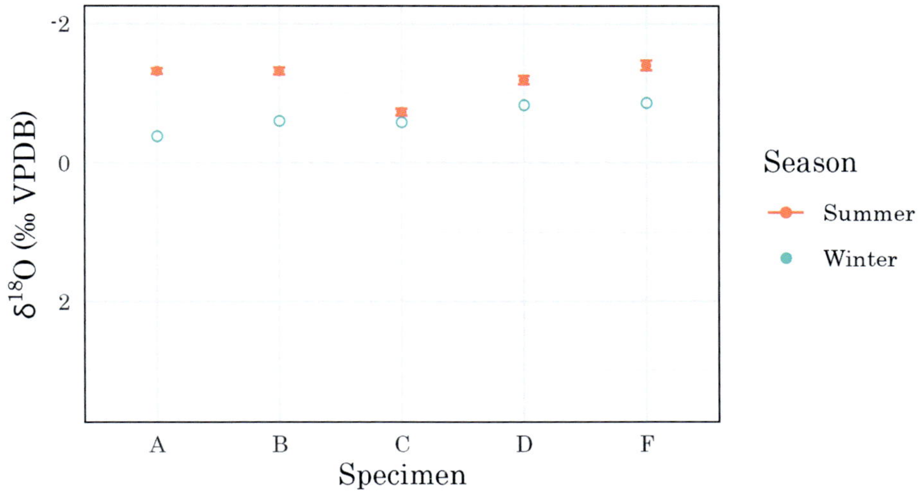

Fig. 100 Seasonal δ18O values for unit FAN 127.

whole extent with the exception of the last increment of growth that indicates a summer collection. The shell was less than 1 year old. Two oxygen isotope samples were extracted from this shell: The summer sample had a value of −1.32 ‰, while the winter −0.60 ‰. Interestingly, this summer increment yielded the exact same value as the summer isotope sample of FAN_127_A. Furthermore, as was the case with specimen FAN_127_A, this low »winter« value could potentially reflect a spring rather than a winter value.

Specimen FAN_127_C presented a very similar picture with the previous specimen: Low Mg/Ca ratio with the exception of the last increment of growth. The shell was collected during the summer and was less than 1 year old. The summer oxygen isotope sample had a value of −0.73 ‰, while the winter −0.58 ‰.

Specimen FAN_127_D recorded seasonal variability patterns that were difficult to interpret. Two winters and a summer period are probably reflected in the increments of this sample, which indicate that the shell was collected during winter and that the shell was 1.5 years old. Two oxygen isotope samples were collected: The summer sample had a value of −1.19 ‰, while the winter −0.83 ‰. The low »winter« value may reflect a spring.

Specimen FAN_127_E presented homogeneous and unclear elemental results. Thus, no season of collection could be determined. The mollusc was less than 1 year old.

Specimen FAN_127_F is similar to the shells of this unit having seasonal increments that are not well-defined. It probably reflects a winter followed by a summer period. The declining trend of the line scan shows that the shell was probably collected

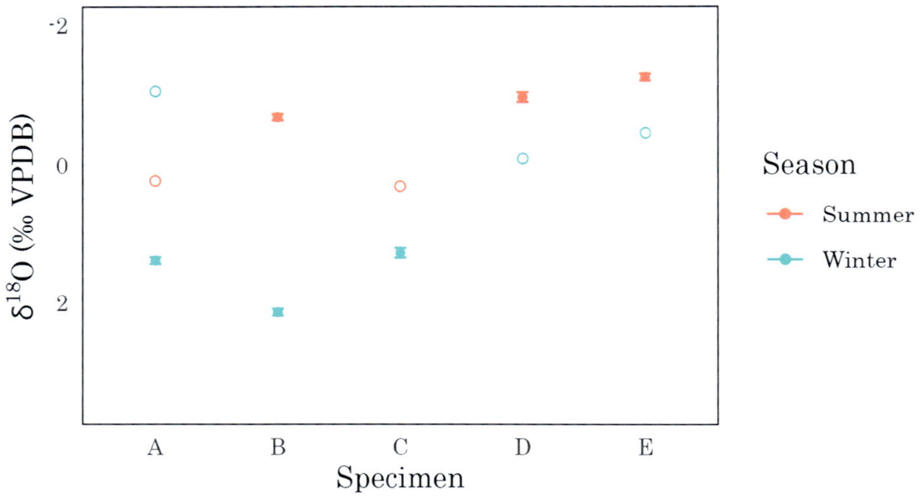

Fig. 101 Seasonal $\delta^{18}O$ values for unit FAN 124.

during autumn. The mollusc was less than 1 year old. Two oxygen isotope samples were extracted: The summer sample had a value of −1.40 ‰, which is very close to the summer oxygen isotope samples of specimens A and B. The »winter« sample had a value of −0.86 ‰, which is very low and may reflect a spring season rather than a winter one.

Specimen FAN_127_G also presents a similar picture to the aforementioned specimens of this unit: Seasonal increments are unclear, probably because the specimen was less than 1 year old. No season of collection could be determined.

FAN 124

Excavation unit FAN 124 dates to the Middle Neolithic period and contains five specimens. The excavation unit winter increment isotope average is 1.56 ‰ and the summer average is −0.88 ‰ (**fig. 101**).

Specimen FAN_124_A recorded homogeneous seasonal patterns, which are due to the specimen's young age (less than 1 year old). No season of collection could be attributed to the specimen. One oxygen isotope sample was collected which had a value of −1.06 ‰. It could potentially represent a spring value.

Specimen FAN_124_B reflects two summers and one winter. The specimen was 1.5 years old and collected during summer. Three oxygen isotope samples were collected: The summer samples had values of −0.47 ‰ and −0.82 ‰, while the winter sample had a value of 1.84 ‰

Specimen FAN_124_C reflects a summer and a winter period. However, the edge increment of growth is not particularly clear and for this reason no season of collection was determined, nor an age estimation was made. Two oxygen isotope samples were retrieved: The winter increment yielded a value of 1.28 ‰, while the summer a value of 0.31 ‰. The winter value is warmer than the period average, while the summer value is relatively high.

Specimen FAN_124_D recorded a winter or spring and a summer season in its increments. It was collected during the summer and was less than 1 year old. Two oxygen isotope samples were extracted from this shell: The summer sample had a value of −0.97 ‰ and the »winter« sample had a value of −0.09 ‰, which is a quite elevated value which may be indicative of a spring rather than a winter season, or may be due to the specimen's young age.

Specimen FAN_124_E has a similar image to that of the previous specimen: It recorded a winter or a spring and a summer. It was collected during the late summer or autumn and was less than 1 year old. Two oxygen isotope samples were collected: The summer sample had a value of −1.26 ‰ and the »winter« sample −0.46 ‰. The latter value could probably represent a spring increment.

FAN 123

Excavation unit FAN 123 dates to the Middle Neolithic period. Specimen FAN_123_A reflects a winter and a summer period. The abrupt drop in the Mg/Ca ratio at the edge of the shell indicates that it was collected during autumn and was approximately 1 year old. Two oxygen isotope samples were extracted: The summer increment oxygen isotope sample yielded

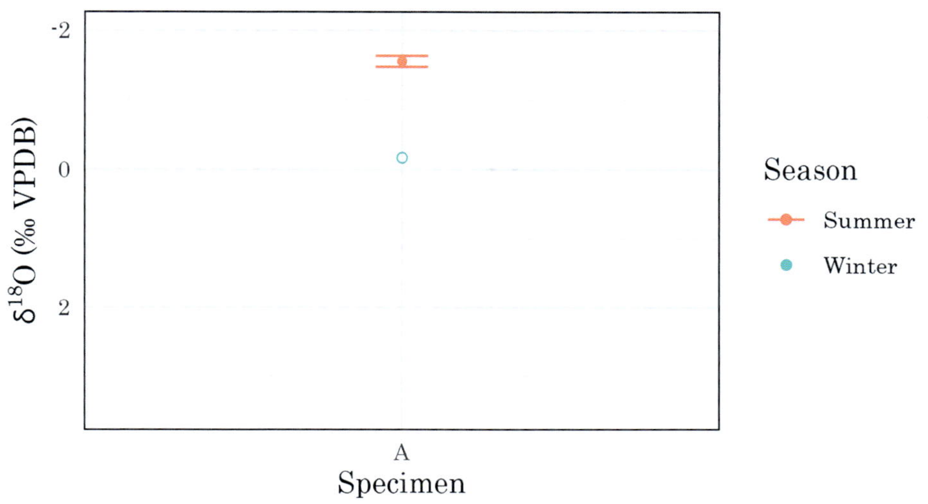

Fig. 102 Seasonal δ^{18}O values for unit FAN 123.

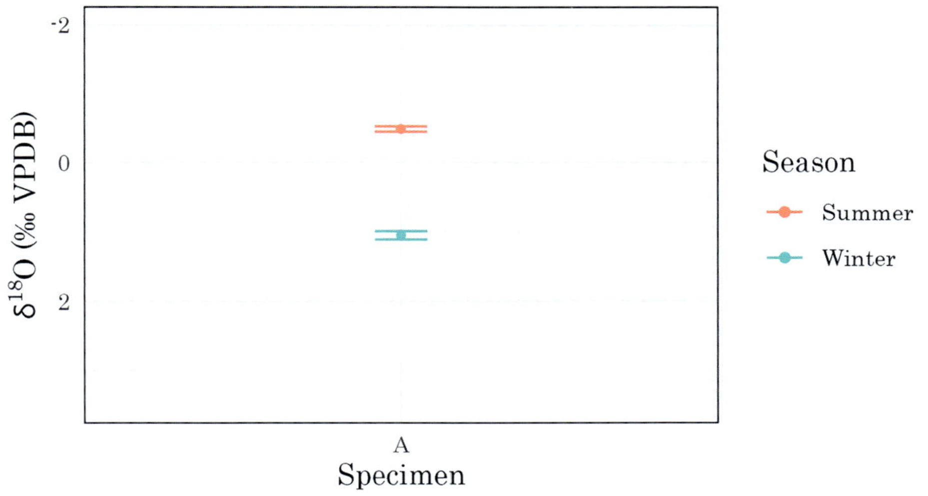

Fig. 103 Seasonal δ^{18}O values for unit L5 37.

a value of −1.56 ‰ and the winter sample −0.17 ‰ (**fig. 102**). The latter follows the »warm winter« tendency of the winters of the previous excavation unit.

4.2.1.3. Trench L5

L5 37

Excavation unit L5 37 belongs to the Paralia area and dates to the Early Neolithic period. Even though elemental data are not particularly clear in this case, specimen L5_37_A reflects a summer season towards the apex and a winter season towards the edge. Thus, it was collected during winter and was 1 year old. Two oxygen isotope samples were collected: The winter sample yielded a value of 1.05 ‰, which is slightly warmer than the period average (1.24 ‰) and the summer sample −0.49 ‰ (**fig. 103**).

L5 33

Excavation unit L5 33 also dates to the Early Neolithic period. Specimen L5_33_A recorded one or two summers and one winter. It was collected during spring/summer and was probably 1.5 years old when it was collected. Three oxygen isotope samples were collected: The summer increments had values of −0.67 ‰ (latest) and −0.98 ‰ (penultimate), with an average of −0.83 ‰, while the broad winter increment had a value of 1.16 ‰, which is very close to the period average (1.24 ‰) (**fig. 104**).

4.2.1.4. Trench L5NE

L5NE 64

Excavation unit L5NE 64 is part of the Paralia trenches and trench L5. It dates to the Early Neo-

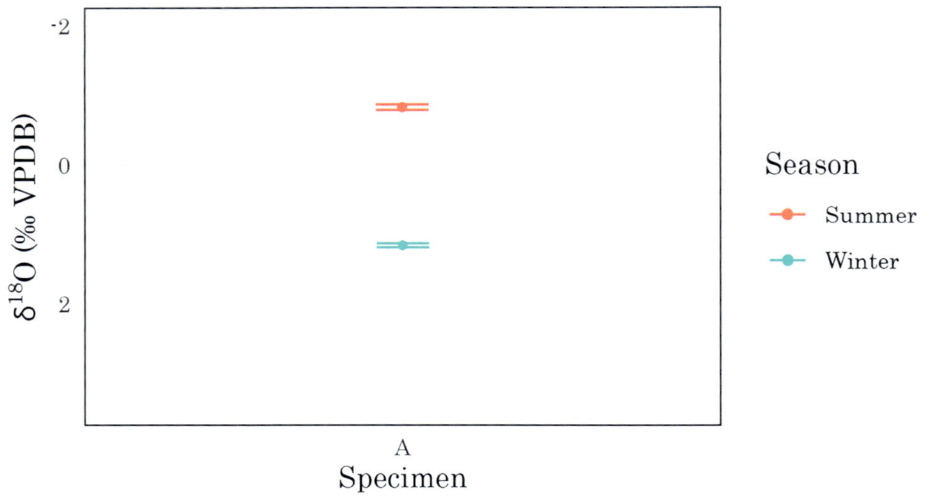

Fig. 104 Seasonal $\delta^{18}O$ values for unit L5 33.

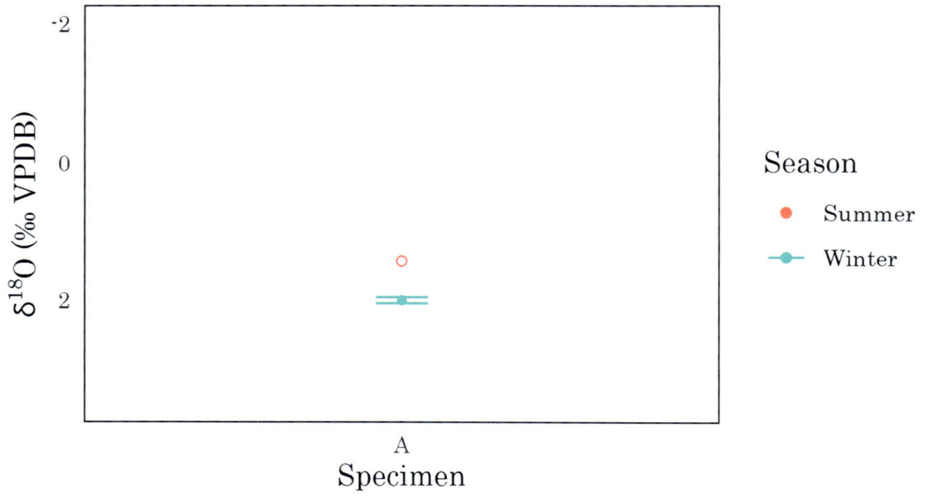

Fig. 105 Seasonal $\delta^{18}O$ values for unit L5NE 64.

lithic period. Specimen L5NE_64_A presents broad winter and thin summer increments. It recorded two or three summers and two winters, however the edge increment is not well-preserved in order to identify the season of collection. Three oxygen isotope samples were collected from the specimen: The winter samples had values of 2.21 ‰ and 1.79 ‰ (winter average 2.00 ‰), indicating cold winters, significantly higher than the period average (1.24 ‰) (**fig. 105**). The summer sample yielded also a relatively high value (1.43 ‰) which could be due to time-averaging while sampling as the increment was particularly thin and thus difficult to accurately sample.

L5NE 56

Excavation unit L5NE 56 dates to the Early Neolithic period and contains two samples. The excavation unit winter increment isotope average is 1.35 ‰, while the summer average is −0.91 ‰ (**fig. 106**).

Specimen L5NE_56_A recorded two winters and one summer and was thus approximately 1.5 years old. Based on the last increment of growth, the mollusc was collected during winter. Two oxygen isotope samples were extracted: The winter sample had a value of 1.53 ‰ and the summer sample −0.98 ‰.

L5NE_56_B recorded two summers and one winter in its increments. The mollusc was collected during autumn as there is a drop in the Mg/Ca ratio following the last summer at the edge. The mollusc was approximately 1.5 years old. Three oxygen isotope samples were collected from this sample: The summer samples had values of −0.52 ‰ and −1.24 ‰, and the winter sample 1.16 ‰, which is close to the average winter value of this chronological period (1.24 ‰).

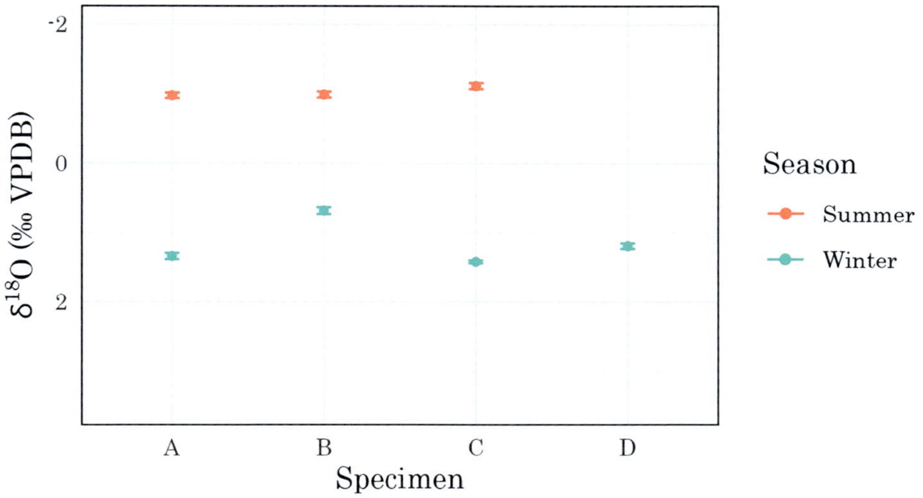

Fig. 106 Seasonal δ^{18}O values for unit L5NE 56.

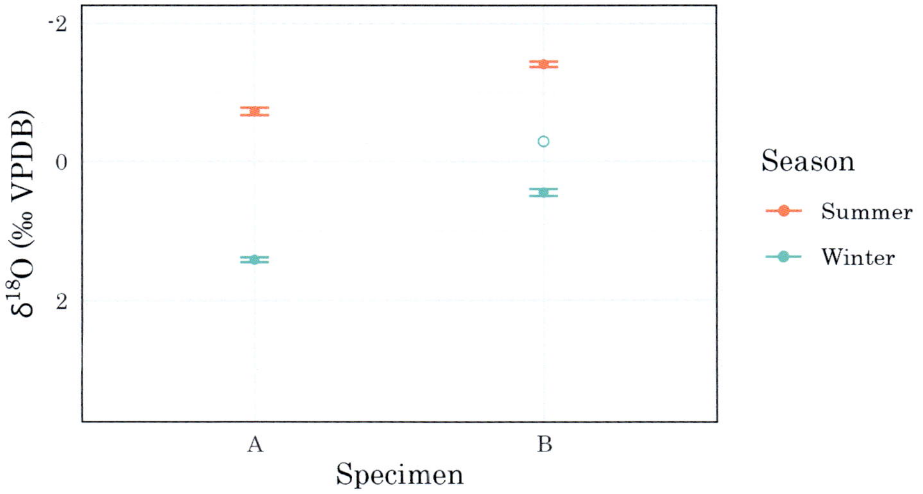

Fig. 107 Seasonal δ^{18}O values for unit L5NE 55.

L5NE 55

Excavation unit L5NE 55 also dates to the Early Neolithic period and contains specimen L5NE_55_A. The anterior side of the shell reflects one winter and one summer, while it shows signs of a decline in the Mg/Ca ratio towards the edge. This indicates that the shell was collected during autumn and was 1 year old. The high Mg/Ca ratio near the edge visible in the line scan is due to noise in the laser signal. Two oxygen isotope samples were retrieved: The summer sample yielded a value of −1.48 ‰, while the winter sample 1.46 ‰ (**fig. 107**).

L5NE 54

Excavation unit L5NE 54 dates to the Early Neolithic period and contains two specimens. Specimen L5NE_54_A has two broad winter increments and a shorter and thinner summer increment. The warming trend visible in the line scan at the edge of the shell indicates that the mollusc was collected during late winter or early spring and was 1.5 years old. Two oxygen isotope samples were collected: The winter sample had a value of 1.83 ‰, which is significantly colder than the period average (**fig. 108**). The summer sample was sampled twice in order to best capture the thin summer increment, which resulted in −1.25 ‰.

The elemental data of the fragmented specimen L5NE_54_B presented a lot of noise and were difficult to interpret. As seasonal cyclicity patterns were not clear enough, no season of collection was identified and the shell was not targeted for isotope analysis.

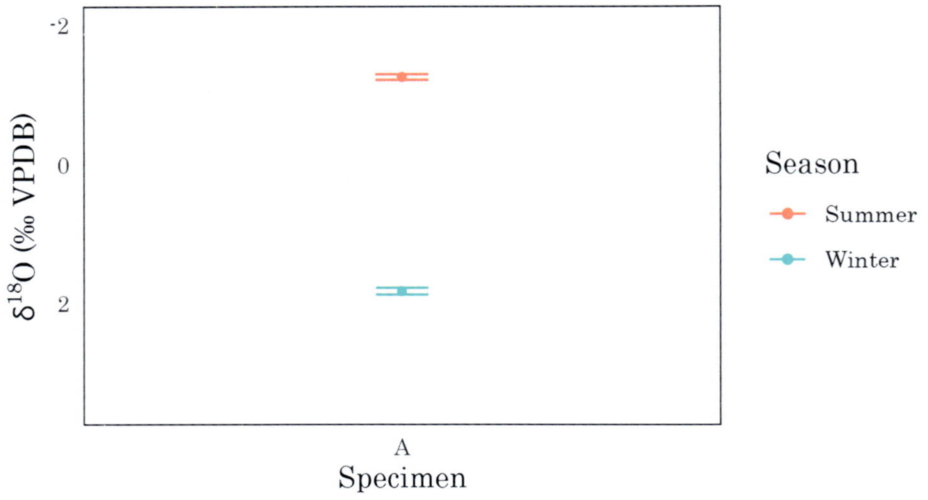

Fig. 108 Seasonal $\delta^{18}O$ values for unit L5NE 54.

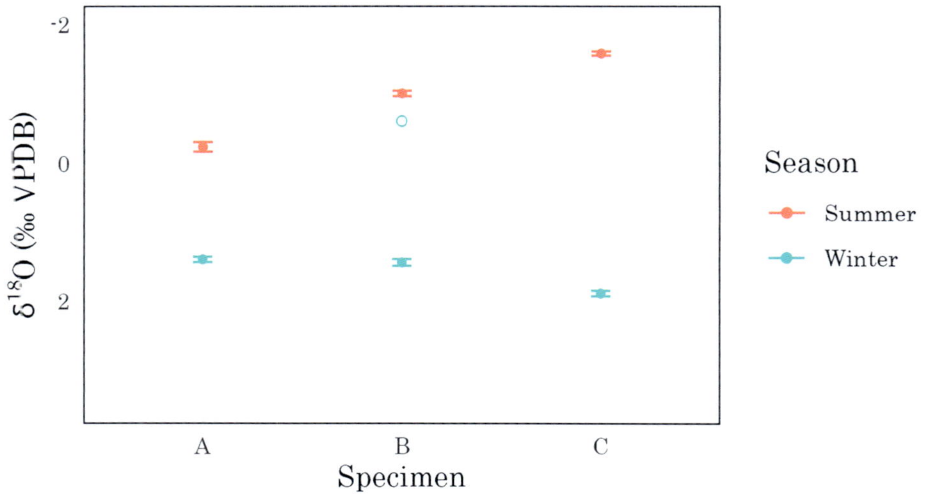

Fig. 109 Seasonal $\delta^{18}O$ values for unit L5NE 48.

L5NE 48

Excavation unit L5NE 48 dates to the Early Neolithic period and contains four shells. The excavation unit winter increment isotope average is 1.57 ‰ and the summer average is −0.95 ‰ (**fig. 109**).

Specimen L5NE_48_A reflects three winters and three summers. The winter increments are broad, while the summer ones are thin. The last increment of growth shows high Mg/Ca ratio indicating that the shell was collected during summer and was 3 years old. Two oxygen isotope samples were collected: The winter sample had a value of 1.39 ‰ and the summer sample a value of −0.24 ‰.

Specimen L5NE_48_B recorded two winters and one summer. Thus, the specimen was 1.5 years old and was collected during winter. The high Mg/Ca ratio at the edge of the line scan represents noise. Three oxygen isotope samples were collected: The winter sample of the last increment had a value of 1.44 ‰, whereas the penultimate winter had a value of −0.61 ‰, which is too low for a winter increment and may be due to time-averaging while sampling. The summer increment sample yielded a value of −1.01 ‰.

Specimen L5NE_48_C recorded two summers and one winter. This seasonal cyclicity is very well visible in the quasi-sinusoidal curve of the line scan. The shell was collected during the summer and was 1.5 years old at that time. Two oxygen isotope samples were retrieved from this sample: The summer sample yielded a value of −1.59 ‰, whereas the winter had a value of 1.89 ‰. The winter measurement indicates a colder winter from the period average (1.24 ‰).

The elemental data of the fragmented specimen L5NE_48_D were difficult to interpret. Seasonal pat-

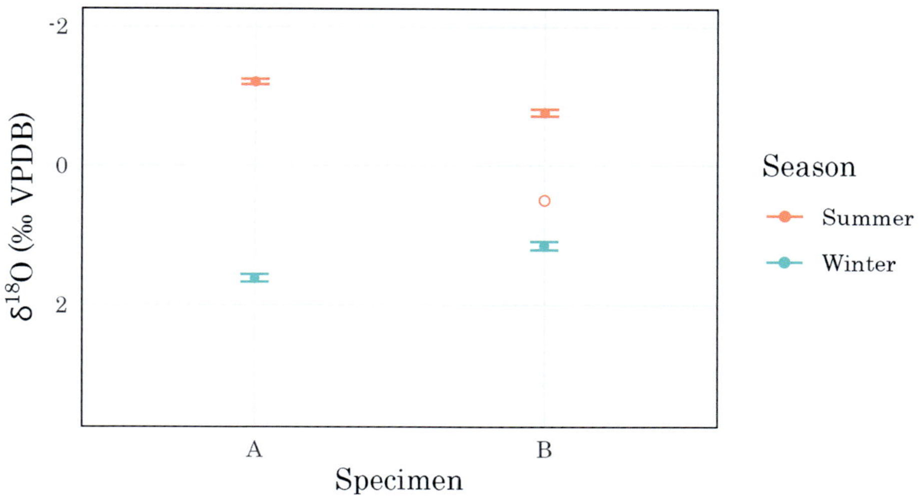

Fig. 110 Seasonal $\delta^{18}O$ values for unit L5NE 45.

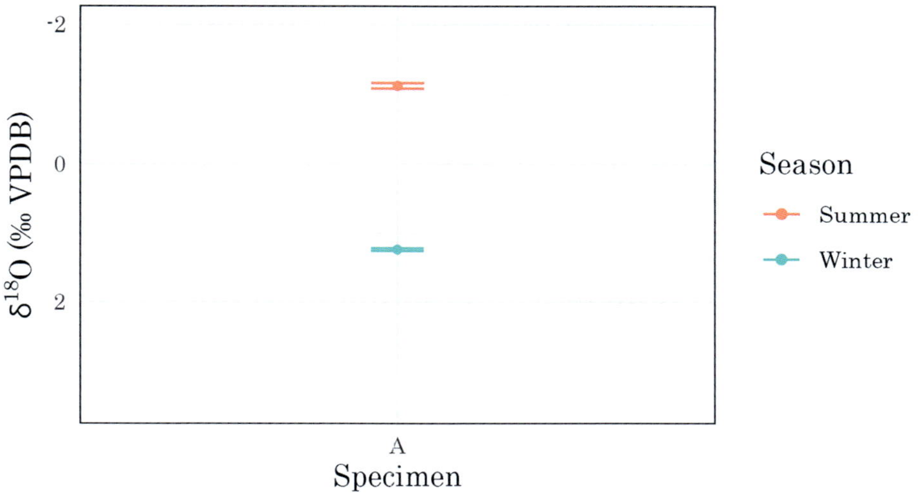

Fig. 111 Seasonal $\delta^{18}O$ values for unit L5NE 43.

terns were not clearly defined and the edge is probably not preserved. Thus, no season of collection was identified with certainty.

L5NE 45

Excavation unit L5NE 45 dates to the Early Neolithic period. Specimen L5NE_45_A reflects in its increments two winters and one summer. The last increment of growth points towards a winter period, while the line scan has an upward trend, which points towards a winter or a spring collection. The mollusc was 1.5 years old. The summer increment is unusually extensive indicating that the mollusc grew significantly during that time. Two oxygen isotope samples were analysed: The winter sample had a value of 1.81 ‰, and the summer had a value of −1.07 ‰. The winter measurement indicates a colder season than the average winter (1.24 ‰) (**fig. 110**).

L5NE 43

Excavation unit L5NE 43 dates to the Early Neolithic period. Sample L5NE_43_A recorded three summers and two winters. Thus, the mollusc was 2.5 years old and was collected during summer/autumn as the elemental 30 µm scan of the anterior shows low Mg/Ca ratio at the edge of the sample. The line scan of the anterior side shows abrupt seasonal alternation, which may represent growth stops. Three oxygen isotope samples were collected from this sample: The summer samples had values of −1.49 ‰ and −0.76 ‰ (summer average −1.13 ‰) and the winter sample 1.24 ‰. The latter indicates a winter that is very similar to the average of this period (1.24 ‰) (**fig. 111**).

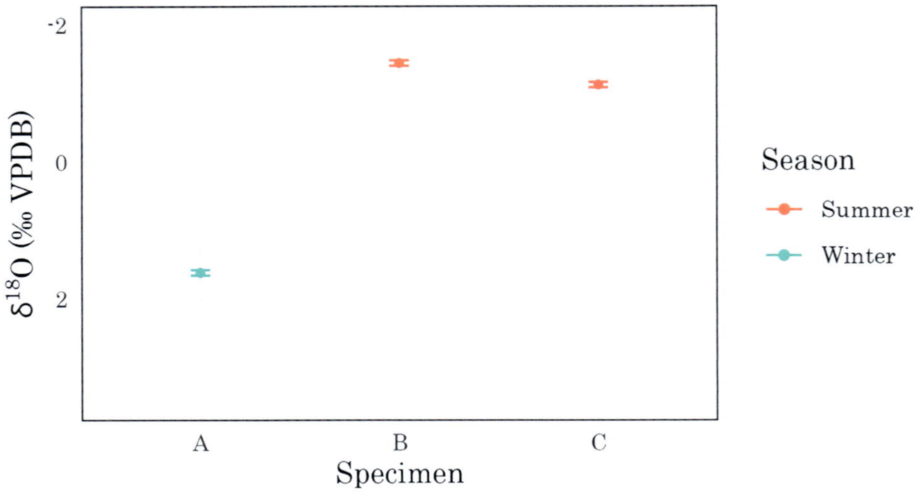

Fig. 112 Seasonal $\delta^{18}O$ values for unit Q5N 65.

4.2.1.5. Trench O5

O5 81

Excavation unit O5 81 is part of the Paralia trench O5 and dates to the Early Neolithic period. Specimen O5_81_A has two summer increments and one or two winters. The latest winter which is particularly extended. The mollusc was at least 1.5 years old and the increased Mg/Ca ratio towards the edge shows that the specimen was collected during the summer. No oxygen isotope samples were extracted from this specimen.

4.2.1.6. Trench Q5N

Q5N 65

Trench Q5N belongs to the Paralia site and dates to the Early Neolithic period. Excavation unit Q5N 65 had three shell specimens. The excavation unit winter increment isotope average is 1.63 ‰ and the summer average is −1.28 ‰ (**fig. 112**).

Specimen Q5N_65_A recorded three winters and two summers. The last increment of growth has a low Mg/Ca ratio indicating that the shell was collected during winter. Moreover, the mollusc was 2.5 years old when it was collected. Two oxygen isotope samples were collected from the two winter increments. The samples yielded values of 1.54 ‰ (last winter) and 1.71 ‰ (penultimate winter).

Specimen Q5N_65_B reflected the presence of two winters and one summer. The shell was collected during winter and was 1.5 years old. One oxygen isotope sample was collected from the summer increment which had a value of −1.44 ‰.

Specimen Q5N_65_C is a relatively young specimen and reflects potentially a summer and a winter period, although seasonal cyclicity patterns are not particularly clear. One oxygen isotope sample was extracted from the increments with higher Mg/Ca ratio, which yielded a value of −1.12 ‰.

Q5N 56

Excavation unit Q5N 56 contains four shell specimens and dates to the Early Neolithic period. The excavation unit winter increment isotope average is 1.00 ‰ and the summer average is −1.15 ‰ (**fig. 113**).

Specimen Q5N_56_A presented an unusual picture of seasonal cyclicity: Approximately six thin summer increments are visible in the elemental maps of the shell, which are also well-illustrated in the line scan of the posterior. Thus, the mollusc was 6 years old and is the oldest specimen of the archaeological shells that were analysed in terms of years of age. The shell was collected during autumn as there is a cooling trend present after the last summer increment. By looking further at the line scan, seasonal alternation is abrupt as the curves are angular. A winter increment oxygen isotope value yielded a value of 1.16 ‰.

Specimen Q5N_56_B recorded two summers and one winter in its increments and was thus 1.5 years old. The last increment of growth is a summer which shows signs of cooling. For this reason, the shell was collected during summer/autumn. Two oxygen isotope samples were extracted from the specimen: The summer sample had a value of −1.22 ‰, while the winter had a value of 0.21 ‰,which is warmer than the period average (1.24 ‰).

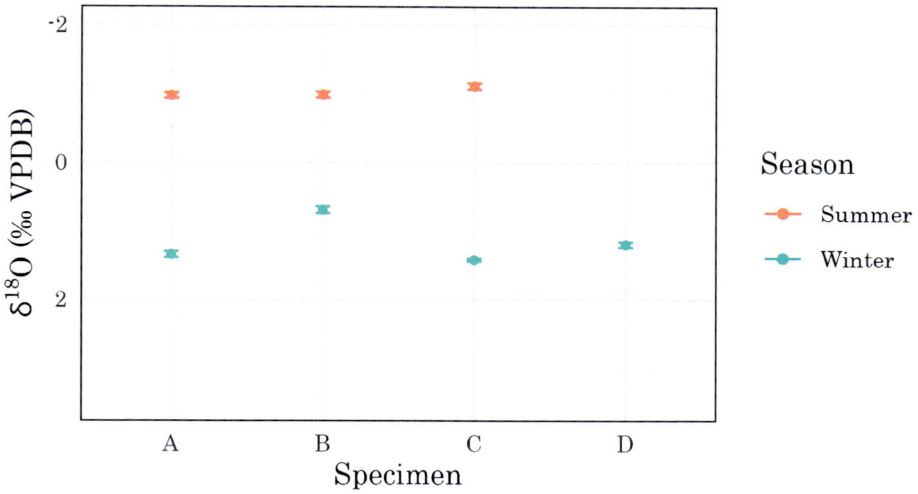

Fig. 113 Seasonal $\delta^{18}O$ values for unit Q5N 56.

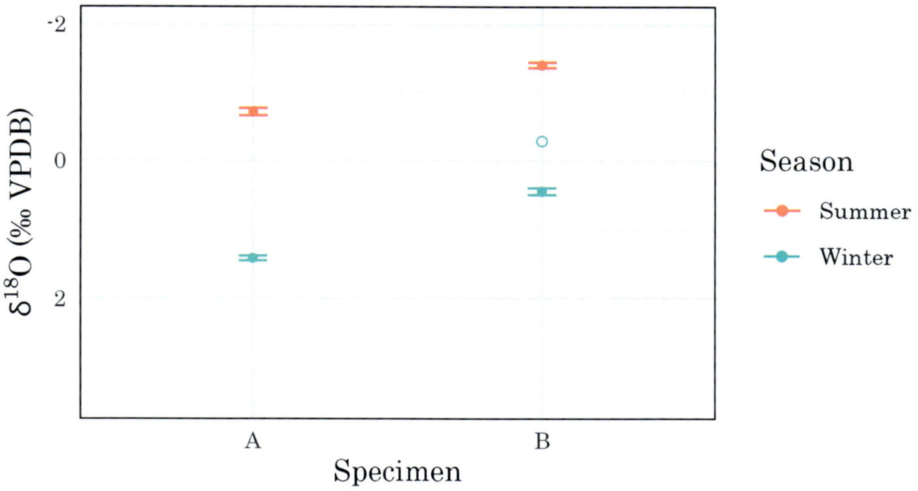

Fig. 114 Seasonal $\delta^{18}O$ values for unit Q5N 55.

Specimen Q5N_56_C presented a similar picture with specimen B: Two summers and one winter are visible in its increments. The cooling trend of the quasi-sinusoidal curve of the line scan immediately after the last summer increment indicates that the shell was collected during autumn, while the specimen was 1.5 years old. Three oxygen isotope samples were collected: The summer increments had values of −1.18 ‰ and −1.06 ‰, while the winter increment had a value of 1.42 ‰. The two summer increments are very consistent among them.

Specimen Q5N_56_D presented limited cyclicity patterns and appears to have been less than 1 year old. In its larger extent the shell has a low Mg/Ca ratio and was thus collected during winter. One oxygen isotope sample was collected from the shell's winter increments which yielded a value of 1.19 ‰.

Q5N 55

Excavation unit Q5N 55 contains two specimens and dates to the Early Neolithic period. The excavation unit winter increment isotope average is 0.92 ‰ and the summer average is −0.69 ‰ (**fig. 114**).

Specimen Q5N_55_A reflects two winters and one summer. The thin summer line indicates a short summer, while the line scan recorded a lot of noise in the area of the penultimate winter. The mollusc must have been 1.5 years old and was collected winter, based on the last increment of growth that presents a low Mg/Ca ratio. Two oxygen isotope samples were retrieved: The winter sample yielded a value of 1.38 ‰, while the summer sample yielded a value of 0.03 ‰. The summer increment was sampled twice in order to best capture the thin summer increment. Nonetheless, the result is relatively high which could be due to time-averaging while sampling.

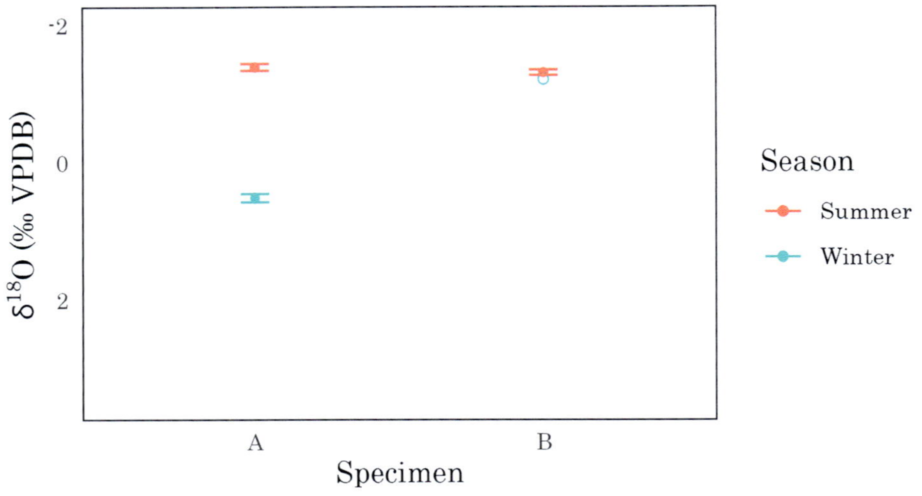

Fig. 115 Seasonal $\delta^{18}O$ values for unit Q5N 47.

Specimen Q5N_55_B recorded three winters and two summers and was thus 2.5 years old. The last increment of growth indicates a winter collection. The line scan presents a lot of noise and could be due to the focusing of the sample while scanning. Three oxygen isotope samples were extracted: The winter increments were both relatively elevated and had values of 0.45 ‰ and −0.29 ‰, while the summer increment had a value of −1.41 ‰.

Q5N 47

Excavation unit Q5N 47 contains three specimens and dates to the Early Neolithic period. The excavation unit summer increment isotope average is −1.35 ‰ (**fig. 115**).

Specimen Q5N_47_A recorded one winter and one summer and was 1 year old. The last increment of growth indicates that the shell was collected during the summer. Two oxygen isotope samples were retrieved: The summer sample yielded a value of −1.39 ‰, while the winter sample yielded a value of 0.51 ‰. The winter value indicates a warmer than the period average winter (1.24 ‰).

Specimen Q5N_47_B presented unclear patterns of seasonal cyclicity, probably because of its young age (less than 1 year old). Both the elemental maps and the line scan did not contribute to a season of collection estimation. Two oxygen isotope samples were collected from the sample: The winter sample had a value of −1.21 ‰, which is particularly low and contradicts the estimation that this was actually a winter increment. The summer sample yielded a value of −1.31 ‰.

Specimen Q5N_47_C is a fragment and did not preserve any calcitic material. No further analyses were run.

Q5N 45

Excavation unit Q5N 45 has two specimens and dates to the Early Neolithic period. The excavation unit winter average is 1.28 ‰, while the summer is −1.05 ‰ (**fig. 116**).

A season of collection could not be determined with certainty for specimen Q5N_45_A, as its seasonal increments are not distinct nor well-defined. Two oxygen isotope samples were collected: The summer sample resulted in a value of −1.34 ‰, while the winter sample had a value of 1.41 ‰. The latter value is colder than the Early Neolithic period average (1.24 ‰).

Specimen Q5N_45_B reflected three winters and two thin summers and was thus 2.5 years old. Both summer increments appear gradually and have distinct lines of high Mg/Ca ratio. The last increment of growth indicates a winter collection. Three oxygen isotope samples were extracted: The summer increments had values of 0.50 ‰ (last summer) and −0.76 ‰ (penultimate summer), while the winter increment had a value of 1.15 ‰. The last summer increment's value is relatively cold for a summer period and may indicate that the thin summer increment was not sampled accurately enough.

Q5N 26

Excavation unit Q5N 26 also dates to the Early Neolithic period. Specimen Q5N_26_A recorded one

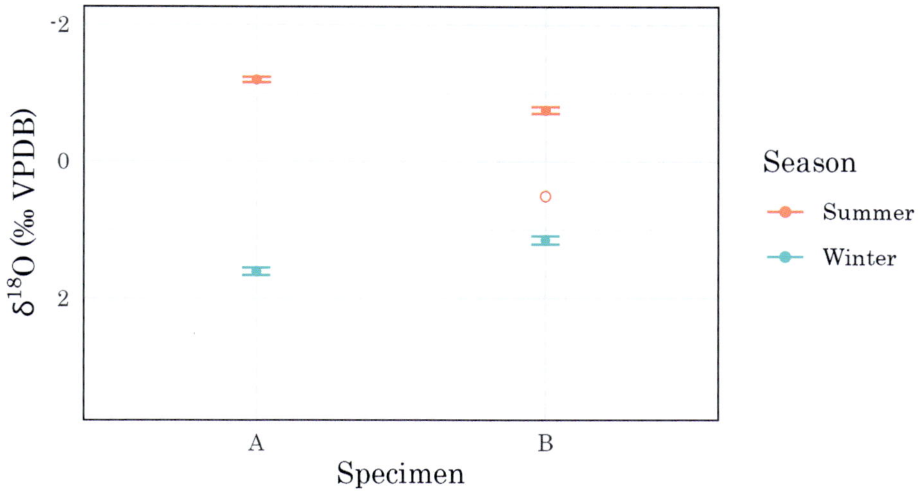

Fig. 116 Seasonal δ18O values for unit Q5N 45.

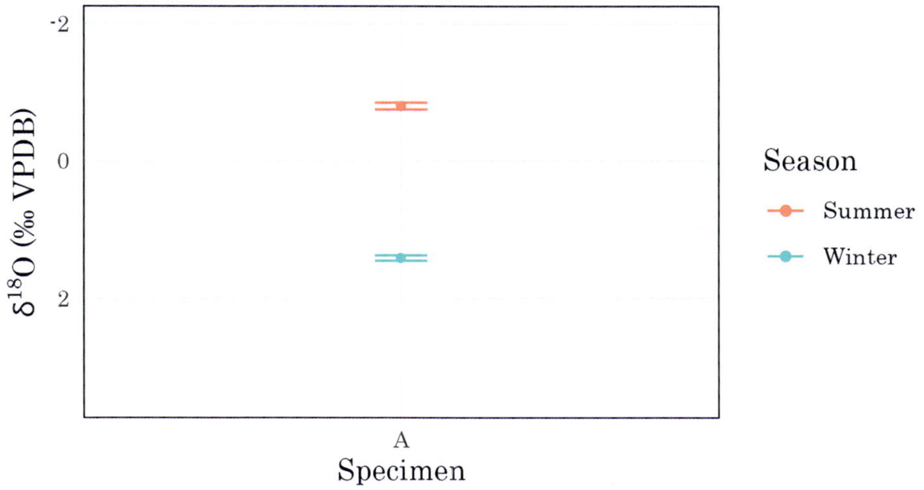

Fig. 117 Seasonal δ18O values for unit Q5N 26.

winter and one summer and the beginning of a second winter. Thus, the mollusc was at least 1 year old and was collected during early winter. The extended summer increment had an isotope value of −0.81 ‰, while the winter increment 1.41 ‰, which is colder than the period average (1.24 ‰) (**fig. 117**).

Q5N 25

Excavation unit Q5N 25 contains two specimens and dates to the Early Neolithic period. Specimen Q5N_25_A presented homogeneous patterns of seasonal cyclicity which could not be interpreted in order to estimate a season of collection. The line scan fluctuating at the low Mg/Ca ratios, has some peaks that could be interpreted as summers.

Specimen Q5N_25_B is a fragment recorded one winter and one summer increment. The shell was collected during summer. No age determination could be made due its inadequate preservation. Two oxygen isotope samples were extracted from this shell: The summer sample resulted in a value of −1.38 ‰, while the winter sample had a value of 1.29 ‰ (**fig. 118**).

Q5N 23

Excavation unit Q5N 23 contains two specimens and dates to the Early Neolithic period. The excavation unit winter increment isotope average is 1.35 ‰ (**fig. 119**).

Shell Q5N_23_A recorded in its increments four winters and three or four summers. Thus, the shell was at least 3.5 years old, while the last increment of growth seems to be characterised by high Mg/Ca ratio indicating a summer collection. Two winter oxygen isotope samples were collected and had consistent values of 1.15 ‰ and 1.24 ‰. The angular

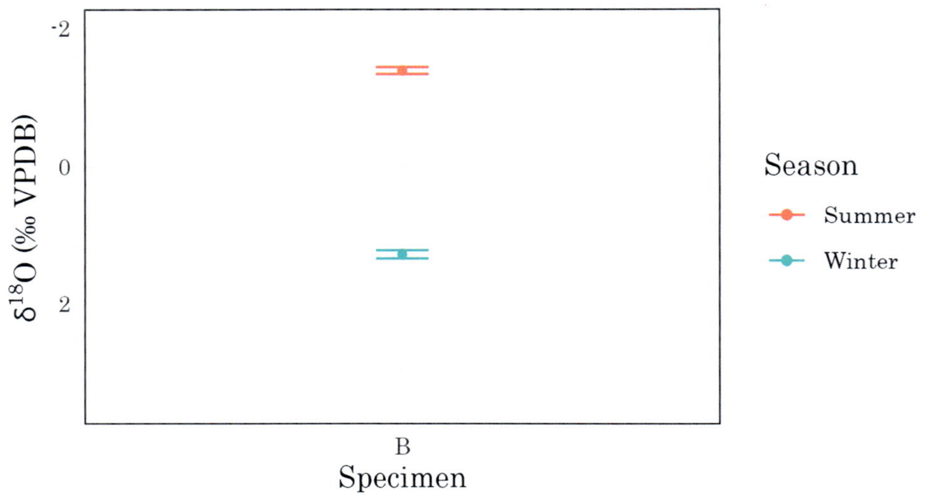

Fig. 118 Seasonal $\delta^{18}O$ values for unit Q5N 25.

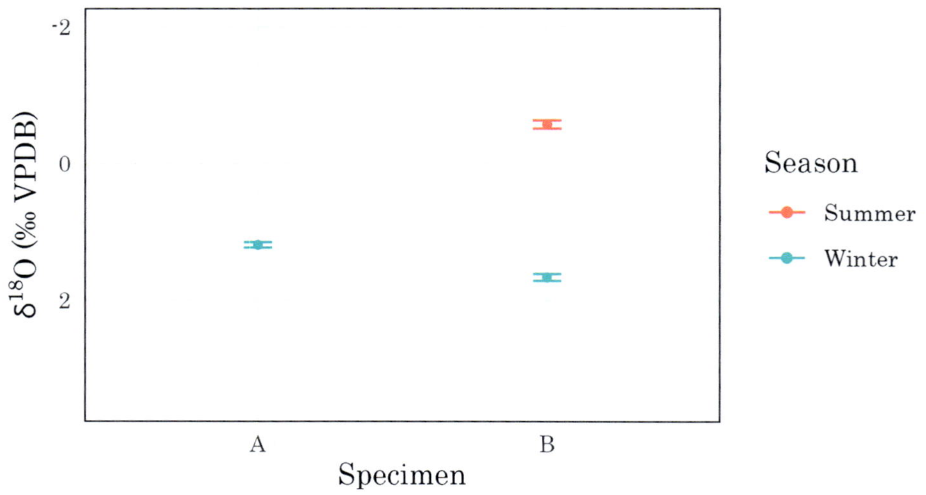

Fig. 119 Seasonal $\delta^{18}O$ values for unit Q5N 23.

curves of the line scan show abrupt changes from one season to another, which could be indicative of growth stops.

Specimen Q5N_23_B reflects two summers and one winter. The mollusc was 1.5 years old and was collected during autumn as indicated by the cooling tendency of the elemental data following the last summer. All seasonal increments were broad, while seasonal alternation was abrupt from the last winter towards the last summer recorded in the shell. Two oxygen isotope samples were retrieved: The summer sample resulted in a value of −0.57 ‰, while the winter sample had a value of 1.67 ‰.

4.2.1.7. Trench Q5S

Q5S 230

Trench Q5S belongs to the Paralia area and dates to the Early Neolithic period. The excavation unit winter isotope average is 1.18 ‰, while the summer is −0.57 ‰ (**fig. 120**).

Excavation unit Q5S 230 contains 27 shell specimens and it is the unit with the largest number of shells. Specimen Q5S_230_A recorded two or three winters and two summers. The mollusc was at least 2 years old and was collected during winter. The last summer increment is a thin and distinct line and its peak in the line scan is abrupt and angular. Three oxygen isotope samples were extracted: The two samples collected from the winter increments had values of 1.39 ‰ and 1.68 ‰, while the summer increment had a value of −0.51 ‰.

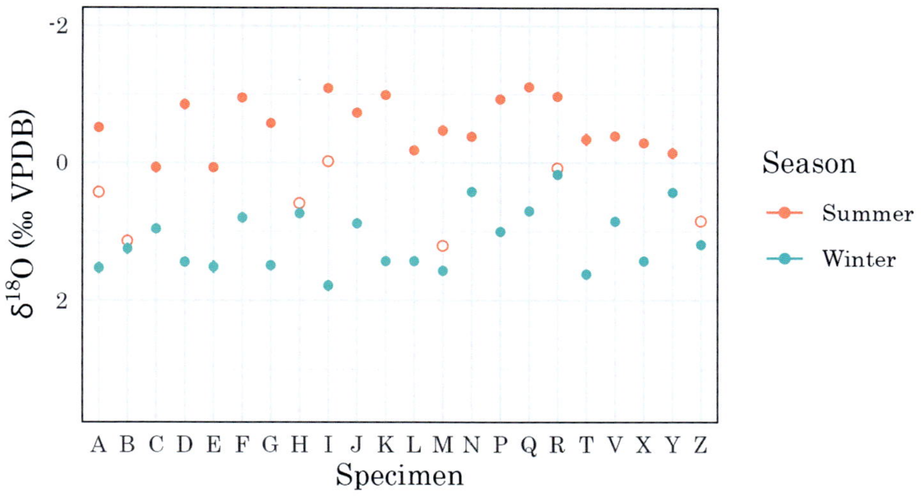

Fig. 120 Seasonal $\delta^{18}O$ values for unit Q5S 230.

Specimen Q5S_230_B has a very similar picture to that of specimen A: It recorded two winters and two summers, with the latest summer being particularly thin. The mollusc was 2 years old and was collected during winter. Two oxygen isotope samples were collected: The summer sample had a relatively high value (1.14 ‰), which may be due to time-averaging due to the thinness of the increment, while the winter increment had a value of 1.25 ‰.

Specimen Q5S_230_C has a very similar picture to that of the previous specimens and was also collected during winter. It records two winters and two summers. The mollusc was 2 years old. Two oxygen isotope samples were retrieved: The summer increment sample had a value of 0.07 ‰ and the winter one had a value of 0.96 ‰.

Specimen Q5S_230_D recorded a relatively broad summer and potentially two winters. The last increment of growth is a winter and the shell was approximately 1.5 years old. Two oxygen isotope samples were collected: The winter sample had a value of 1.44 ‰ and the winter sample had a value of −0.85 ‰.

Specimen Q5S_230_E reflects two winters and one summer, which has a distinct characteristic summer line. The mollusc was collected during winter and was 1.5 years old. Two oxygen isotope samples were retrieved: The summer increment sample had a value of 0.07 ‰ and the winter one had a value of 1.51 ‰.

Specimen Q5S_230_F also follows the same pattern, although the summer increment is not particularly thin. The mollusc was collected during winter and three oxygen isotope samples were extracted: The two winter samples had values of 1.14 ‰ and 0.45 ‰, while the summer increment had a value of −0.95 ‰.

Q5S_230_G is different from the above specimens. It reflects in its increments two winters and two summers. The cooling trend in the line scan visible after the last summer indicates that the shell was collected during autumn. The mollusc was 2 years old when it was collected. The line scan shows gradual warming (winter towards summer) trends and abrupt cooling trends (summer towards winter). Three oxygen isotope samples were collected: The summer samples had values of −0.30 ‰ and −0.85 ‰, while the winter sample had a value of 1.49 ‰.

Specimen Q5S_230_H reflects four winters and three summers. The mollusc was 3.5 years old. The elemental maps show broad winter increments and thin summer increments, while the line scan shows angular summer peaks, which may be suggestive of growth stops. Avoiding time-averaging effects while sampling for oxygen isotopes in this case was difficult because of the thinness of the summer increments. The two summer samples had values of 0.52 ‰ and 0.66 ‰, which are relatively cold, while the two winter samples had values of 0.68 ‰ and 0.79 ‰.

Specimen Q5S_230_I recorded two winters and two summers. Thus, it was 2 years old at the time of death. The declining trend in the Mg/Ca ratio as well as the low ratio of the last increment of growth after the last summer indicates that the shell was collected during autumn or during winter. Three oxygen isotope samples were collected: The summer samples had values of −0.02 ‰ (last summer increment) and −1.09 ‰ (penultimate summer increment), while the winter sample had a value of 1.78 ‰.

Specimen Q5S_230_J reflects two winters and one summer. Thus, the mollusc was 1.5 years old when it was collected, while the season of collection was winter. The end of the line scan has noise and is not to be confused with a summer signal. Three oxygen isotope samples were collected: The two winter samples had values of 0.96 ‰ and 0.80 ‰, while the summer sample had a value of −0.73 ‰.

Specimen Q5S_230_K also reflects one or two winters and one summer. It was at least 1 year old when it was collected. The low Mg/Ca ratio at the last increment of growth of the anterior edge indicates that the shell was collected during winter. Two oxygen isotope samples were extracted: The winter sample yielded a value of 1.43 ‰ and the summer sample a value of −0.99 ‰.

Specimen Q5S_230_L is similar to specimens A, B, C, and E. There are two winter increments and one or two summers, the last of which is thin and distinct. The specimen was at least 1.5 years old and was collected during winter. Two oxygen isotope samples were collected: The summer sample had a value of −0.18 ‰ and the winter sample 1.43 ‰.

Specimen Q5S_230_M recorded two or three winters and two summers. It was thus at least 2 years old when it was collected, while the last increment of growth indicates a winter collection. The summer increments are relatively broad and the warming trends towards the summer seasons are gradual while the cooling trends are abrupt based on the line scan. Three oxygen isotope samples were collected: The two summer samples had values of −0.47 ‰ and 1.21 ‰, while the latter value is particularly high and may be due to inaccurate sampling. The winter isotope sample yielded a value of 1.57 ‰.

Sample Q5S_230_N recorded two winters and one summer and was thus 1.5 years old. The low Mg/Ca ratio at the last increment of growth indicates that the shell was collected during winter. The line scan presented a lot of noise. Two oxygen isotope samples were extracted: The summer sample had a value of −0.38 ‰, while the winter sample 0.42 ‰.

Specimen Q5S_230_O did not present interpretable seasonal cyclicity patterns. The elemental maps show a rather homogeneous distribution of low Mg/Ca ratio across the shell, which is probably due to the molluscs' young age (less than 1 year old). The line scan presents a lot of noise. Nonetheless, the shell was probably collected during winter. No isotope analysis was run on this sample.

Seasonal alternation is not particularly clear in specimen Q5S_230_P as it is unclear how many winters and summers are present. At least two winters and one or two summers may be present in this shell, while the last increment reflects an autumn or winter collection. Two oxygen isotope samples were collected: The summer sample yielded a value of −0.93 ‰, while the winter sample 1.00 ‰.

Specimen Q5S_230_Q potentially reflects a winter and a summer increment. However, the edge is not preserved and no season of collection was determined nor an age estimation made. The line scan has a significant amount of noise. Two oxygen isotope samples were collected: The summer sample yielded a value of −1.11 ‰, while the winter sample 0.70 ‰.

Specimen Q5S_230_R recorded three winters and two summers and was 2.5 years old. The last increment of growth has a low Mg/Ca ratio and indicates that the shell was collected during winter. Three oxygen isotope samples were extracted: The summer samples yielded values of 0.08 ‰ (last summer) and −0.97 ‰ (penultimate summer), while the winter sample had a value of 0.17 ‰. The winter value is particularly warm.

Specimens Q5S_230_S and Q5S_230_U and Q5S_230_W did not preserve enough calcitic material nor any edges, thus no further analyses were done, nor any further conclusions were drawn.

Specimen Q5S_230_T recorded two winters and one summer. It was 1.5 years old and collected during winter. Two oxygen isotope samples were retrieved: The winter sample had a value of 1.62 ‰ and the summer sample a value of −0.34 ‰.

Specimen Q5S_230_V reflects a winter season and potentially a summer season. As no seasonal cyclicity is present it appears that the mollusc was less than 1 year old. The low Mg/Ca ratio indicates that the shell was collected during winter. Two oxygen

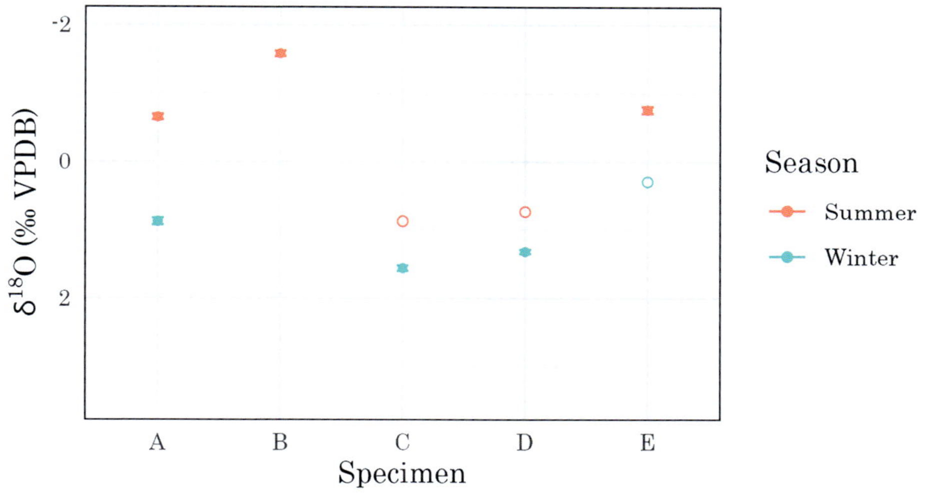

Fig. 121 Seasonal $\delta^{18}O$ values for unit Q5S 222.

isotope samples were extracted: The winter sample had a value of 0.85 ‰ and the summer sample a value of −0.39 ‰.

Specimen Q5S_230_X is a fragment and recorded two winters and one summer. The summer signal is very similar to that of specimens A, B, C, E and L. As it is a fragment an age estimation could not be made, while the warming trend after the last winter indicates a winter/spring season of collection. Two oxygen isotope samples were collected: The winter sample yielded a value of 1.43 ‰ and a summer value of −0.29 ‰.

Specimen Q5S_230_Y is also a fragment and has a very similar elemental picture with that of specimen X. Two winters and one summer are present in the shell, whereas the summer increment is very thin and distinct. Two oxygen isotope samples were retrieved: The summer sample had a value of −0.14 ‰ and the winter sample a value of 0.43 ‰.

Specimen Q5S_230_Z is a fragment and recorded two winters and almost two summers. As it is a fragment an age estimation could not be made. The last increment of growth indicates a spring/summer collection, which is best visible in the elemental scan. Two oxygen isotope samples were extracted: The winter sample had a value of 1.19 ‰ and the summer sample a value of 0.85 ‰. The summer sample is particularly high and could be due to time-averaging while sampling as the summer increment sampled is very thin.

Specimen Q5S_230_AA is also a fragment and recorded two summers and two winters. The penultimate winter is particularly extended, while both summers are short. The shell was collected during late autumn or early winter. The line scan presents a clear quasi-sinusoidal curve, which illustrates seasonal cyclicity well. Two oxygen isotope samples were collected from this specimen. The winter increment gave a value of 1.50 ‰, while the summer increment a value of 0.43 ‰.

Q5S 222

Excavation unit Q5S 222 dates to the Early Neolithic period and contains five shell specimens. The excavation unit winter isotope average is 1.25 ‰, while the summer is −0.99 ‰ (**fig. 121**).

Specimen Q5S_222_A recorded two winters and one summer and thus was 1.5 years old when it was collected. The summer increment is quite broad for a summer increment, while it appears that the shell was collected during winter. The high Mg/Ca values at the end of the line scan of the posterior represent noise and not a certain season of collection. Two oxygen isotope samples were extracted from the shell: The summer sample yielded a value of −0.65 ‰ and the winter sample a value of 0.88 ‰. The latter value indicates a warmer than the period average winter (1.24 ‰).

Specimen Q5S_222_B reflects two winters and a very thin summer, even though seasonal cyclicity patterns are not particularly distinct in this case. The mollusc was 1.5 years old. Furthermore, the last increment of growth appears to be a winter, while the line scan shows a very high degree of noise. One oxygen isotope sample was collected from the thin summer increment which yielded a value of −1.58 ‰.

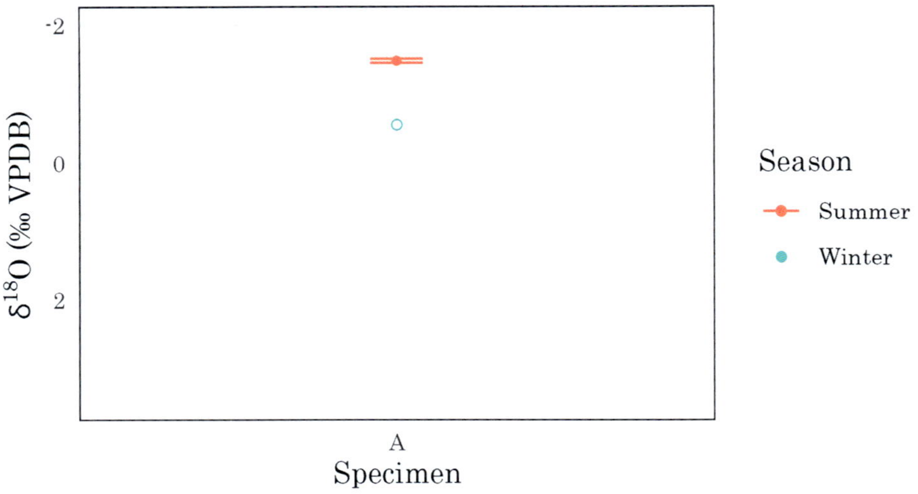

Fig. 122 Seasonal $\delta^{18}O$ values for unit Q5S 191.

Specimen Q5S_222_C recorded two winters and one summer. Thus, the mollusc was 1.5 years old and was collected during winter. Two oxygen isotope samples were collected from the specimen: The winter sample had a value of 1.56 ‰, while the summer sample had a value of 0.88 ‰. The latter had a particularly high for a summer sample and could be due to inaccurate sampling and time-averaging.

Specimen Q5S_222_D reflected also two winters and one thin summer. The mollusc was 1.5 years old and was collected during winter. The high value at the edge of the line scan is noise and not a summer signal. Two oxygen isotope samples were extracted: The summer sample yielded a value of 0.74 ‰, while the winter sample yielded a value of 1.32 ‰. The summer value is relatively high for a summer signal and this could be attributed to the thinness of the increment and subsequent time-averaging.

Specimen Q5S_222_E is a fragment and recorded two winters and two summers. Both summers are characteristically thin. The shell was collected during late summer or early autumn, while due to its inadequate preservation no age determination could be made. The line scan presents a clear quasi-sinusoidal curve, exhibiting repetitive seasonal cyclicity. Two oxygen isotope samples were collected from this specimen: The winter increment gave a value of 0.29 ‰, while the summer increment a value of −0.76 ‰. The winter increment is relatively warmer than expected and could be due to inaccurate sampling as the mollusc did not grow significantly during that period and growth was limited.

Q5S 191

Excavation unit Q5S 191 also dates to the Early Neolithic period. In the case of specimen Q5S_191_A seasonal variability is not decipherable. Potentially, a lower Mg/Ca ratio area is visible closer to the apex, and a higher Mg/Ca ratio area towards the edge, however no season of collection conclusion could be drawn. The shell appears to have been less than 1 year old. Two oxygen isotope samples were extracted: The summer sample yielded a value of −1.48 ‰, while the winter sample yielded a value of −0.56 ‰ (**fig. 122**). The latter is quite more elevated than expected and could be due to the shell's young age, which did not allow for the adequate targeting of seasonal extremes.

Q5S 189

Excavation unit Q5S 189 dates to the Early Neolithic period and contains five shells. The excavation unit winter isotope average is 1.36 ‰, while the summer is −1.01 ‰ (**fig. 123**).

Specimen Q5S_189_A recorded two winters and one summer and was thus 1.5 years old. The penultimate winter is quite extended, while the season of collection appears to be late autumn or early winter. Two oxygen isotope samples were collected from this shell: The summer sample had a value of −0.56 ‰, while the winter sample had a value of −0.38 ‰. The latter is quite elevated for a winter value.

Specimen Q5S_189_B recorded one summer and one winter and was approximately 1 year old. The line scan is not particularly interpretable, however illustrates the overall trends. The season of collection

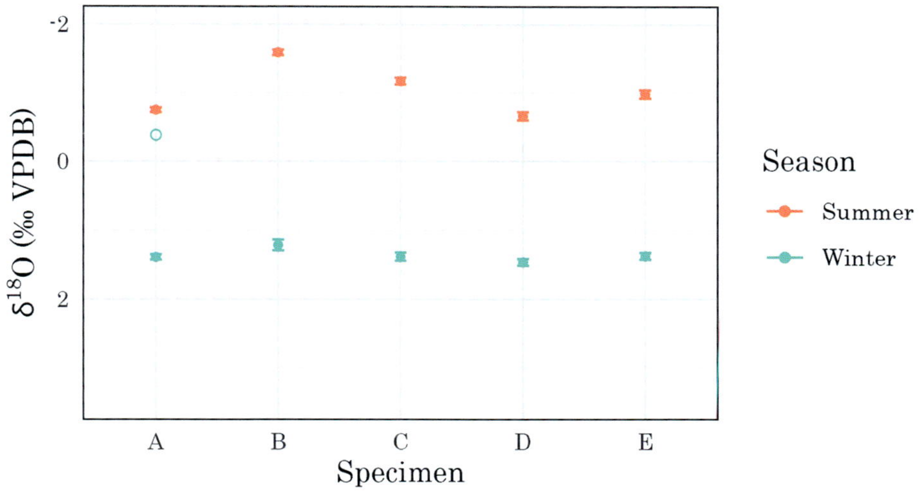

Fig. 123 Seasonal $\delta^{18}O$ values for unit Q5S 189.

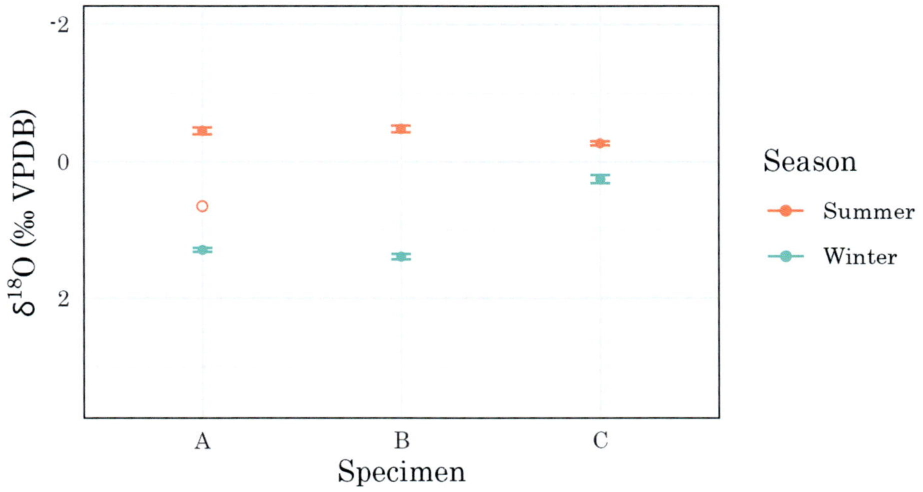

Fig. 124 Seasonal $\delta^{18}O$ values for unit Q5S 188.

was probably winter. Two oxygen isotope samples were collected from this specimen: The winter increment yielded a value of 1.21 ‰, which is very close to the period average, while the summer increment yielded a value of −1.59 ‰.

Specimen Q5S_189_C reflects a spring/summer season and one winter season and was thus almost 1 year old. The last increment of growth has a low Mg/Ca ratio and indicates a winter collection. Two oxygen isotope samples were extracted: The winter increment gave a value of 1.38 ‰, while the summer increment a value of −1.17 ‰.

Specimen Q5S_189_D presents a similar picture to that of the previous specimen, having however a less pronounced summer. Seasonal cyclicity is not well-defined in this specimen as seasonal increments are not particularly distinct. The shell recorded one summer and one or two winters and was thus at least 1 year old. Based on the elemental map of the anterior, the shell was collected during winter. Two oxygen isotope samples were extracted: The winter sample had a value of 1.46 ‰, while the summer sample had a value of −0.66 ‰.

Specimen Q5S_189_E recorded two winters and one thin and distinct summer. The last increment of growth is an extended broad winter increment, which indicates that the shell was collected during winter. As the shell is fragmented an age estimation could be not made. Two oxygen isotope samples were extracted: The winter increment gave a value of 1.37 ‰, while the summer increment a value of −0.98 ‰.

Q5S 188

Excavation unit Q5S 188 dates to the Early Neolithic period and contains three shells. The excavation unit

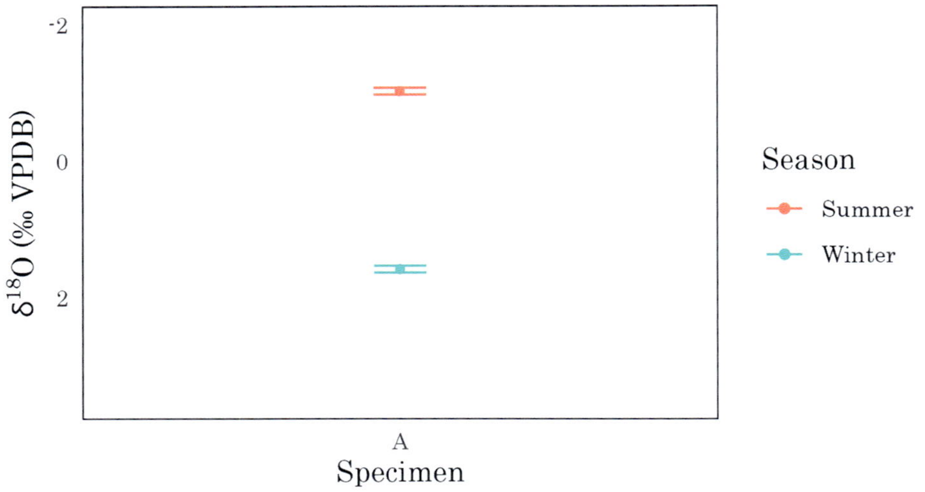

Fig. 125 Seasonal δ18O values for unit Q5S 187.

winter isotope average is 0.98 ‰, while the summer is −0.40 ‰ (**fig. 124**).

Specimen Q5S_188_A recorded two winters and two summers. The summer increments, better visible in the anterior edge, are very thin. The mollusc was 2 years old and was collected during autumn as attested by the cooling trend at the last growth increment. Three oxygen isotope samples were collected in this case: The summer samples had values of −0.45 ‰ (last summer) and 0.65 ‰ (penultimate summer), while the winter sample had a value of 1.29 ‰. The penultimate summer value is relatively high for a summer increment and could be due to the thinness of the increment.

Specimen Q5S_188_B reflects two winters and one summer. Thus, the mollusc was 1.5 years old and was collected during winter as attested by the low Mg/Ca ratio at the last increment of growth. Two oxygen isotope samples were collected: The summer sample had a value of −0.48 ‰ and the penultimate winter sample a value of 1.39 ‰.

Specimen Q5S_188_C recorded a winter and a spring or summer season. The mollusc was probably less than 1 year old as there is no seasonal cyclicity observed rather than a downward trend visible in the line scan. The season of collection appears to have been winter. Two oxygen isotope samples were collected: The winter sample had a value of 0.25 ‰ and the summer sample −0.27 ‰. Both isotope results are quite close and indicate that the specimen was short-lived and that seasonal extremes are not yet developed.

Q5S 187

Excavation unit Q5S 187 also dates to the Early Neolithic period. Specimen Q5S_187_A recorded two winters and two summers. The mollusc was 2 years old at the time of death and was collected during winter. The last summer is relatively broad and transitions from one season to another are gradual. Three oxygen isotope samples were extracted: The winter samples had consistent values of 1.64 ‰ and 1.54 ‰ (average of 1.59 ‰), while the summer sample had a value of −1.02 ‰ (**fig. 125**).

Q5S 185

Excavation unit Q5S 185 dates to the Early Neolithic period and contains four shells. The average winter isotope value of this excavation unit is 0.44 ‰ and the summer −0.84 ‰ (**fig. 126**).

Specimen Q5S_185_A reflects two winters and two summers as seen on the anterior edge and its line scan. Thus, the specimen was 2 years old. The shell was collected during autumn as a cooling trend is present after the last summer. Three oxygen isotope samples were collected from this shell: The summer samples had consistent values of −0.45 ‰ and 0.27 ‰, while the winter sample had a value of 0.51 ‰. The latter is relatively low for a winter increment.

Specimen Q5S_185_B recorded two winters and one summer as visible on the anterior edge. The mollusc was 1.5 years old and was collected during winter. Two oxygen isotope samples were extracted: The winter sample had a value of −0.68 ‰, while the summer sample had a value of −1.38 ‰. The winter

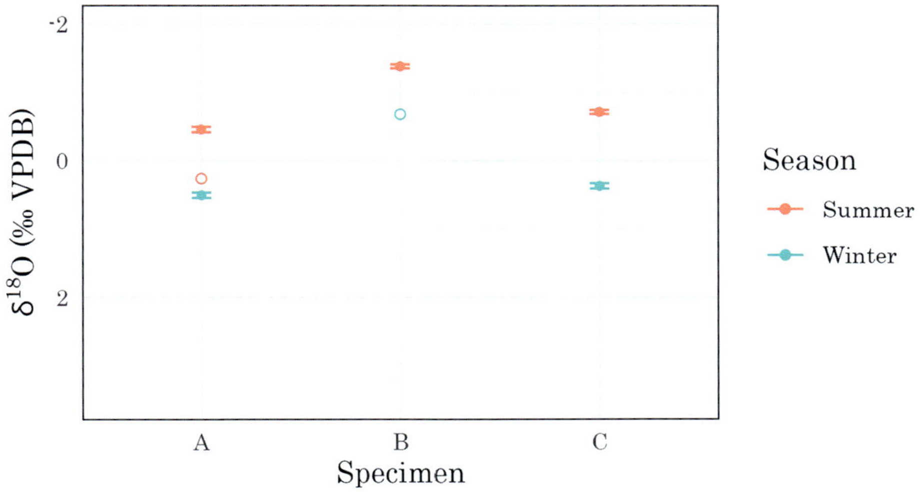

Fig. 126 Seasonal $\delta^{18}O$ values for unit Q5S 185.

value is particularly low and may represent an autumn rather than a colder season.

Specimen Q5S_185_C also reflects two winters and one summer. The mollusc was 1.5 years old and was collected during winter. Two oxygen isotope samples were collected: The winter sample yielded a value of 0.37 ‰, while the summer sample yielded a value of −0.71 ‰.

Specimen Q5S_185_D is quite fragmented and only a small amount of the calcitic layers is preserved. None of the edges were preserved and for this reason no season of collection was identified and no age determination was made. Due to the specimen's fragmented state, no oxygen isotope samples were retrieved.

4.2.1.8. Summary of Elemental Mapping Results

The aim of the elemental analysis was to determine the general growth structure of *Patella* specimens based on the Mg/Ca distributions linked to Sea Surface Temperature (SST). As described in Chapter 3, by examining the Mg/Ca data the instances of summer and winter are revealed. The counting of the seasonal increments allows the estimation of the biological age of the mollusc, while the identification of the last increment of growth reflects the season of death, capture, consumption or collection.

Mg/Ca elemental mapping of the archaeological shell sections revealed in 93 % of the cases (167 specimens) seasonally-resolved growth patterns. 13 specimens produced results that could not be interpreted in terms of seasonal cyclicity and in all of these cases the season of collection could not be inferred. In comparison with the modern shell specimens, the archaeological ones were not intact, as described in the previous chapter, and did not always preserve their edges. Given that the preservation of the shell edge was essential to estimate the season of collection, samples with no edge (17 %, 30 specimens) were excluded from the season of collection estimates. In addition, the age of the specimens was not determined in the aforementioned cases as the last growth increments were missing.

In the cases where an edge was preserved, the season of collection estimate of 13 % of the samples (24 specimens) was not considered to be certain and for 15 of these (8 %) the season of collection was considered to be unknown. Thus, only 9 of the 24 samples were assigned to a specific season of collection with a degree of uncertainty. These 9 »uncertain« samples were excluded from the summary graphs and overall result considerations together with specimens without an edge, while the unknown season of collection specimens are distinctly indicated as such. In total, of the 180 samples after subtracting the specimens without an edge and the uncertain season of collection, 141 specimens are included in the season of collection results. The season of collection results are further discussed in Chapter 5. All these estimates are based on the *Patella* species that were selected for this study, while as shown in Chapter 2, limpets are not the dominant species at the site (Shackleton 1988, 12–19).

Overall seasonal growth increments were relatively regular and consistent. More, specifically

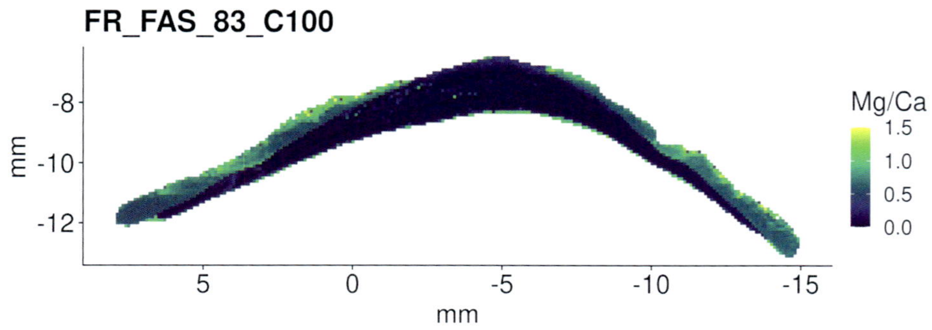

Fig. 127 Example of a shell (FR_FAS_83_C) that did not preserve sufficient calcitic material and was not selected for isotope analysis.

visible growth stops were observed only on eight samples (4 %) (FAN_134_C, FAN_162_A and B, FAN_163_B, FAN_172_B, FAS_124_A, L5NE_56_A and L5NE_64_A). Broader than usual summer increments were noticed on twelve specimens, that is on the 7 % of the assemblage (FAS_83_A, FAS_91_B, FAS_145_A, FAS_148_A, FAS_186_A, FAS_195_B, FAS_207_A, L5NE_45_A, L5NE_48_B, Q5N_23_B, Q5N_25_B, Q5S_185_B), while very thin summer increments were observed on 26 specimens (14 %).

4.2.2. Oxygen Isotope Analysis

As explained in Chapter 3, calcium carbonate samples were sampled from positions where according to the elemental data the seasonal minima and the maxima are situated. This enables the analysis of larger assemblages by extracting fewer isotope samples per specimen. Thus, 381 carbonate powder samples were extracted in total from the 153 of the 180 archaeological shell specimens which revealed seasonally-resolved cyclicity patterns. The rest of the 27 archaeological shell specimens did not undergo isotope analysis as it was not deemed meaningful to proceed further in their analysis. These were shell specimens that either did not yield seasonally-resolved elemental results (e. g. FR_FAS_74_C), or that did not preserve sufficient material of their calcitic parts (e. g. FR_FAS_83_C, FR_FAN_163, FR_FAN_175_B) (**fig. 127**).

From the 381 carbonate powder samples extracted 50 were not taken into consideration (13 %) in the subsequent analysis as their results were not considered to be trustworthy (**tab. A.9**). In some of the cases, this was because the results did not satisfactorily match the elemental data and the accompanying isotope results in the same specimen. In other cases, specimen size and seasonal increments were too small to be adequately captured. To tackle the problem of the very thin summer increments, two samples were taken from the area in order to best sample the increment and the one with the best result was kept. Finally, in some cases shell specimens were too young and the seasonal increments could not be adequately distinguished and attributed to a certain season, something that was made further obvious by the oxygen isotope results.

Figure 128 depicts the isotope ranges by excavation unit and secondly by chronological period. Ranges consist of the minimum and maximum $\delta^{18}O$ values of each excavation unit that may contain multiple shell specimens, while the minimum values represent the summer increments and the maximum winter ones. The amplitude between the minimum and maximum values represents the seasonality of temperature (Kwiecien 2022, Box 1), while this amplitude constitutes an aspect of climate variability. In this case seasonal amplitude appears to be variable through time, while being particularly smaller during the Lower and the Final Mesolithic, as well as the Initial and the Final Neolithic period.

More specifically, during the Final Palaeolithic period, isotope ranges reflect lower temperatures both during summer and winter (**fig. 128**). The coldest winter of all the samples analysed is recorded in unit FAS 207 of the Final Palaeolithic (3.22 ‰). Winter increments during this period vary from 2.15 to 3.22 ‰ and summers from −0.85 to 0.50 ‰. During the Mesolithic period, winters become slightly warmer compared to the previous period, ranging from 0.67 to 2.35 ‰ and summers do not exceed -1.09 ‰, while having a maximum value of 0.29 ‰.

From the transitional period of the Final Mesolithic to the Initial Neolithic a warming trend of the summer increments becomes more pronounced

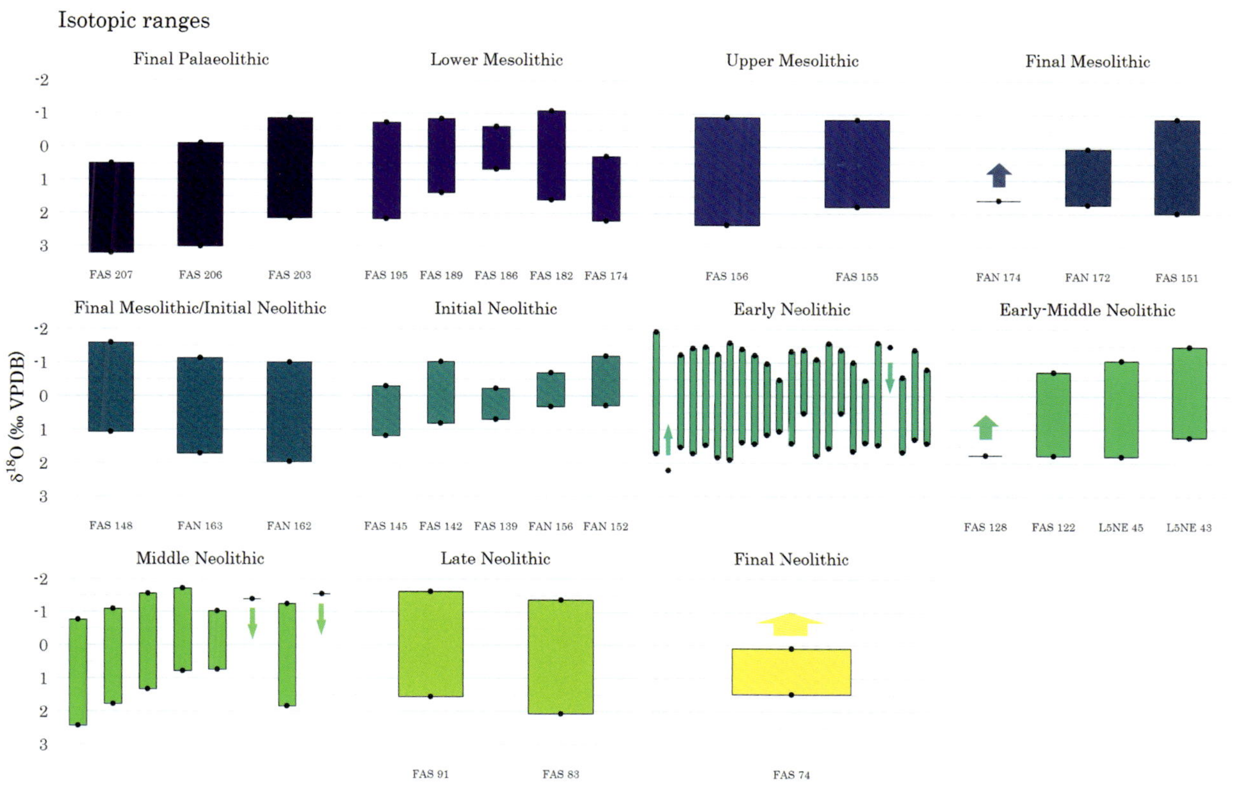

Fig. 128 Isotopic ranges by excavation unit (n = 59). Top points of each bar (representing an excavation unit) represent summer increment values and the bottom point winter increments values.

(from −1.6 to −1.00 ‰), while winters are still relatively cold (from 1.06 to 1.96 ‰). During the Initial Neolithic period, summer values range from −1.2 to −0.22 ‰ and winter ones from 0.29 to 1.18 ‰. Interestingly, isotope ranges are quite short during this time.

The warming trend of the Neolithic will result in the warmest recorded summer value among all the samples analysed, as evidenced by a shell specimen from unit FAN 144 (−1.92 ‰), dated to the Early Neolithic period. Furthermore, summer values during the Early Neolithic period range from −1.92 to −0.48 ‰. In addition, winter values during the Early Neolithic range from 0.51 to 2.21 ‰. One can notice extended seasonal variability during the Early Neolithic period, that is there are significant differences between annual minima and maxima in a large number of excavation units.

Shell samples from four excavation units of the Early-Middle Neolithic period yielded summer increment values that range from −1.49 to −0.72 ‰. Winter increment values appear to be more consistent and range from 1.24 to 1.81 ‰. During the Middle Neolithic, summer increment values range from −1.73 to −0.76 ‰, while winter increments range from 0.74 to 2.44 ‰. Winter increments of the Middle Neolithic shells are quite variable, due to their young age (average age 1.5 years). The same is not the case for the two excavation units belonging to the Late Neolithic period, as the isotope ranges were more consistent. Summer increment values range from −1.63 to −1.39 ‰, while winter ones from 1.55 to 2.07 ‰. Lastly, only one excavation unit was dated to the Final Neolithic period. The summer value is significantly low at 0.09 ‰ and the winter value at 1.47 ‰. The latter summer value comes only from one specimen and does not appear to be quite reliable.

The differences between the minimum and the maximum isotope values were calculated for each unit and then averaged to reflect the mean difference for each chronological period. The amplitude of the seasonal changes is reflected in **table 14** and represents the mean annual ranges by period.

The sample standard deviation was calculated separately for winter and summer oxygen isotope values for every excavation unit where more than

Chronological period	Mean annual range (‰)	Minimum δ18O (‰)	Maximum δ18O (‰)
Final Palaeolithic	2.95	-0.85	3.22
Lower Mesolithic	2.20	-1.09	2.23
Upper Mesolithic	2.95	-0.91	2.35
Final Mesolithic	2.27 [1.66]	-0.87	1.98
Final Mesolithic/Initial Neolithic	2.82	-1.60	1.96
Initial Neolithic	1.35	-1.20	1.18
Early Neolithic	2.64 [2.48]	-1.92	2.21
Early-Middle Neolithic	2.70 [2.19]	-1.49	1.81
Middle Neolithic	2.73 [2.35]	-1.73	2.44
Late Neolithic	3.32	-1.63	2.07
Final Neolithic (only 1 sample)	1.38	0.09	1.47

Tab. 14 Mean annual ranges by chronological period. The ranges represent mean values of the annual ranges (difference between δ18Omin and δ18Omax within one specimen) across one period. Values in brackets represent calculations where all values (also the uncertain) are considered.

one same-season measurement had been made. The same was done for every chronological period. The two sets of standard deviations (winter and summer) are displayed by excavation unit and by chronological period (see **figs 129–130; tabs A.12–A.15**). Standard deviation calculations were run twice, with and without the less reliable values. The less reliable measurements were certain oxygen isotope measurements (n = 50) that were excluded from the standard deviation calculations as unreliable either because the seasonal increments boundaries were considered to be unclear and the specimens too young, or because the certain summer increments were too thin to be accurately sampled and captured. Here, I discuss the more reliable results, while the graphs that also include the less reliable results are shown in **Appendix I**. Notably, there is no substantial difference between the two sets of results, suggesting that the exclusion of the less reliable measurements had minimal impact on the overall variability patterns. Regarding the reliable sample results, in general, on the excavation unit level the summer standard deviation is on average slightly higher (0.44 ‰) in comparison to the winter standard deviation (0.37 ‰) (**fig. 130**), while the same is the case on the chronology level where the summer increments standard deviations average at 0.54 ‰ and the winter ones at 0.45 ‰ (**fig. 129**).

More specifically, on the wider chronological period level certain groups of specimens stand out: Firstly, the highest standard deviation is present at the summer increments of the Late Neolithic specimens (1.41 ‰). These calculations were based on specimens FAS_83_A and B and FAS_91_A and B. This may be partially due to the very thin summer increments sampled in the case of specimen FAS_83_B that yielded a comparatively high value, while the winter increments of the same set of samples presented a small standard deviation. Subsequently, the Middle Neolithic period had relatively high winter increments standard deviation values (0.61 ‰) coupled with less high summer increments standard deviation (0.40 ‰). The Middle Neolithic specimens are a quite young group of specimens (mean 1.9 years of age) and seasonal cyclicity is not well-defined on the elemental maps. The high standard deviation of the winter increments of these specimens may be related to the ambiguous cyclicity patterns where seasonal increments are not easily distinguishable on the elemental data (e. g. FAN_124_A, D and E, FAN_127_A, B, C, D, E, F and G, FAN_129_A, FAN_131_A and D, FAN_132_B, C and E, FAN_134_E).

Subsequently, the Final Palaeolithic and the Lower Mesolithic winter increments present relatively high standard deviation summer values (0.61 and 0.57 ‰ respectively), whereas the respective sum-

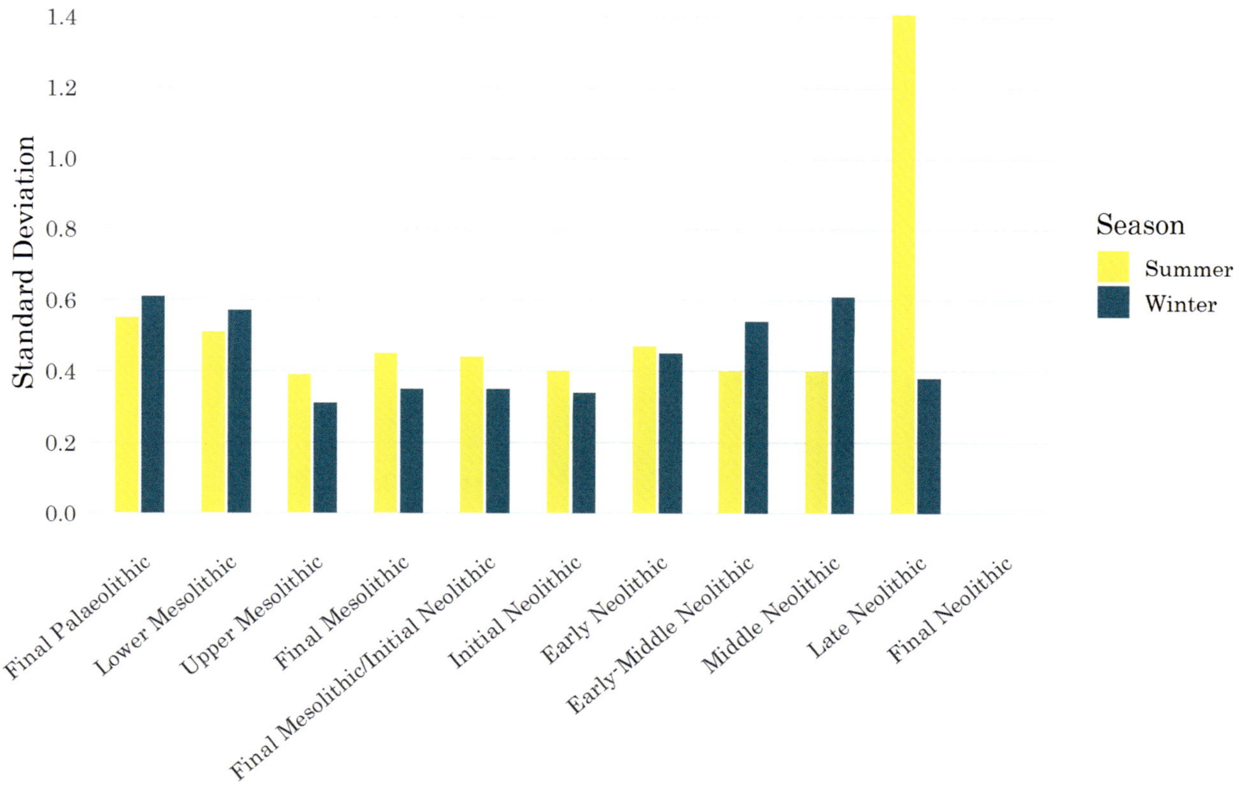

Fig. 129 Winter and summer standard deviation of oxygen isotope values by chronological period.

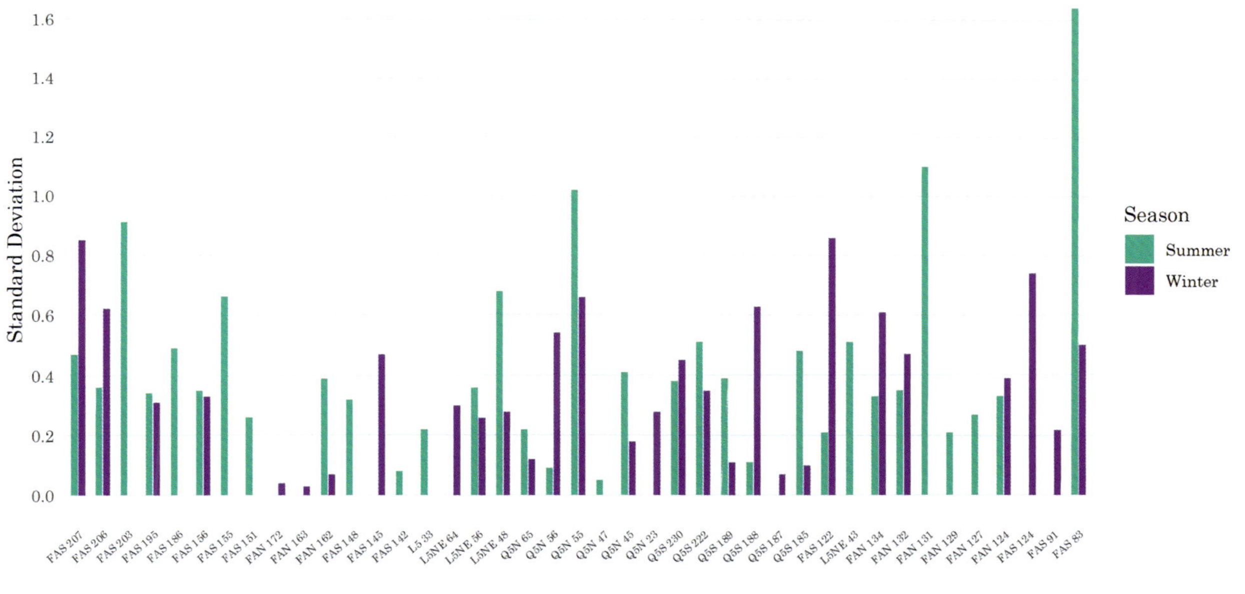

Fig. 130 Winter and summer standard deviation of oxygen isotope values by excavation unit in chronological order, starting from the left with the older specimens.

mer increments were consistent with similar standard deviation values of 0.55 and 0.51 ‰ (**fig. 129**). The Early-Middle Neolithic had a winter standard deviation of 0.54 ‰ and a summer standard deviation of 0.40 ‰. The Early Neolithic had a winter standard deviation of 0.45 ‰ and a summer standard deviation of 0.47 ‰. Moreover, the winter increments of the rest of the periods provided consistent oxygen isotope values with a standard deviation of 0.45 ‰ and below.

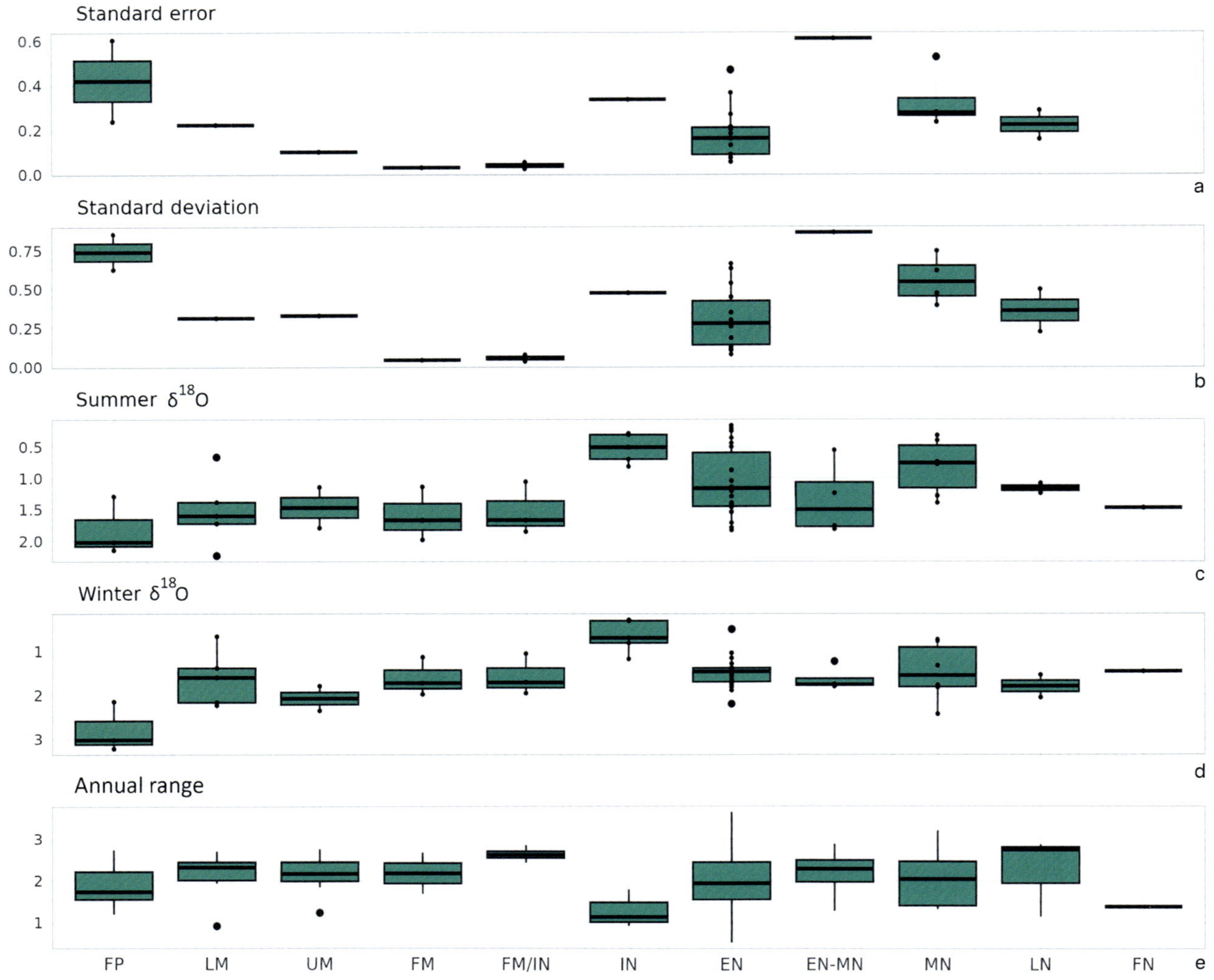

Fig. 131 Graph displaying summary statistics for the standard error (**a**), the standard deviation (**b**), the summer $\delta^{18}O$ (**c**), the winter $\delta^{18}O$ (**d**) and the annual range (**e**). – Chronology legend: FP: Final Palaeolithic; LM: Lower Mesolithic; UM: Upper Mesolithic; FM: Final Mesolithic; FM/IN: Final Mesolithic/Initial Neolithic; EN: Early Neolithic; MN: Middle Neolithic; LN: Late Neolithic; FN: Final Neolithic.

When examining the standard deviation by excavation unit, the picture is similar to that of the chronological period standard deviation. The highest standard deviation (1.64 ‰) is found in the summer increments of the Late Neolithic FAS 83 excavation unit. Subsequently, the summer increments of the Middle Neolithic unit FAN 131 is also significantly high (1.1 ‰), which could be due to the very thin summer increments of specimen FAN_131_B. Subsequently, the summer increments of the Early Neolithic unit Q5N 55 is also significantly high (1.02 ‰). On the other hand, among the lowest standard deviation values are those of the Final Mesolithic/Initial Neolithic winter increments of FAN 163 (0.03 ‰) and the Final Mesolithic winter increments of FAN 172 (0.04 ‰), as well as the winter increments of the Early Neolithic excavation units Q5SN 65 (0.12 ‰), Q5S 185 (0.1 ‰), Q5S 187 (0.07 ‰), Q5S 189 (0.11 ‰), L5NE 56 (0.26 ‰). The summer increments standard deviation of the Early Neolithic is higher with an average of 0.38 ‰, while the winter standard deviation of the Early Neolithic averages at 0.31 ‰.

The following summary statistics of the overall distribution of the standard deviation, standard error and the $\delta^{18}O$ values is shown in **figure 131**. The overall distribution of all four variables shows that higher variability is present during the Final Palaeolithic, the Early Neolithic and the Middle Neolithic period. The more scattered values in all four categories are detected in the Early Neolithic period. The Initial Neolithic period presents the higher $\delta^{18}O$ val-

Fig. 132 Scatterplot of the oxygen isotope values of the last winter of growth oxygen isotope of each specimen grouped by chronological period. The average of the winter values was calculated separately for each period and subsequently set to zero (dashed black line) while the isotope values (points) were adjusted in relation to each average value.

ues both during the summer and during the winter. This is further examined in section 5.3.1.

In order to detect seasonal extremes, I examined the specimens' last winter increments in comparison to the average of winter isotope values of each unit and period. In order to examine the winter increments' variability, the average of the winter values of each chronological period and the isotope ratios in comparison to that average were plotted (**fig. 132**). Only the winter increment values that represented the last winter recorded in the shell are considered in this case. These sequences give a glimpse of winter variability through the stratigraphy as they show how colder or warmer winters alternated across time.

Furthermore, in most chronological periods, isotope values are distributed around the average while certain cases stand out: The more populated groups of values of the Early Neolithic (n = 43) and the Middle Neolithic period (n = 13) present more variable last winters of growth with more values deviating from the period average. More specifically, in the case of the Early Neolithic samples, it appears that a large number of winters (n = 29) are below, that is colder than the period average (1.24 ‰, set at 0 in **fig. 133**). During the Middle Neolithic, oxygen isotope values of the last winter increment are less clustered around the average, while six specimens presented warmer last winters than the period average (1.27 ‰ set at 0 in the graph) and seven winter increments that were colder than the average. On the other hand, little winter variability can be seen during the Final Palaeolithic or the Late Neolithic period.

When examining these datasets by excavation unit and comparing them with the previous graph (**fig. 133**), it appears that the large amount of colder than average winters of the Early Neolithic period are more prevalent in unit Q5S 230, where 9 winter values out of 15 are below the unit average. As for the Middle Neolithic period, winters are more variable in units FAS 122, FAS 156, FAN 132 and FAN 134. Furthermore, when splitting the datasets between the cave and the Paralia site, no further differences become visible (**figs A.47–A.48**).

The overall plotting of isotope results reveals an interesting picture (**fig. 134**). All isotope values were placed in a scatter plot where the x axis represents

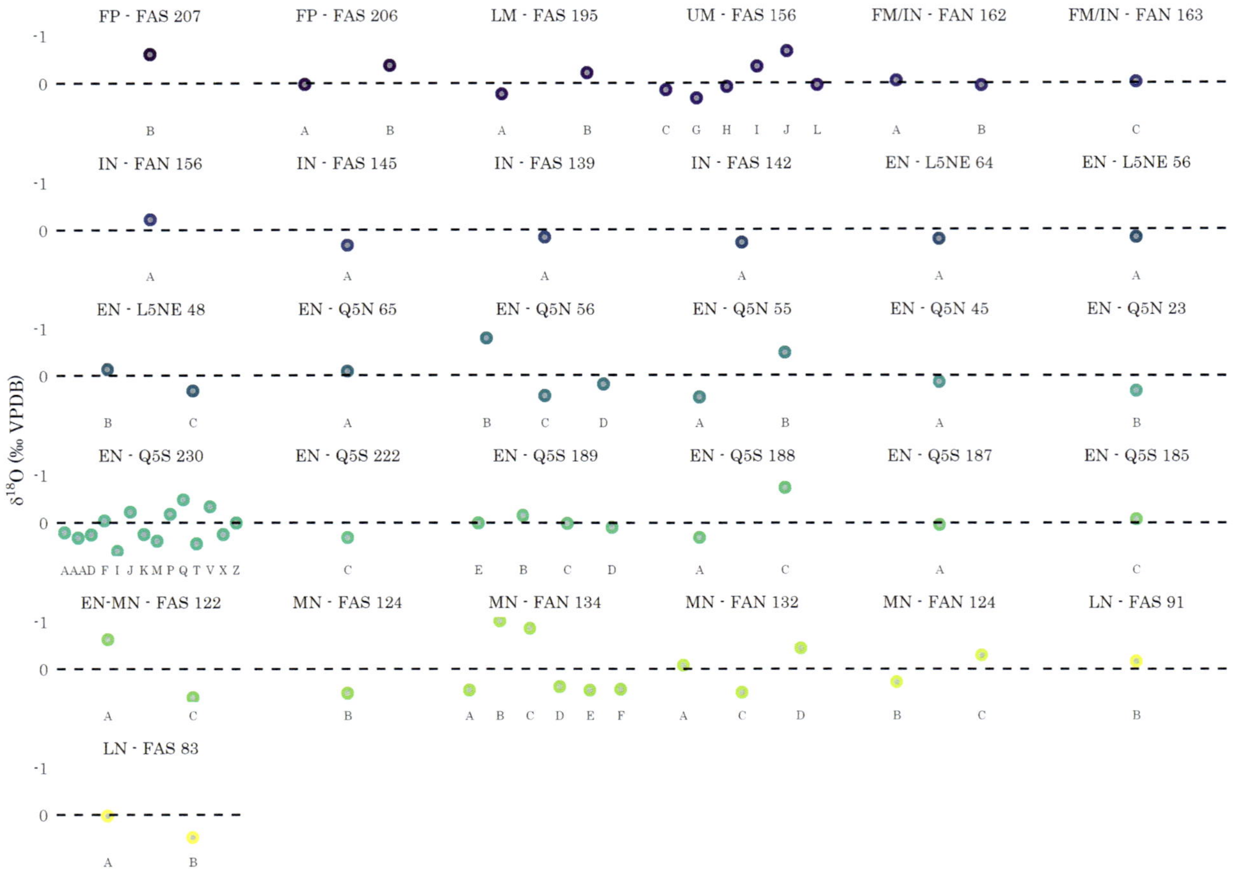

Fig. 133 Scatterplot of the oxygen isotope values of the last winter of growth oxygen isotope of each specimen grouped by excavation unit. On the x axis of each graph the specimen ID is indicated. The average of the winter values was calculated separately for each excavation unit and subsequently set to zero (dashed black line) while the isotope values (points) were adjusted in relation to each average value.

time and a Locally Estimated Scatterplot Smoothing (LOESS) curve was applied in order to best visualise data trends. As described in section 4.2.1, the higher seasonal isotope values are dated to the Final Palaeolithic period. The highest winter value is recorded in a shell of this period. During the Mesolithic period, there is a gradual overall decline of oxygen isotope values that culminates in the Early Neolithic period, where the group of isotope values are at their lowest peak. The winter values show a more intense peak. The Early-Middle Neolithic group of samples indicate a rise which is more pronounced in the winter values. In addition, the isotope values of the Middle Neolithic shells show an increase in temperature, which is again more intense in the winter set of values. Finally, the Late Neolithic isotope values indicate a decrease of temperature, nonetheless these samples are very few to clearly display any meaningful trends.

To summarise, the 329 carbonate powder samples extracted from 153 archaeological shells yielded results that shed light on the archaeological samples on various levels. The isotope results on the specimen, the excavation unit and the wider chronological level allow us to reconstruct relative changes in sea surface temperatures in the past. In the larger scale, this provides a high-resolution picture of changes in the climate at the time of human occupation.

More specifically, the plotting of secure isotope ratios across time revealed highest seasonal values during the Final Palaeolithic. The coldest winter of the analysed assemblage is recorded in a unit dated to the Final Palaeolithic period (3.22 ‰). During the Mesolithic, relative temperatures progressively rose leading to the Early Neolithic where the warmest summer value (–1.92 ‰) belongs to. During the Early-Middle Neolithic period isotope ratios drop and increase again during the Middle Neolithic.

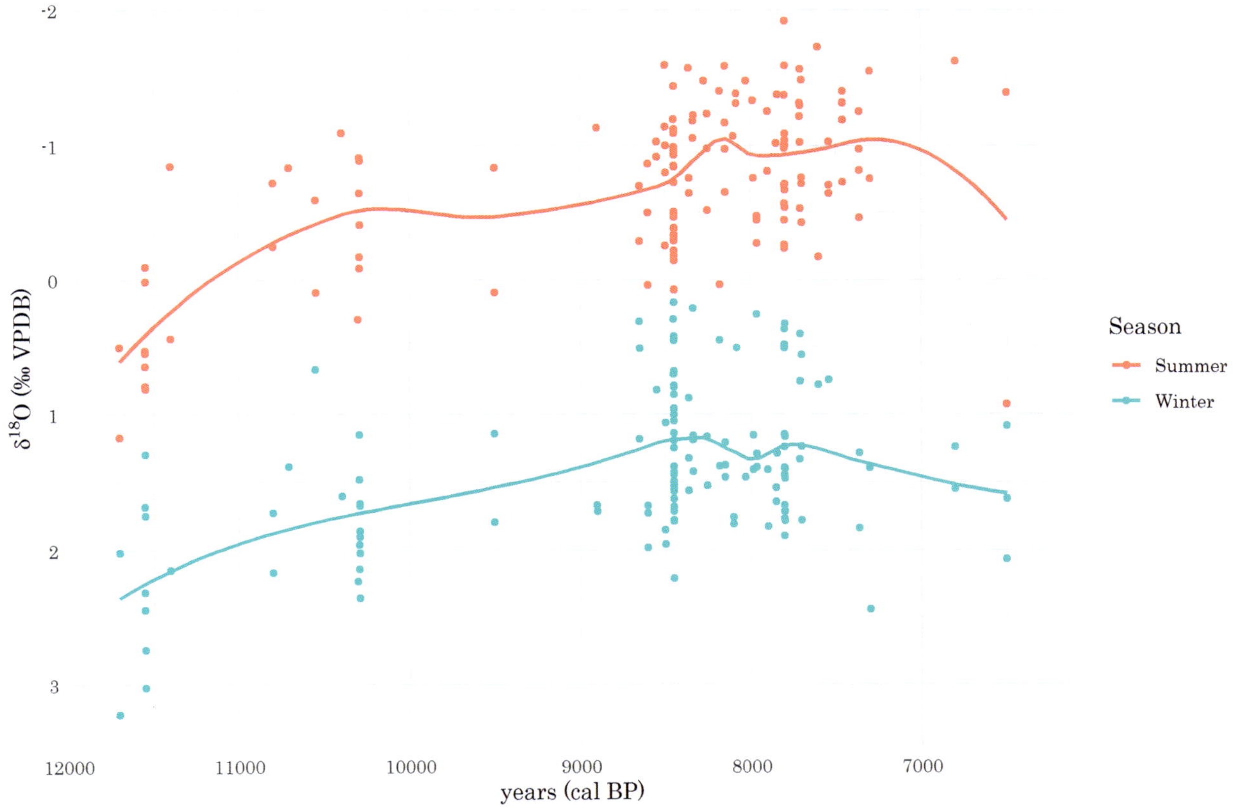

Fig. 134 Summary plot of certain seasonal isotope values (n = 279) both from Franchthi Cave and the Paralia.

Furthermore, the minimum (representing summer increments) and maximum $\delta^{18}O$ values (representing winter increments) were detected in each specimen, excavation unit and period and from these, the mean annual ranges were calculated. This seasonal amplitude is variable through time, while being smaller during the Lower Mesolithic, the Initial and the Final Neolithic period. Large seasonal amplitude is present during the Late Neolithic, the Final Palaeolithic and the Upper Mesolithic period. The last winter comparison to the average of winter isotope ratios illustrates the increased winter variability recorded in the shells across the stratigraphy. In most of the cases, isotope values were close to the average, while during the Early Neolithic and the Middle Neolithic period reflected more variable winters with more isotope ratios deviating from the period average.

4.2.3. AAR Analysis

The results of the 66 samples selected for Amino Acid Racemisation (AAR) analysis were not as satisfactory as the previous groups of results. Even though the technique is well-established, in this case the samples yielded values that were not in accordance with the relative chronologies of the shells. The examination of the D/L ratio of four amino acids presented here, aspartic acid, glutamic acid, alanine and valine, did not contribute to the better understanding of the Franchthi stratigraphy as initially aimed.

As seen in **figures 135–138**, the samples that belong to the Final Palaeolithic do not appear to be older, that is closer to a D/L ratio of 1, as initially expected. Furthermore, samples dating to the Neolithic period are widely spread in all of the four amino acids. More specifically, the Middle Neolithic FAN_132_B shell followed by the Early Neolithic shell Q5N_65_A appear to be the oldest according to the results from all amino acids, while in the alanine results, the Early Neolithic L5NE_55_A is the oldest.

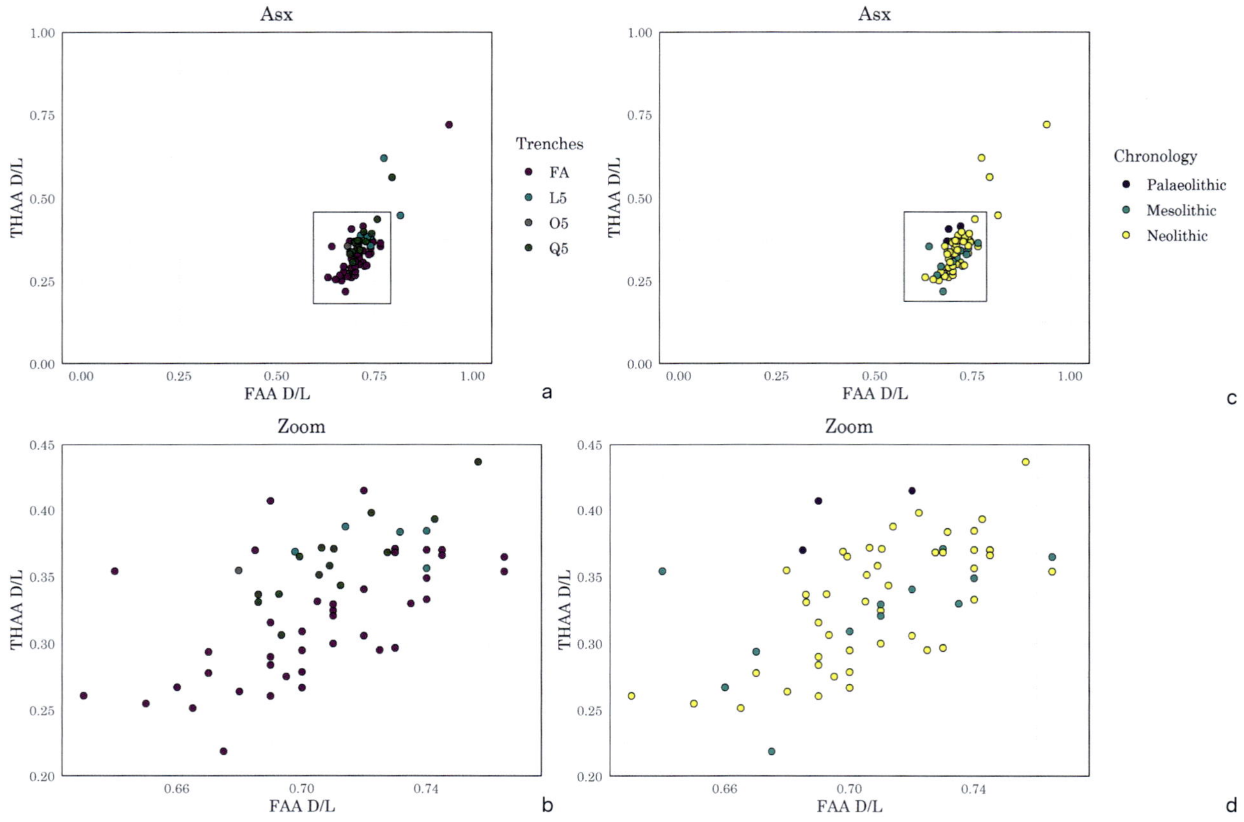

Fig. 135 Aspartic acid hydrolysable amino acids (THAA) D/L ratio against free amino acid fraction (FAA) D/L ratio: **a** Colours indicate the trench that the sample belongs to. – **b** Detail of the previous graph. – **c** Colours indicate the relative chronology. – **d** Detail of the previous graph, indicated by the grey rectangular above.

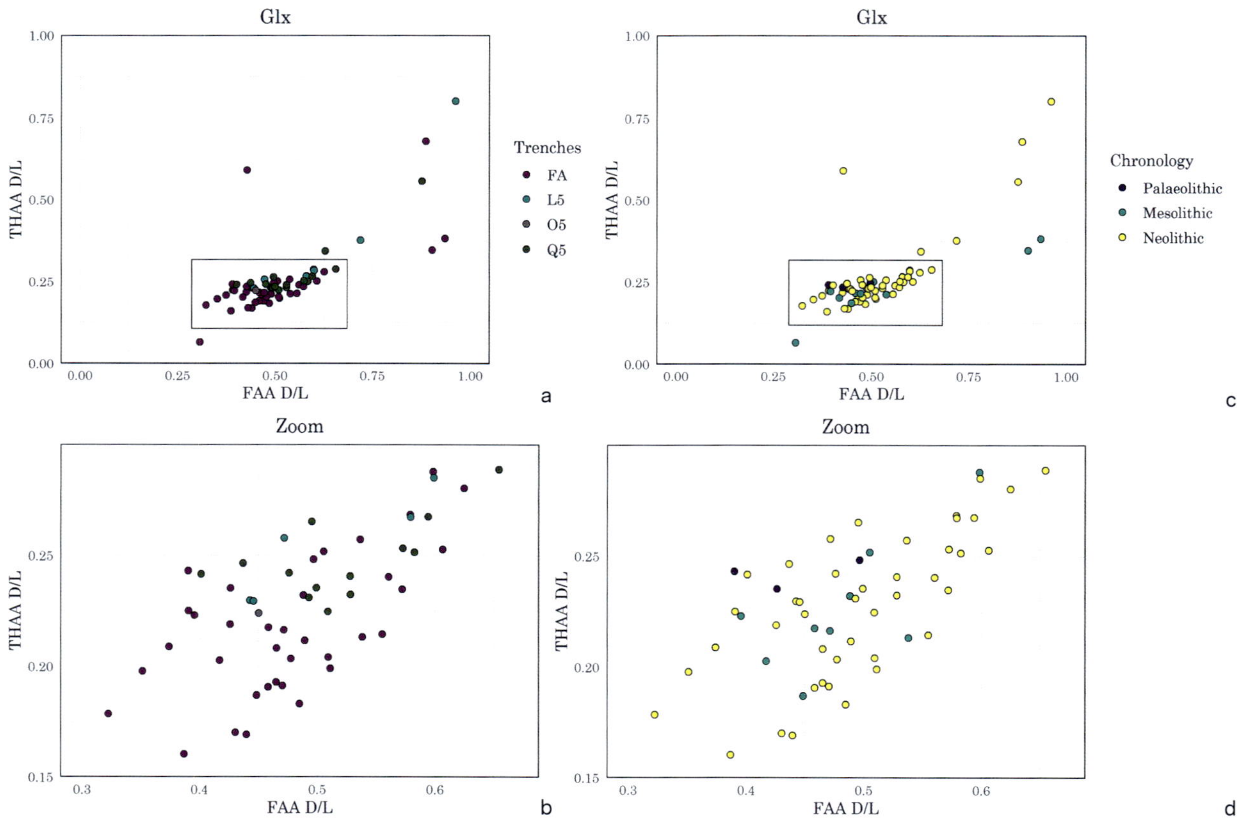

Fig. 136 Glutamic acid hydrolysable amino acids (THAA) D/L ratio against free amino acid fraction (FAA) D/L ratio: **a** Colours indicate the trench that the sample belongs to. – **b** Detail of the previous graph. – **c** Colours indicate the relative chronology. – **d** Detail of the previous graph, indicated by the grey rectangular above.

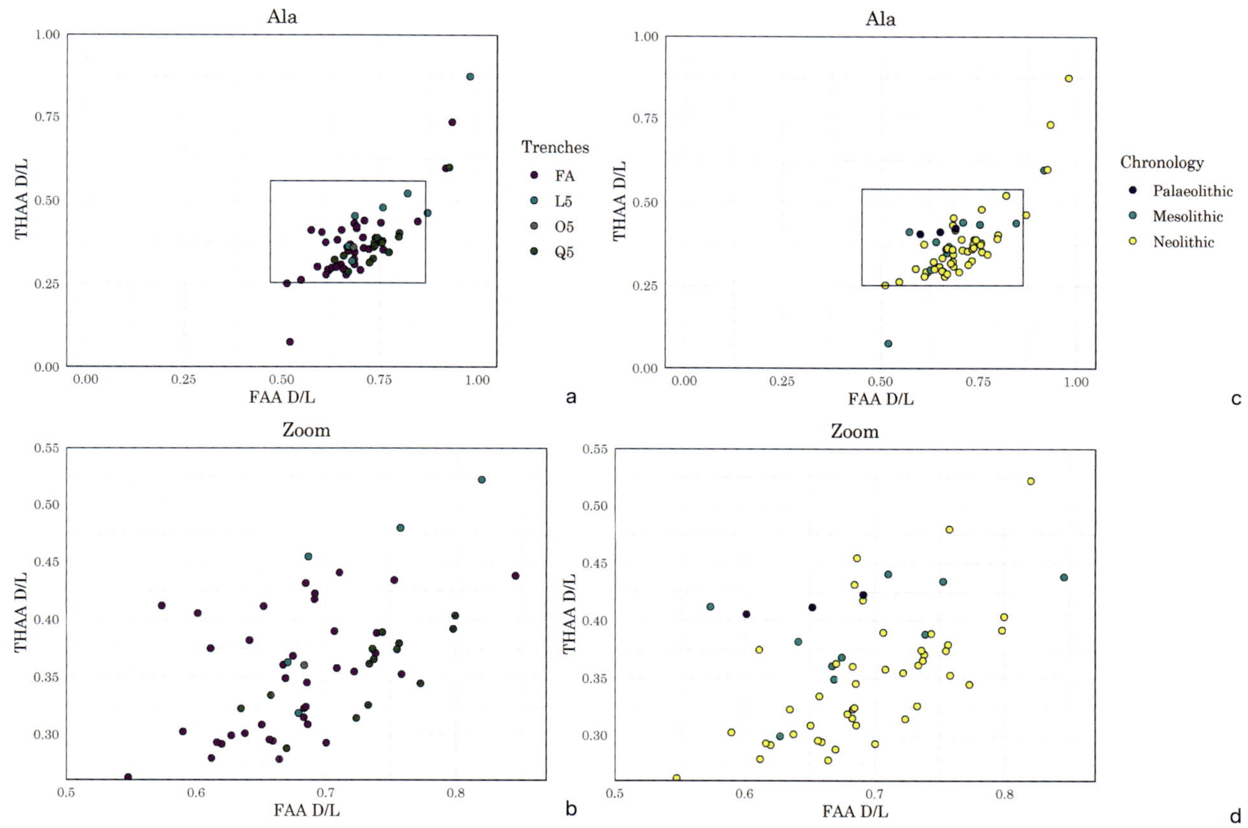

Fig. 137 Alanine hydrolysable amino acids (THAA) D/L ratio against free amino acid fraction (FAA) D/L ratio: **a** Colours indicate the trench that the sample belongs to. – **b** Detail of the previous graph. – **c** Colours indicate the relative chronology. – **d** Detail of the previous graph, indicated by the grey rectangular above.

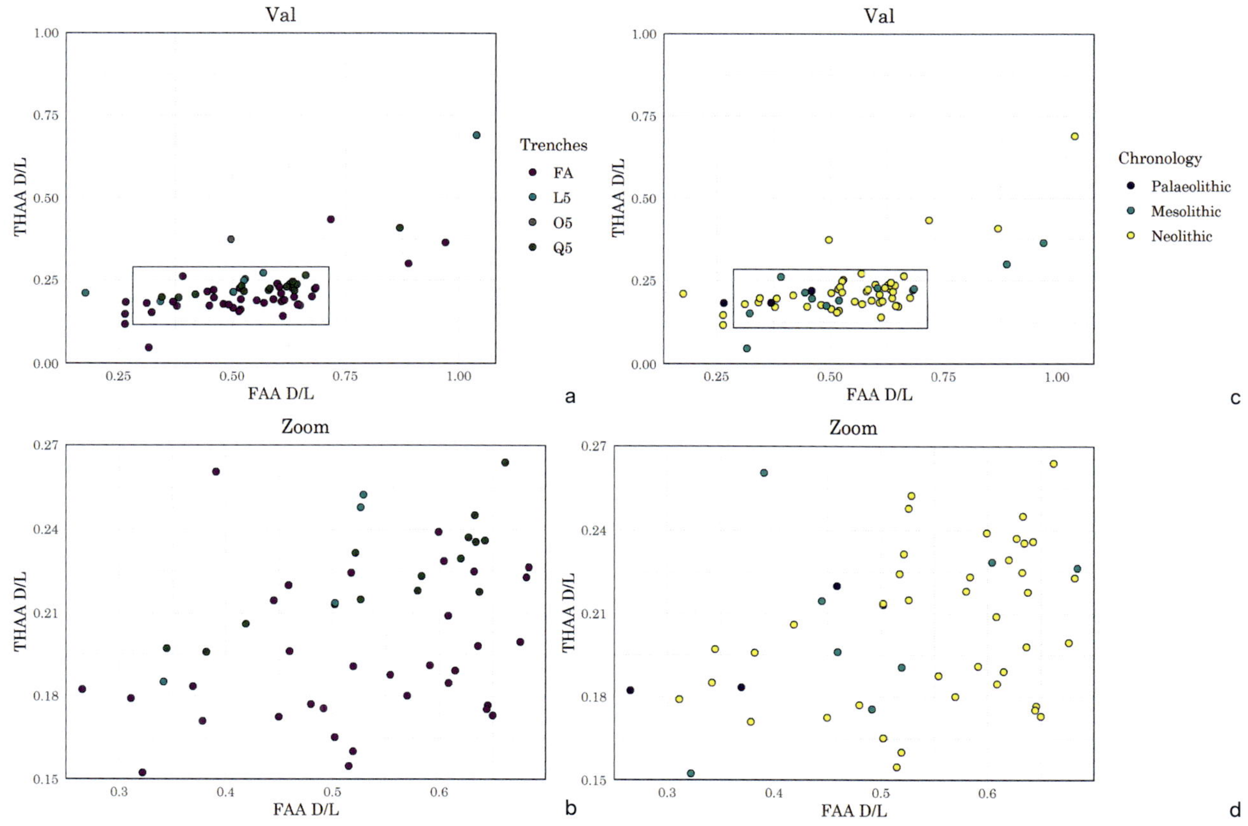

Fig. 138 Valine hydrolysable amino acids (THAA) D/L ratio against free amino acid fraction (FAA) D/L ratio: **a** Colours indicate the trench that the sample belongs to. – **b** Detail of the previous graph. – **c** Colours indicate the relative chronology. – **d** Detail of the previous graph, indicated by the grey rectangular above.

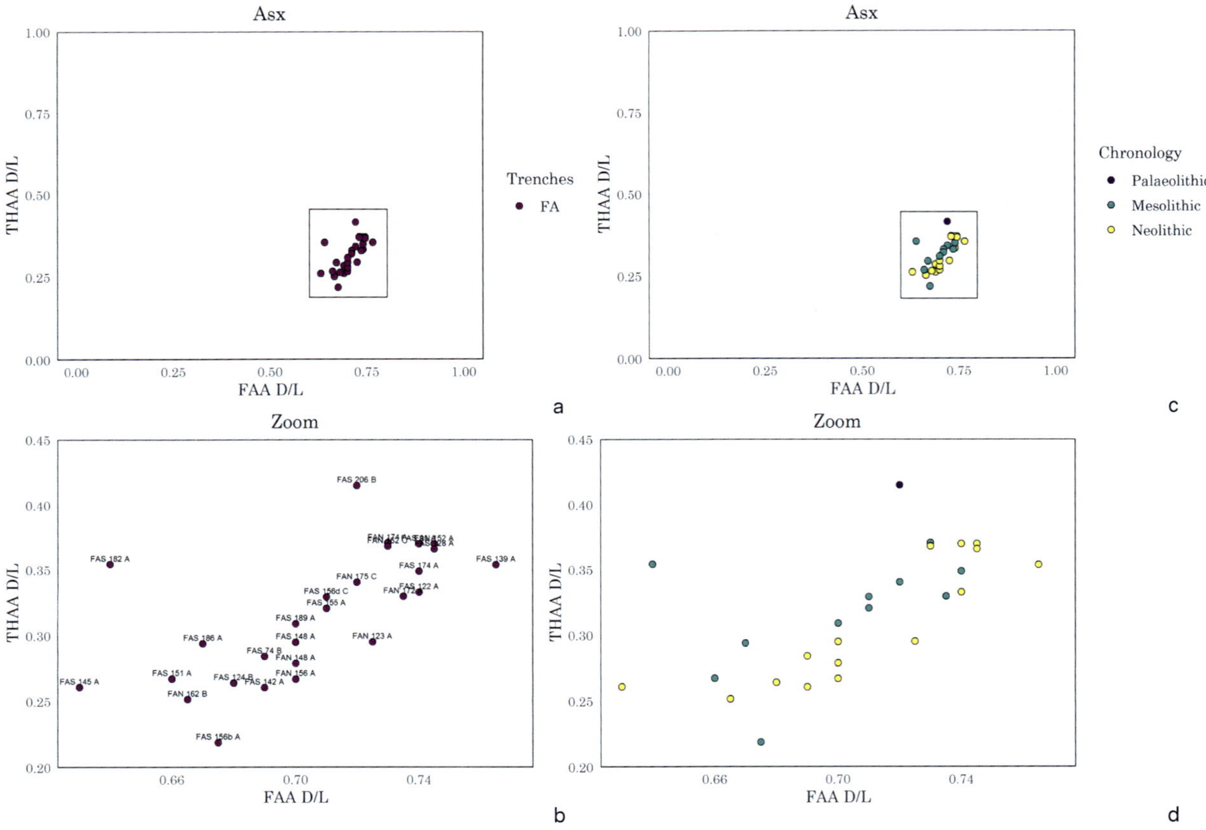

Fig. 139 Aspartic acid hydrolysable amino acids (THAA) D/L ratio against free amino acid fraction (FAA) D/L ratio. The Paralia samples and the samples with indications of possible burning are filtered out. **a** Colours indicate the trench that the sample belongs to (only FA here). – **b** Detail of the previous graph. – **c** Colours indicate the relative chronology. – **d** Detail of the previous graph, indicated by the grey rectangular above.

Excluding the inaccurately old Neolithic samples, two Lower Mesolithic samples seem to agree with the relative chronology, FAS_182_A and FAS_174_A.

It is possible that the layers from the Paralia trenches may have been more exposed in comparison to those of the cave. Furthermore, exposure to human-induced increased temperatures in *Patella* species increases the rate of the degradation processes and can result in high-racemisation rates that do not correspond to the actual age of the sample (Demarchi et al. 2011, 115; 2013). For this reason, I filtered out the Paralia samples and examined the outcomes of the cave samples (**figs A.50–A.52**). In the case of the aspartic acid results (**fig. 139**), the Final Palaeolithic samples do appear to be older than the other samples. Nonetheless, the overall picture is that samples appear to a large extent clustered together, as for example in the case of aspartic acid, glutamic acid and alanine. This may be due to the fact that the samples are do not have vastly different ages but have a difference of maximum 6,000 years, as well as due to the fact that the technique works better with samples that are dated to the Pleistocene and the Pliocene rather than the Holocene (Bada 1972, 230).

4.2.4. Summary

The results of the 180 archaeological samples analysed via LIBS, oxygen isotope analysis, and amino acid racemisation technique were presented in this chapter. The elemental data and the isotope outcomes of each archaeological specimen was described in detail. A season of collection and a biological age was attributed to each specimen when possible and the isotope ratios were also reported. The Mg/Ca ratio revealed seasonal cyclicity patterns in 167 specimens. After excluding the specimens without an edge and those with an uncertain season of collection, 141 specimens produced reliable season of collection results. These results are further presented in Chapter 5.

The carbonate powder samples extracted from 153 archaeological shells yielded useful isotope results on multiple levels. The minimum and maximum $\delta^{18}O$ values were reported in every specimen, excavation unit and chronological period. The mean annual ranges, the standard deviation and the standard error were calculated for these groups of data. Furthermore, winter variability across the stratigraphy was presented by reporting how colder and warmer winters alternated throughout this limited sequence of shell specimens. Larger temporally-resolved sequences, similar to those of dendrochronology, could not be established given the short-lived *Patella* specimens used in this study. This was not further aided by the AAR dating results, which were not successful in further elaborating the chronological associations among the specimens of the same period or excavation unit. Nonetheless, the oxygen isotope analysis outcomes enabled the reconstruction of past sea surface temperature changes across time providing a fragmented glimpse at the local climate at the time of human occupation.

Chapter 5: Discussion

5.1. Introduction

In this work, I attempt to reconstruct the long-term and seasonally-resolved climatic record of Franchthi and integrate it with the archaeological information on subsistence and mobility. In order to do this, LIBS spectroscopy and stable isotope analysis are employed as explained in Chapter 3. The high degree of correlation (mean $R^2 = 0.89$) among the results of the two techniques retrieved from modern specimens of known environmental conditions provided a robust basis for the combined use of the two methods in an archaeological context. This general correlation was already established (Hausmann et al. 2019a) and was further corroborated in my study of the modern specimens used here. In addition, the combination of the two methods as well as the large number of samples to which they are applied constitutes a novelty in the field, as previous seasonality and palaeoclimatic studies on mollusc shells analysed fewer specimens and/or fewer samples (e. g. Godfrey 1988; Mannino et al. 2003; Wang et al. 2012; Jew et al. 2013; Hausmann/Meredith-Williams 2017a; Parker et al. 2020). This approach provides insights into both seasonal patterns of mollusc exploitation and occupation as well as the palaeoclimatic local record of the site. The different sampling strategies used by other researchers to retrieve seasonal and palaeoclimatic information, include either taking sequential and/or single edge series (see also West et al. 2018, 198). For example, Parker et al. (2020) used only one terminal sample from a large number (n = 100) of *P. candei* from the Canary archipelago, while Hausmann and Meredith-Williams (2017a) sampled approximately 10–15 isotope samples per specimen of *C. fasciatus* to reconstruct seasonality exploitation patterns on the Farasan islands. In addition, the number of specimens analysed also vary in every study. For example, Wang et al. (2012) analysed 10 *P. vulgata* shells of the Roman Warm Period to understand climate variability, while Jew et al. (2013) analysed 40 *Mytilus californianus* shells and Deith (1983) analysed long sequences of five archaeological shells of *M. lineata*.

Importantly, in contrast to studies that only use stable isotope sampling, the combined method employed here offers the fast and efficient scanning of a large number of shells with LIBS, which enables the selection of more reliable specimens for subsequent analysis. In addition, LIBS allows the targeting of specific areas of interest, i. e. the seasonal minima and maxima, which further enables the sampling of a larger number of specimens. Mg/Ca ratio patterns can also contribute to the identification of similar environmental conditions, as it will be explained below in section 5.2.4, and can further enhance our understanding of the nature and composition of the archaeological assemblage. The large number of samples allows to detect such features in the shells assemblage, and thus to better understand the short-term climate variability and the accumulation of the deposits.

In the present chapter, the seasonal exploitation patterns and the palaeoclimatic record are discussed in order to test the hypothesis of the potential climatic impact on subsistence practices at Franchthi. The methodological advances of the employed method to this end, are also summarised and discussed. Finally, based on the above evidence I argue that prehistoric inhabitants of Franchthi demonstrated resilience in response to climate change.

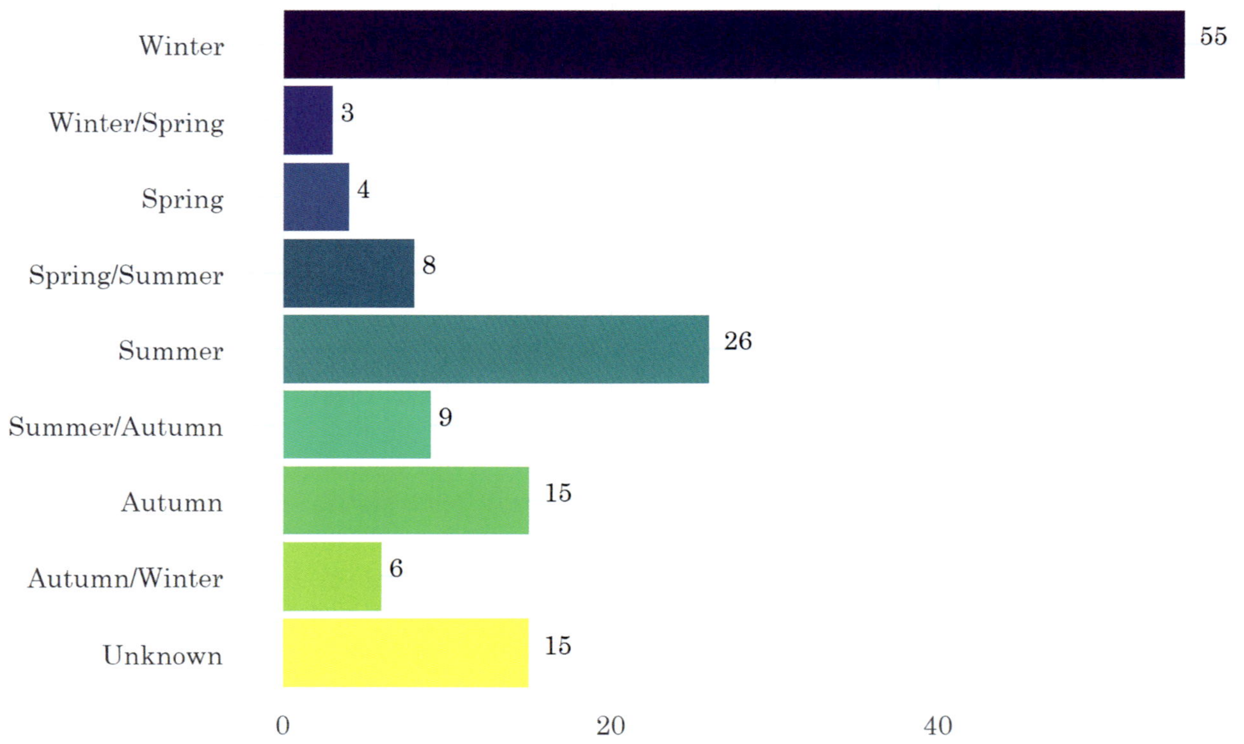

Fig. 140 Summarised season of collection results from samples from both the cave and the Paralia site of Franchthi, as well as from the Final Palaeolithic to the Late Neolithic period (n = 141).

5.2. Seasonality

5.2.1. Season of Collection

The detailed elemental analysis of the archaeological assemblage was described specimen-by-specimen in Chapter 4. Here, I make a synthesis of the season of collection results in order to better understand the seasonality patterns at the site and put these into a wider context of coastal seasonality studies. Across Franchthi's chronology and its various trenches, the analysed shells indicate a wide distribution of collection seasons (**fig. 140**). Of the 141 specimens, 55 were collected during winter, 3 during winter/spring, 4 during spring, 8 during spring/summer, 26 during summer, 9 during summer/autumn, 15 during autumn, 6 during autumn/winter and 15 are of an unknown season of collection. In total, this distribution points towards winter and summer being the main seasons of collection.

In order to further examine these results, I will group them based on their respective spatial and chronological context: When examining the seasonal variability patterns of the Final Palaeolithic and the Mesolithic period in the cave (trenches FAN and FAS), the samples analysed (n = 25) were not numerous, especially when excluding the samples with an unknown season of collection (n = 4), leading to a size for this sub-assemblage of 21 shells. The Final Palaeolithic shells in the case of which a season of collection could be estimated are only six: two shells reflect a winter collection, one a spring/summer collection, one an autumn collection and two an autumn/winter collection (**fig. 141**). Thus, it seems that there is a stronger preference towards the colder months of the year (five out of six samples), even though the sample size is not large enough to draw solid conclusions on the matter.

5.2.1.1. Seasonality during the Mesolithic

Shells in the Mesolithic period (n = 17, with 2 uncertain) are distributed more evenly across the year: four indicated a winter collection, two a spring col-

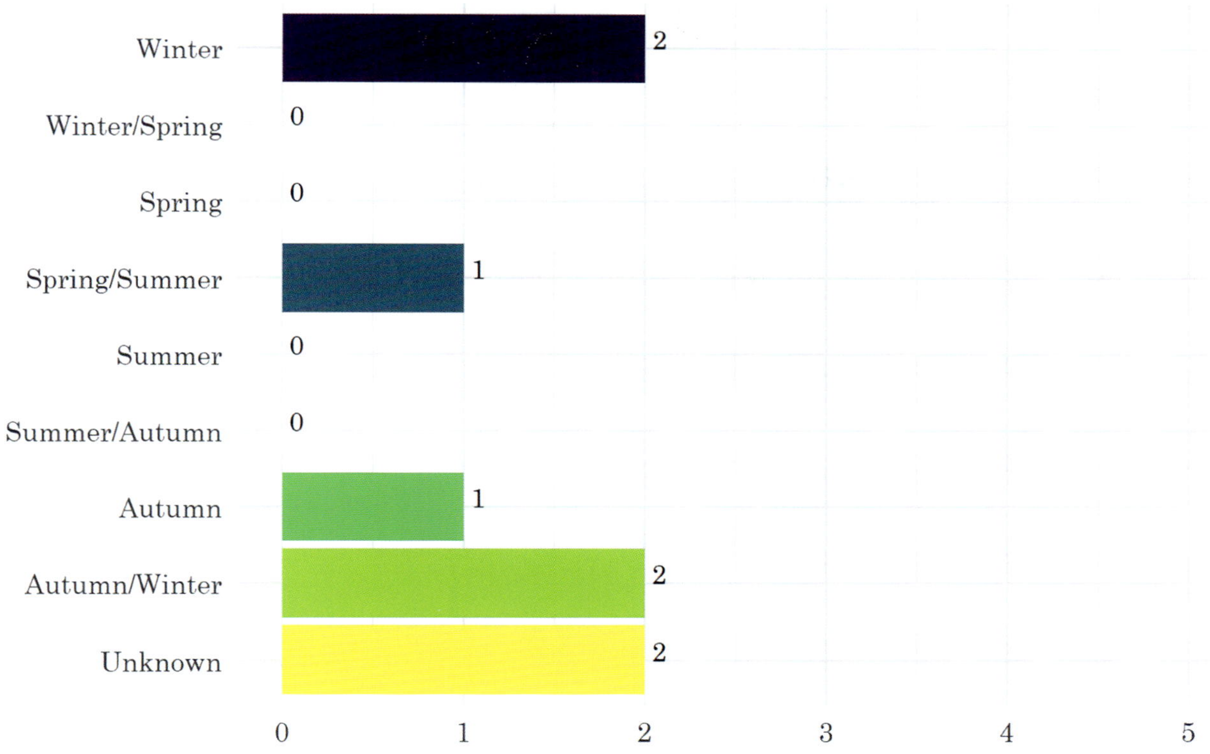

Fig. 141 Season of collection of the Final Palaeolithic shell specimens (n = 8) from the FAS trench.

lection, three a spring/summer collection, three a summer collection and 3 an autumn collection (**fig. 142**). The collection of molluscs during the Mesolithic period does not reflect a specific tendency, even though samples are, also in this case, not numerous. Interestingly, there are no samples collected during liminal periods of the year, that is during winter/spring, spring/summer, summer/autumn and autumn/winter.

When further examining the chronological subdivisions of the Mesolithic period, no obvious patterns emerge. One may notice a preference towards mollusc collection during the warmer months (spring/summer, summer, autumn) during the Lower Mesolithic. During the Upper Mesolithic, samples are distributed across the year, while the same is the case for the Final Mesolithic, during which there are no autumn-collected samples (**fig. 143**).

Thus, when examining the seasonality results (**figs 144–145**), there appears to be a stark contrast between the Paralia winter-collected specimens and the cave summer-collected ones. On a first level, this could be viewed as an indication of intra-site seasonal use of space where people inhabit the different areas during different seasons. Nonetheless, when looking further into the seasonal evidence, these findings are not contemporaneous. The samples selected belong to different areas of the site and also to different time periods. In this manner, a hypothesis cannot stand for specific seasonal use patterns of occupation. In addition, the use and function of the Paralia wall remains are not firmly interpreted by the excavators as they suggest that they may represent dwellings or retaining walls (Jacobsen 1976, 83–84; Vitelli 1993, 31). Even though the intra-site movement hypothesis is rejected, the repeating seasonal exploitation patterns could point towards certain collection and consumption events. This is further analysed in section 5.2.4. Moreover, the above observations constitute an example of how the increased number of samples can provide more refined information on various temporal and spatial scales. Subsequently, I provide the evidence for the shell-derived seasonality of the Neolithic deposits.

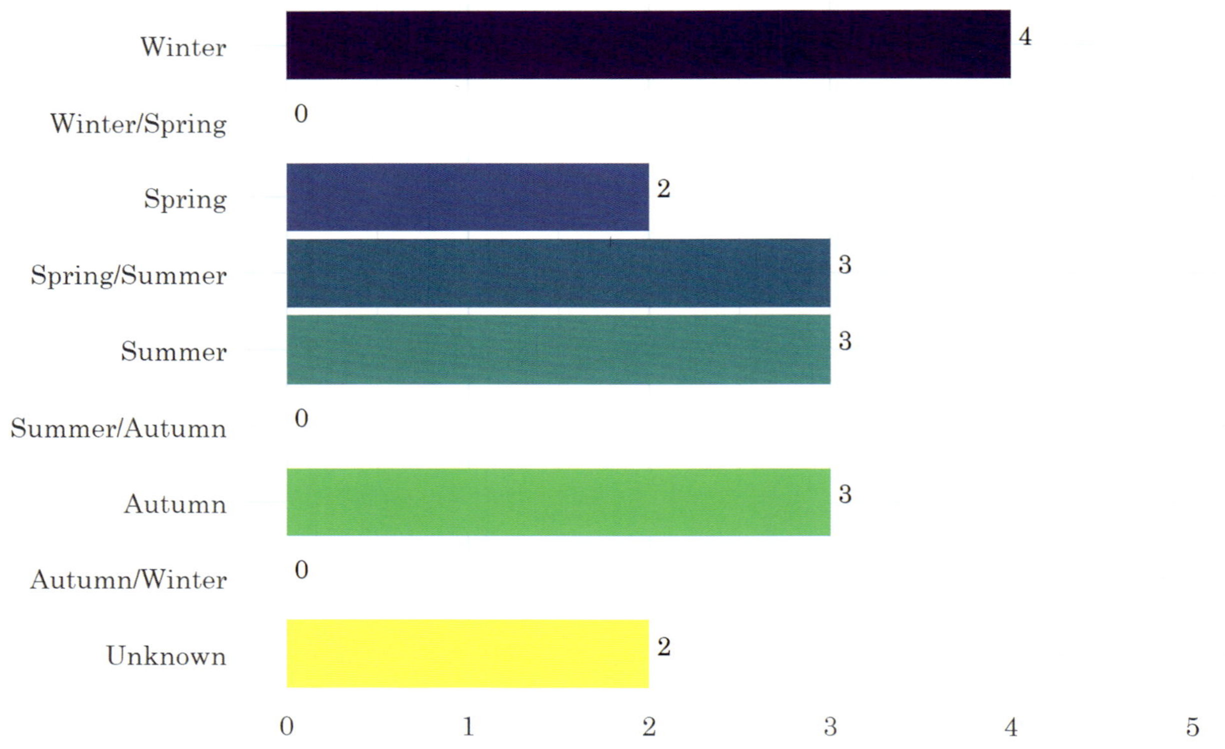

Fig. 142 Season of collection of the Mesolithic shell specimens (n = 17) from the FAS and the FAN trenches.

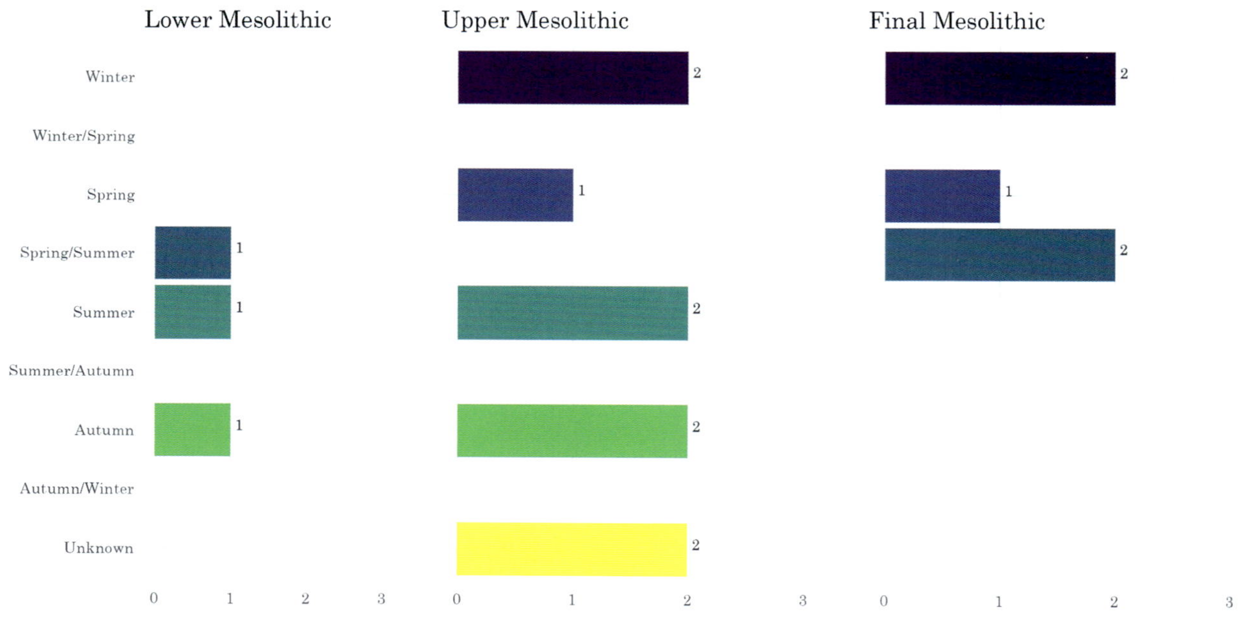

Fig. 143 Season of collection distribution during the Lower (n = 3), the Upper (n = 9) and the Final Mesolithic (n = 5) period.

5.2.1.2. Seasonality during the Neolithic

The Neolithic period was seasonally more variable. Firstly, all seasons are represented, however fewer specimens were collected during winter/spring and spring/summer. The overall comparison of the two areas (cave and Paralia) reveals a stark difference: a strong summer component is attested from the finds at the cave (40 %) and a stronger winter one (56 %) at the Paralia trenches (**figs 144–145**).

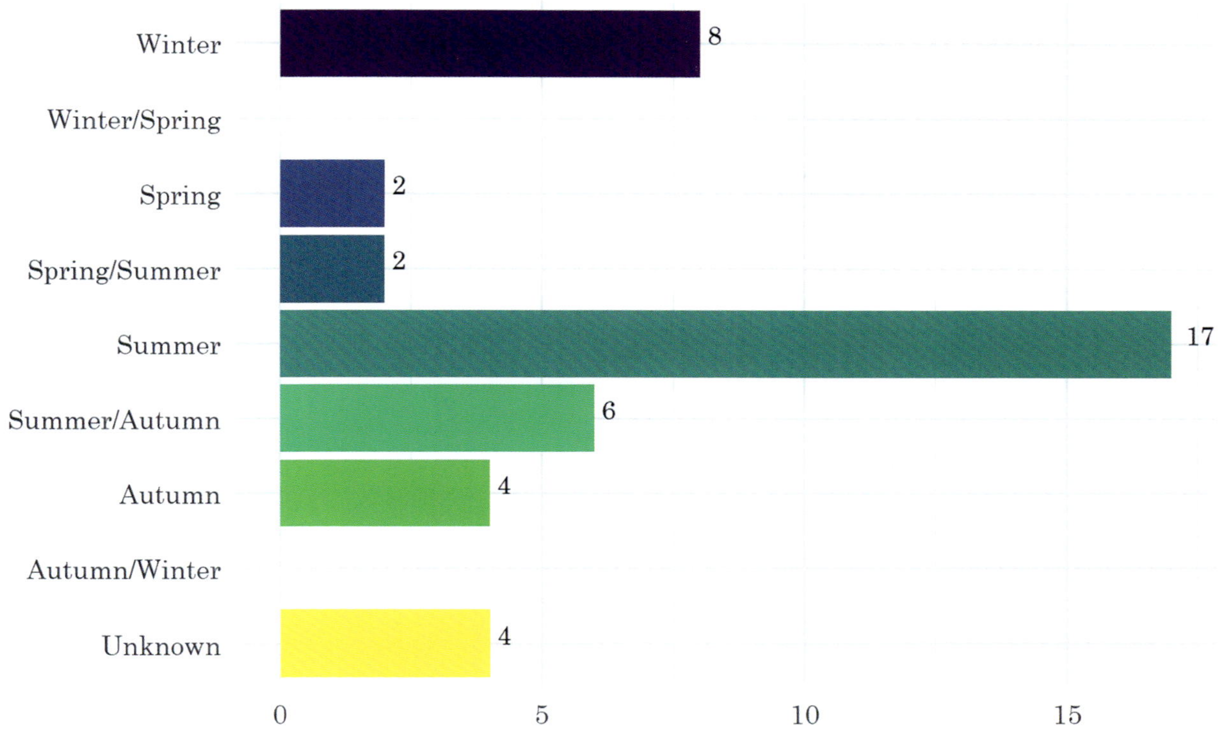

Fig. 144 Summarised season of collection distribution at the cave during the Neolithic period (n = 43).

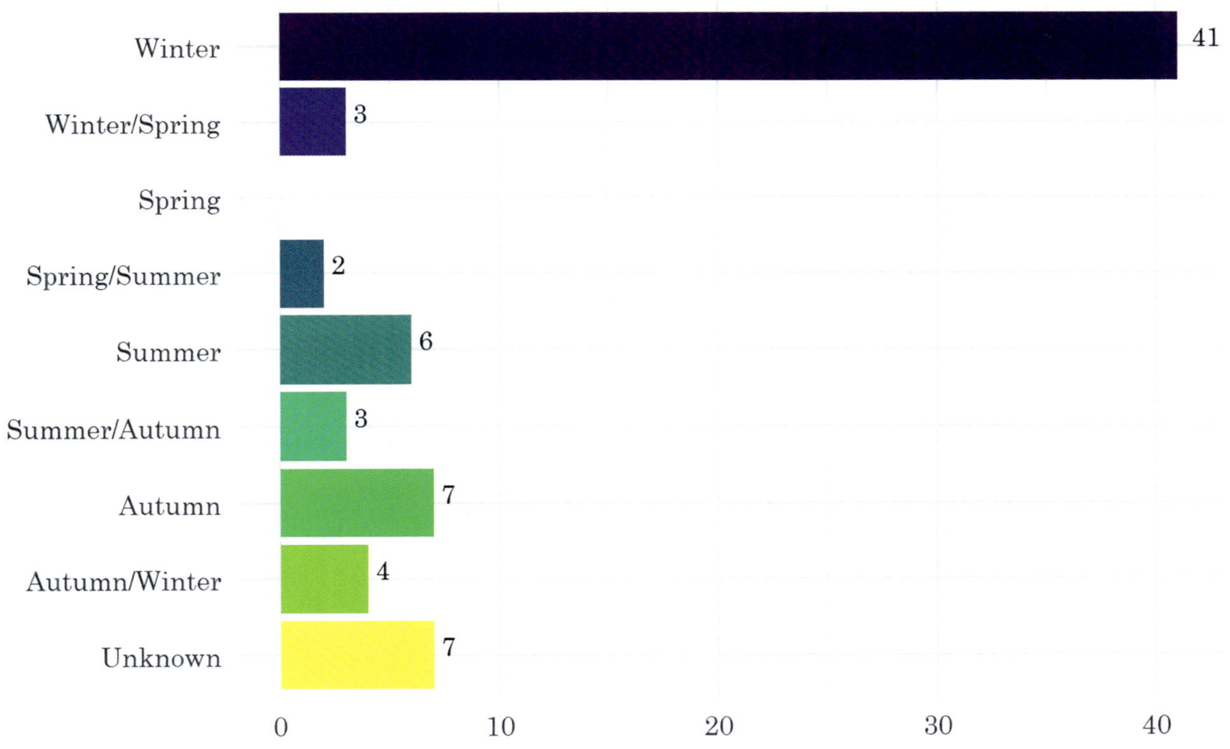

Fig. 145 Summarised season of collection distribution at the Paralia during the Neolithic period (n = 73).

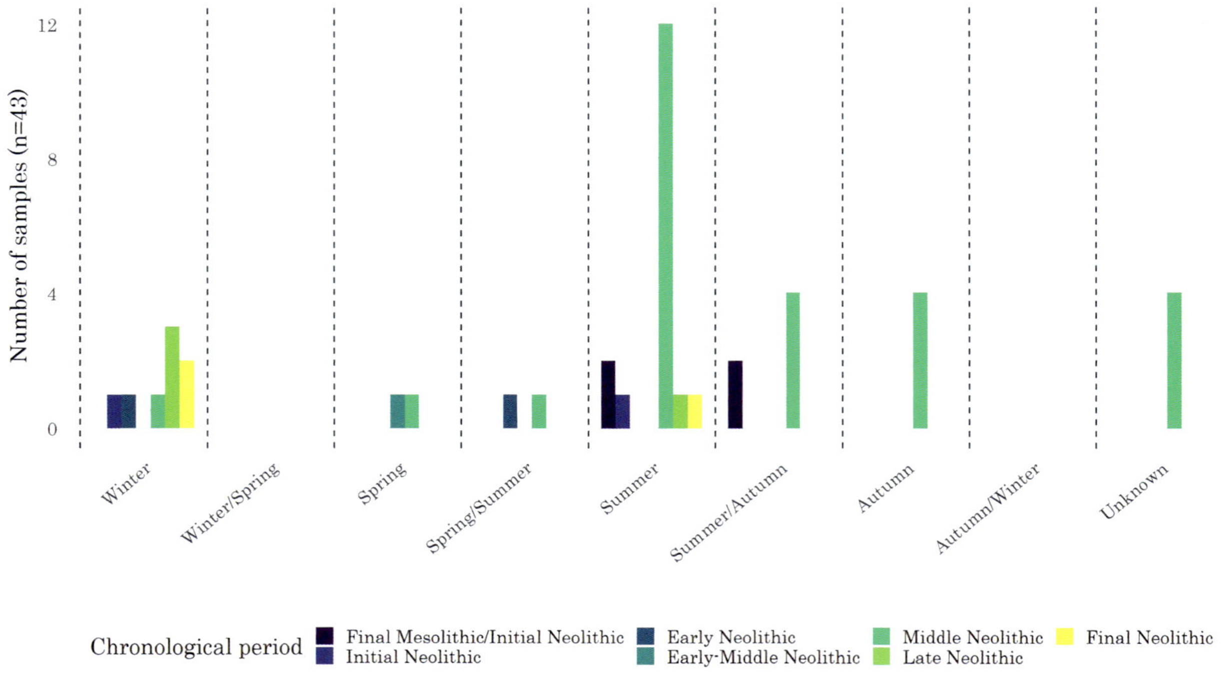

Fig. 146 Season of collection at the cave during the different Neolithic periods (n = 43).

When examining further the distribution of the seasons of collection through the individual periods of the Neolithic, it becomes evident that these specimens are not contemporaneous. The large quantity of summer-collected specimens at the cave is only present during the Middle Neolithic: 44 % (n = 12) of the Middle Neolithic samples were collected during summer, 15 % (n = 4) during summer/autumn and another 15 % (n = 4) during autumn, while the rest of the samples were distributed among winter, spring and spring/summer (n = 3). In addition, this is a pattern that was not identified by the Deith and Shackleton study of 1988, which is discussed in detail in section 5.2.2.

Furthermore, the rest of the Neolithic periods at the cave are represented only by small numbers: Final Mesolithic/Initial Neolithic with summer (n = 2) and summer/autumn (n = 2); Initial Neolithic only one sample collected during the summer; Early Neolithic with two specimens, one collected during the winter and one during the spring/summer period; Early-Middle Neolithic with one shell that was collected during spring; Late and the Final Neolithic are represented by winter (n = 3 and n = 2 respectively) and summer with one shell in each period (**fig. 146**). These results are quite variable and do not point towards specific collection patterns. In addition, the small number of samples, and their uneven distribution, doesn't enable the extraction of any further conclusions.

The Neolithic specimens from the Paralia present an interesting seasonal pattern and show a strong winter component (**fig. 147**). From the 73 analysed specimens that belong to the Paralia trenches, only 3 % (n = 2) are dated to the Early-Middle Neolithic with the rest 97 % dated to the Early Neolithic period (n = 71). A strong winter component is clearly present during the Early Neolithic at the Paralia trenches (58 %, n = 41), 3 % (n = 2) during winter/spring, 3 % (n = 2) during spring/summer, 9 % (n = 6) during summer, 3 % (n = 2) during summer/autumn, 10 % (n = 7) during autumn and 6 % (n = 4) during autumn/winter. The winter and winter/spring element is present in 14 excavation units (out of 25 of the Paralia). Among these, there is a specific one, Q5S 230, in which 17 of 23 samples indicated a winter collection (and 3 more during autumn/winter and one more winter/spring). But also without that specific excavation unit, the cold month collection pattern persists (48 %, n = 24, collected during winter time) at the Paralia.

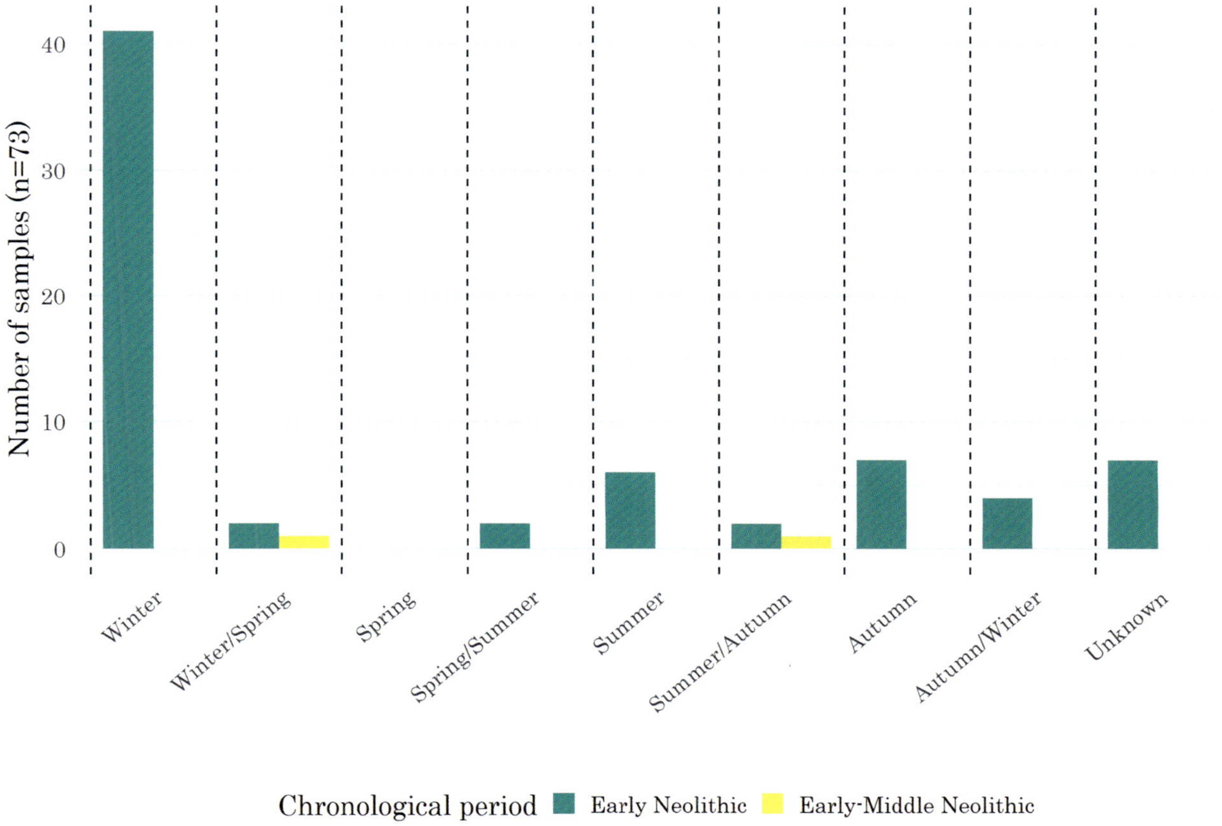

Fig. 147 Season of collection at the Paralia during the different Neolithic periods (n = 73).

5.2.1.3. Summary

Even though there is patchy seasonal evidence between the cave and the Paralia and across Franchthi's occupation, most seasons of the year at the cave are represented during the Final Palaeolithic and the Upper Mesolithic, while year-round collection occurs during the Middle Neolithic at the cave (**fig. 148**). Then, a summer component is prominent, followed by summer/autumn and autumn-collected specimens. In addition, the early Neolithic is best depicted at the Paralia, when specimens collected throughout the year are present, while the winter-collected specimens are the most prominent feature of this group of trenches. In the cave, the early Neolithic is not represented by a sufficient amount of samples to draw any solid conclusions: Only a winter-collected sample and a spring/summer-collected sample are present during the Early Neolithic and the same is the case for the Initial Neolithic which is represented by a single summer-collected specimen.

5.2.2. Past and Current Results Comparisons

M. Deith and N. Shackleton performed isotope analyses on 120 shells that belonged to *Phorcus* (*Monodonta*), *Cerastoderma* and *Cerithium* species (Deith/Shackleton 1988, tab. 5). Here, I compare their results with my own to provide a combined picture of the seasonality of collection at the site. Given that they have analysed different species and I have analysed a large set of samples of *Patella* shells, I expand on their findings and broaden the seasonal scope of the site by combining the two studies.

To begin with, Deith and Shackleton see a shift from a summer and autumn bias of the Palaeolithic and the Mesolithic towards a summer and spring bias during the Neolithic period (1988, 154). Their argument of lack of winter use of the cave during the Final Palaeolithic is weakened by the finds of this study which identified more winter-collected specimens, which nonetheless are not significantly

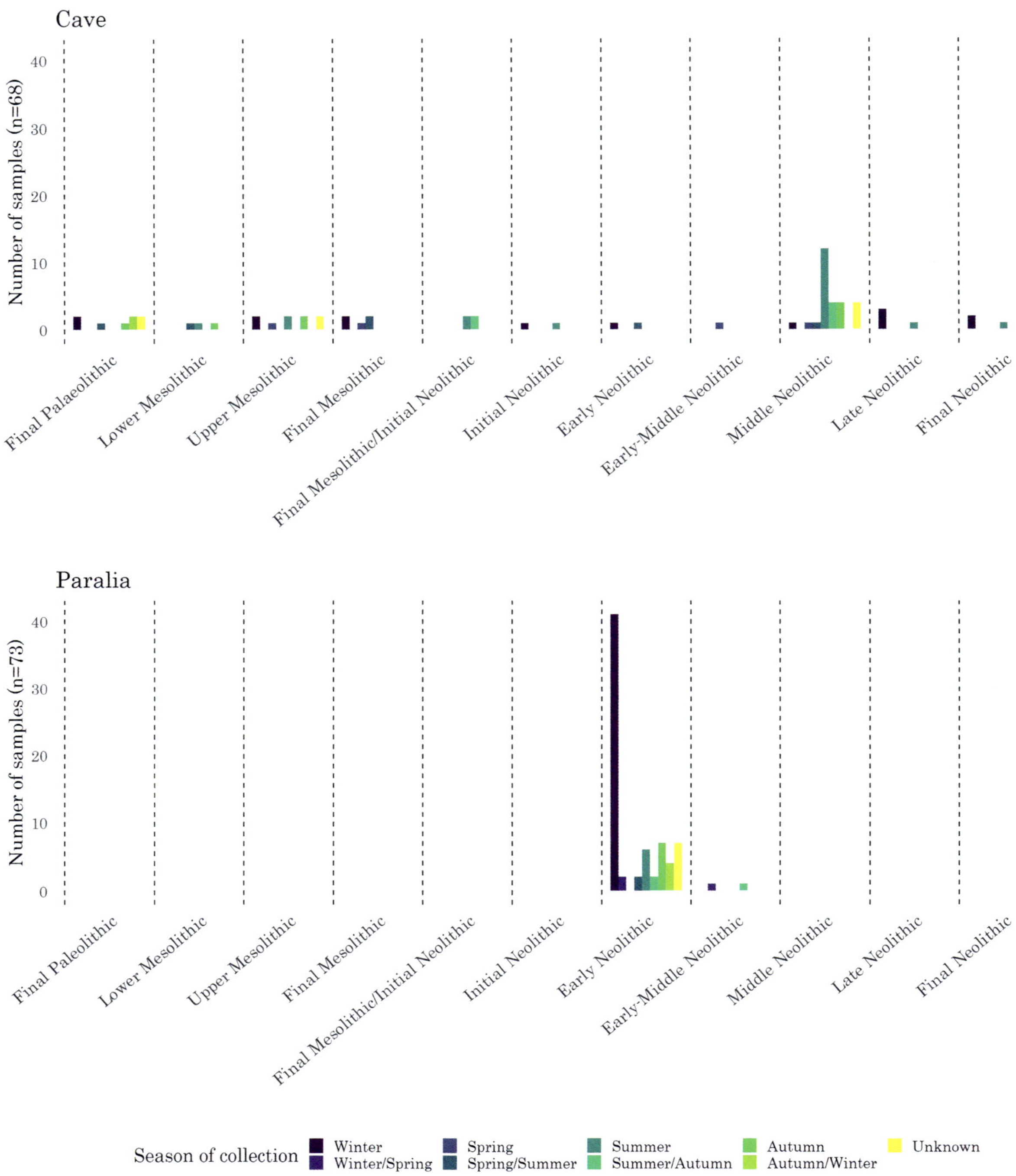

Fig. 148 Diachronic comparison of seasons of collection of the cave (n = 68) and the Paralia (n = 73) by chronological period.

numerous. During the Mesolithic, they suggest that the cave was used as a summer/autumn camp with occasional visits in winter and spring (1988, 154), whereas the present study detected further instances of winter and spring collection. Furthermore, a significant difference between present and past results is noticed during the Early Neolithic: Deith and Shackleton see a strong spring and summer (1988, 154–155), whereas I have identified a large amount of winter-collected specimens. Nonetheless, during the Middle Neolithic the summer element is also strong in my set of samples.

More specifically, for the Final Palaeolithic period, Deith and Shackleton concluded that the site was visited primarily during autumn and secondarily during summer, as they found that the site was vis-

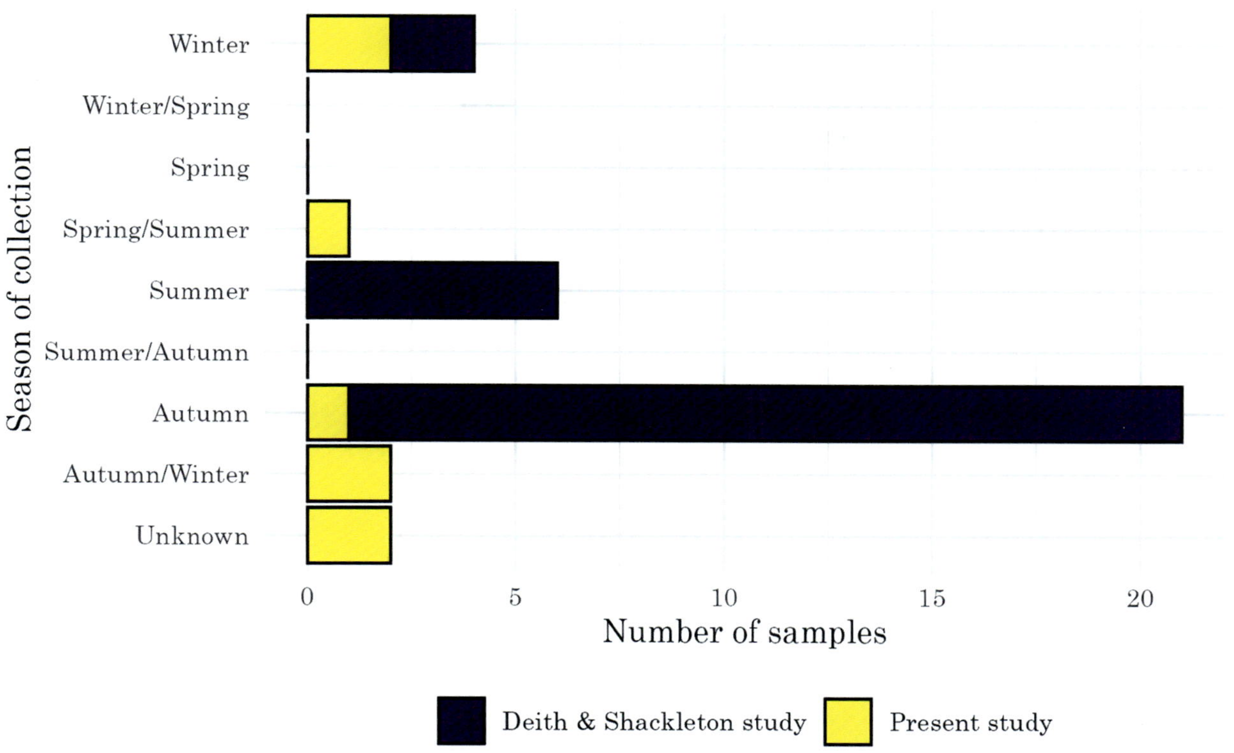

Fig. 149 Combined season of collection results for the Final Palaeolithic period.

ited primarily during autumn (20 out of 28) and secondarily during summer (6 out of 28) (1988, tab. 6). They also assign two shells to a winter collection, while considering that one of them may have also been collected during autumn. Thus, they note that they do not »attach too much significance to single shells as evidence for winter occupation of the site« (1988, 151). All of the Final Palaeolithic specimens analysed by Deith and Shackleton belong to the *Phorcus* (*Monodonta*) species. In the present study, two more winter-collected specimens are added to the overall number of the analysed specimens of this period, strengthening the evidence for winter occupation as well.

The combination of the results of the two studies shows a preference for autumn-collected shells (n = 21), followed by summer- (n = 6) and winter-collected (n = 4) specimens (**fig. 149**).

For the Lower Mesolithic, Deith and Shackleton recognise a broader seasonal distribution: More specifically, they identify seven shells collected during spring, six during summer, ten during autumn and eight during winter. Evidence for year-round collection comes to some extent from *Cerithium* shells, while a large number of *Cerastoderma* shells point towards evidence for winter visits of the cave (1988, 151 tab. 6).

Deith and Shackleton only analysed one Upper Mesolithic *Cerithium* shell from excavation unit FAS 159, which was identified as a summer-collected sample (1988, 153 tab. 6). Furthermore, the Upper Mesolithic period is characterised by the sudden increase of the bluefin tunny (*Thunnus thynnus*), which dominates the faunal assemblage, while during the Final Mesolithic period large-scale fishing ceases. The fishing of the bluefin tuna, a highly seasonal activity due to the fish's migratory movements, takes place during spring and autumn (Desse/Desse-Berset 2002, 88; Theodoropoulou 2007, 61; Mylona 2021, 30).

Regarding the Final Mesolithic period, Deith and Shackleton analysed 18 samples of which one was identified as collected during spring, eight during summer and nine during autumn. Therefore, this group of samples are clustered in summer and autumn (1988, 153 tab. 6).

Overall, Deith and Shackleton suggest that during the Mesolithic the cave was used as a summer/autumn camp with occasional visits in winter and spring (1988, 154). The winter and spring compo-

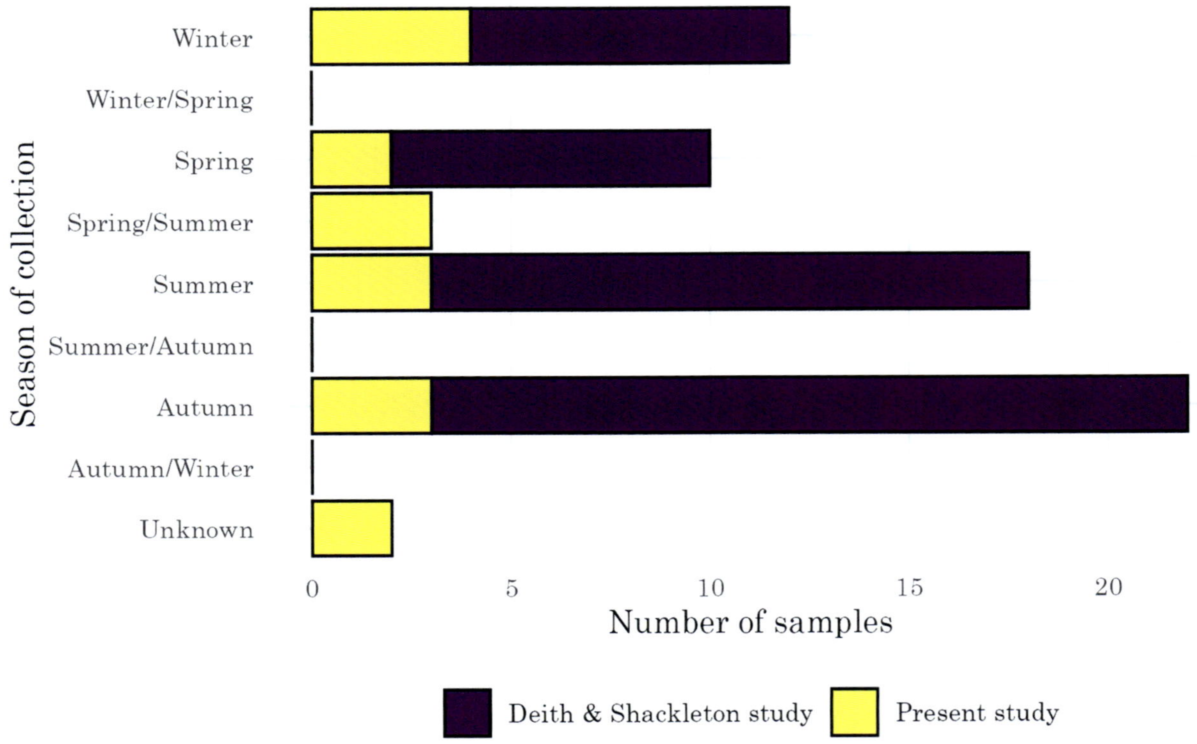

Fig. 150 Combined season of collection results for the Mesolithic period.

nents are further reinforced by the present study, and an almost year-round collection of mollusc shells with an emphasis during the summer/autumn period appears to have taken place during the Mesolithic period (see **fig. 7**). The winter component detected in the present study (n = 4) further corroborates the winter-collected specimens of the Lower Mesolithic (n = 8) of the previous study (**fig. 150**). The differences in the species composition of the specimens analysed by Deith and Shackleton (*Phorcus*, *Cerastoderma*, *Cerithium*) do not show any specific seasonal patterns or biases. A stronger winter component is indicated by the *Cerastoderma* specimens during the Lower Mesolithic, nonetheless the specific species is also collected during spring, summer and autumn (1988, tab. 6).

During the Early and the Middle Neolithic period, grouped as such by Deith and Shackleton the majority of the shells both from the cave and the Paralia were collected during spring (8) and summer (9), while one shell was collected during autumn and one during winter (1988, 153 tab. 6).

In the present study, the Early Neolithic shells belong exclusively to the Paralia area, where winter-collected shells dominate the assemblage, while the Middle Neolithic samples from the cave have a stronger summer and summer/autumn component. This aspect is completely absent from the Early and Middle Neolithic samples studied by Deith and Shackleton.

Regarding the Late Neolithic shells, Deith and Shackleton analysed 13 specimens from the cave. From these, they assigned three to a spring season of collection, five to a summer collection, one to an autumn collection and four to a winter collection (1988, tab. 6), whereas in the present study a winter component is more prevalent during the Late Neolithic, even though the sample size is small. Regarding the species composition of the Deith and Shackleton study, the species composition does not exhibit distinct seasonal patterns; however, there is an uneven distribution of specimens of different species across different periods.

Finally, by taking these interpretations into account, the following assumptions for seasonality of collection and seasonality of occupation at Franchthi can be made. During the Final Palaeolithic, an almost year-round collection of molluscs appears to be the case. Palaeolithic specimens of this study, even

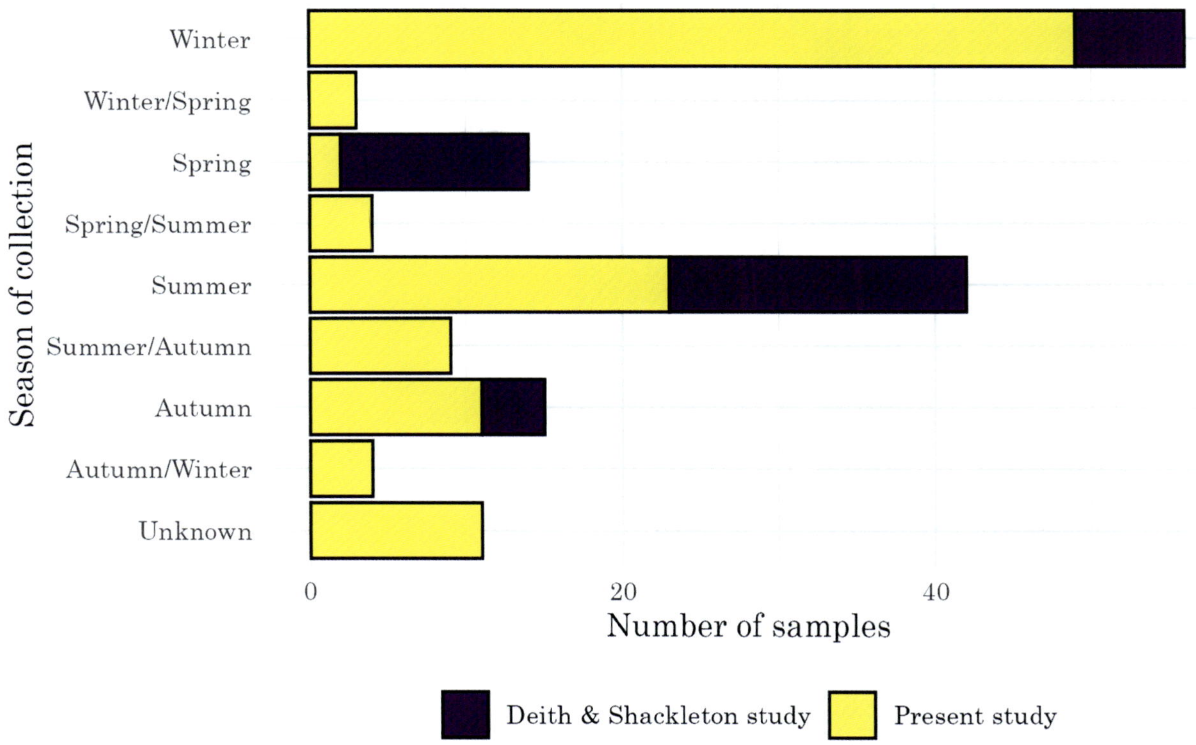

Fig. 151 Combined season of collection results for the Neolithic period.

though of a limited number (n = 6), show most of the seasons are present (two shells collected in winter, one a spring/summer collection, one an autumn and two in autumn/winter). Even though these samples show a preference towards the colder months, Deith and Shackleton see a prevalence of summer and autumn-collected specimens and make a hypothesis for seasonal visits to the cave during autumn and summer (1988, 151).

Deith and Shackleton (1988, 154) conclude that during the Mesolithic period the cave was used as a summer/autumn camp with occasional visits in the winter and spring, although during the Lower Mesolithic they do recognise a broader seasonal distribution. According to the present study, the Mesolithic period contains mollusc shells almost year-round, with a focus on the summer and fall seasons (**fig. 150**).

According to Deith and Shackleton, most of the shells from the cave and the Paralia were collected in the spring and summer of the Early and Middle Neolithic periods (1988, 153 tab. 6). The winter component prevalent in the Early Neolithic results of the present study is not caught at all by the analyses of Deith and Shackleton. While I have identified a significant number (n = 41) of winter-collected specimens from the Early Neolithic Paralia trenches (n = 73), Deith and Shackleton report a stronger spring and summer element (**fig. 151**). In addition, in my Middle Neolithic set of samples the summer aspect is equally prominent.

5.2.3. Analyses on Size and Biological Age

Size and growth patterns can provide valuable information regarding past resource exploitation patterns and aspects such as intensity of exploitation, changes in habitats and marine ecosystems and human-species interactions. Age/size analyses on mollusc life-histories can also reveal shifts in the climate and the environment as these influence significantly the growth rates and size of the specimens (e. g. Jerardino 1997, 1041–1042; Claassen 1998, 136–137; Toniello et al. 2019, 22113). In this case, based on the available data I focus on the specimens' growth rates and how the anthropogenic impact and the environment may have shaped them.

Elemental mapping allowed me also to estimate the biological age of the analysed specimens as

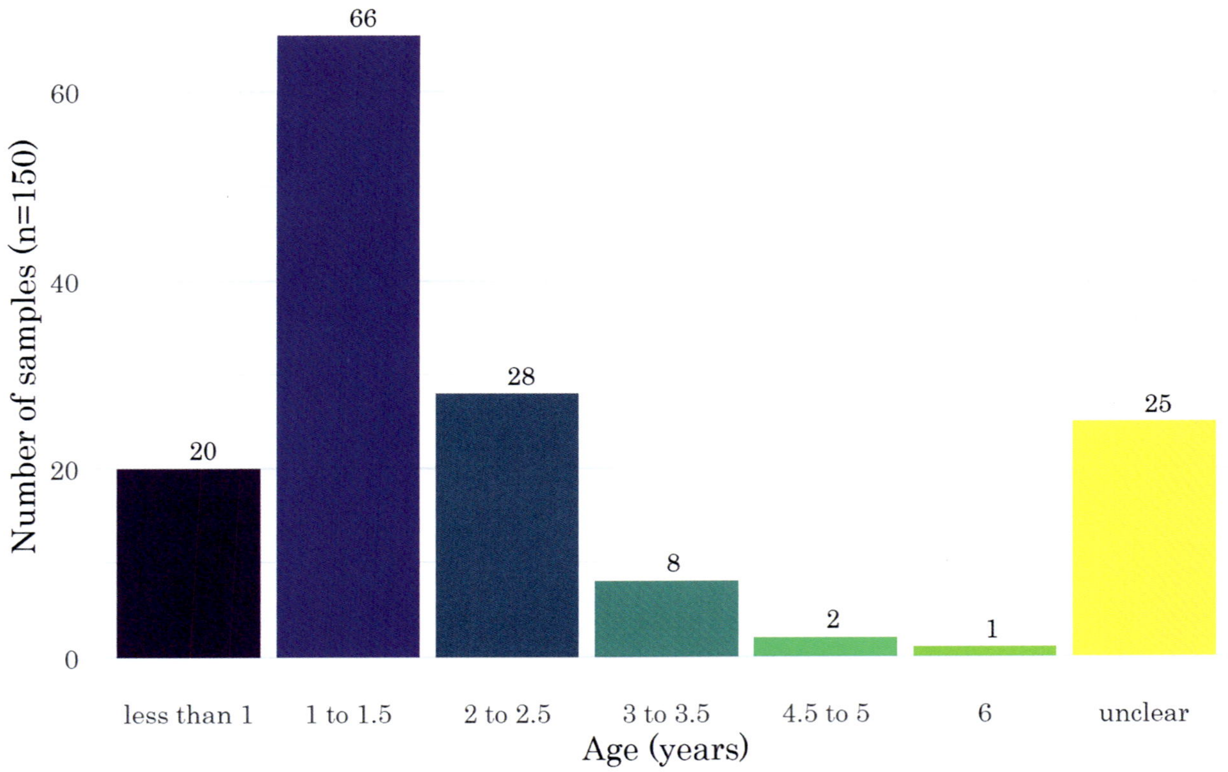

Fig. 152 Estimated years of age among edge preserving specimens (n = 150).

83 % (n = 150) of the shell specimens preserved their edge. In 25 (17 %) of these the age could not be determined and another 20 (13 %) recorded less than a year of growth. As shown in **figure 152**, the largest age group is that of 1 to 1.5 years (44 %), followed by those of 2 to 2.5 years (19 %) and those that were less than 1 year old (13 %). Only eleven specimens in total (7 %) exceeded the three years of age.

Interestingly, the archaeological shells tend to be younger on average (1.8 years) than the modern shells (2.9 years). However, this is not the case for the modern Franchthi samples which had a mean age of 1.6 years and were on average younger than the entire assemblage of the modern specimens.

I compared the age distributions with the seasonal distribution throughout the year to test for any patterns that could point towards specific collection behaviours (e. g. Hallmann et al. 2009; Burchell et al. 2013). Age groups were created based on the distribution of ages found in the dataset to facilitate the analysis and the comparison with size. When examining the age groups of the shells by season of collection, the overall trend is similar to that of **figure 140**, where winter-collected specimens are the most populated category. Nonetheless, it appears that there is a pattern of collecting younger specimens during winter time, while the older ones tend to be collected during the warmer months of the year (**fig. 153**).

The spawning season takes place during the end of the summer (Vafidis et al. 2020, 8), while the summer or autumn component is not particularly strong in any age group. This could imply that the spawning season may have impacted mollusc collection practices, in the sense that their collection was avoided during these times. Ethnographic studies inform on the good ecological knowledge of the collectors of shellfish during certain periods of the year, as well as of the knowledge of the organisms' life-cycles (Waselkov 1987, 111; Claassen 1998, 33–36). The small number of specimens that are less than 1 year old, further corroborates such possible behaviours of »ecological conscience«.

It appears that there is a correlation between the age and size of the archaeological shells. As shown in **figure 154**, there is an upward trend of the length and width of the specimens as the age increases and it appears that the two pairs of variables (length and age and width and age) are defined by a positive cor-

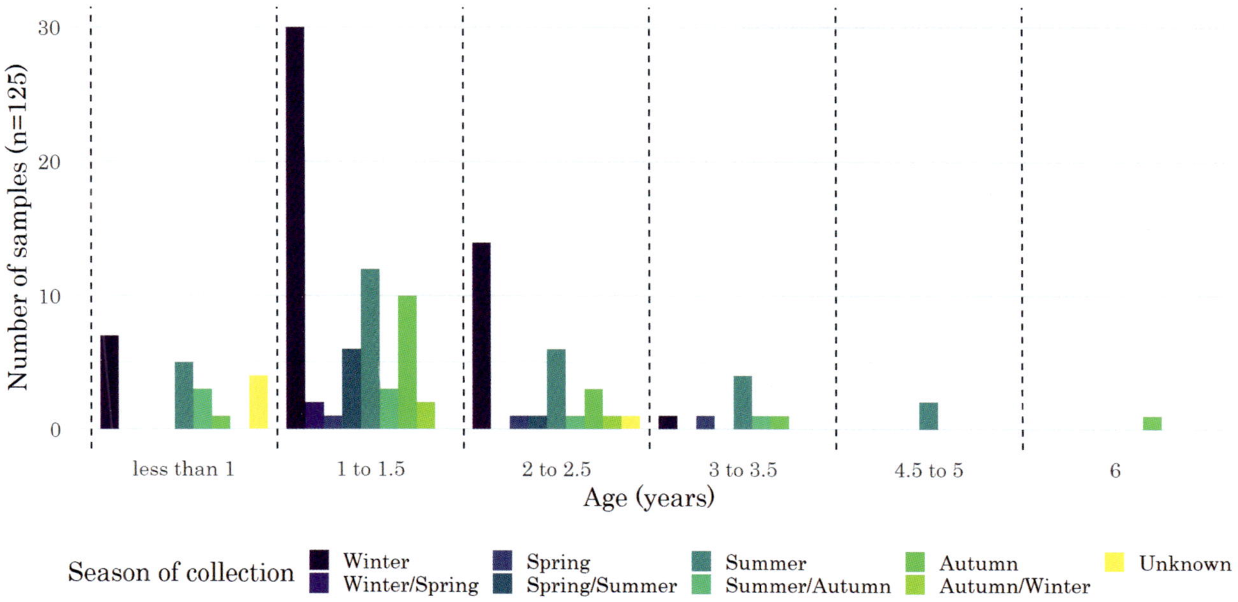

Fig. 153 Estimated years of age by season of collection (n = 125).

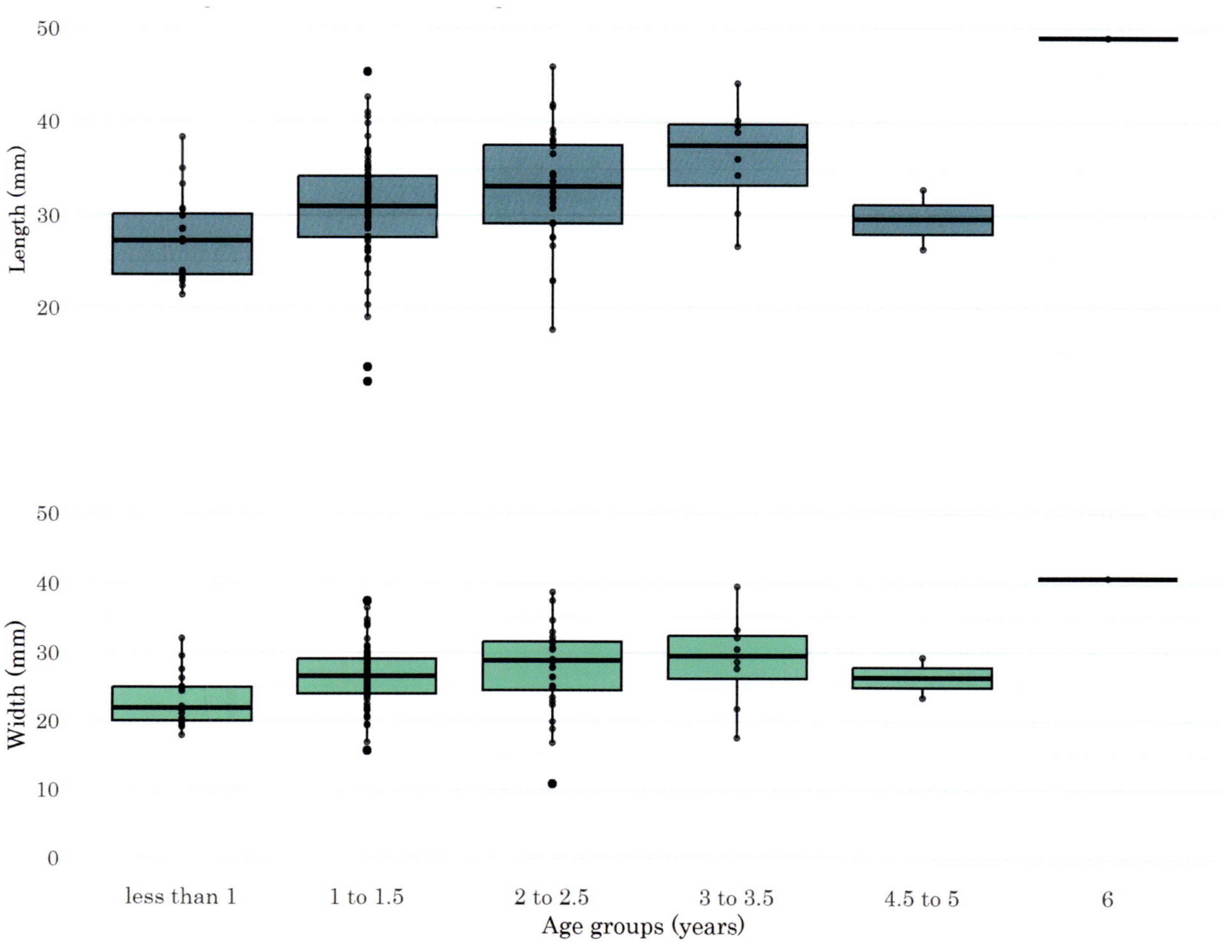

Fig. 154 Barplot comparing the length and width of the archaeological shells with years of age (n = 125).

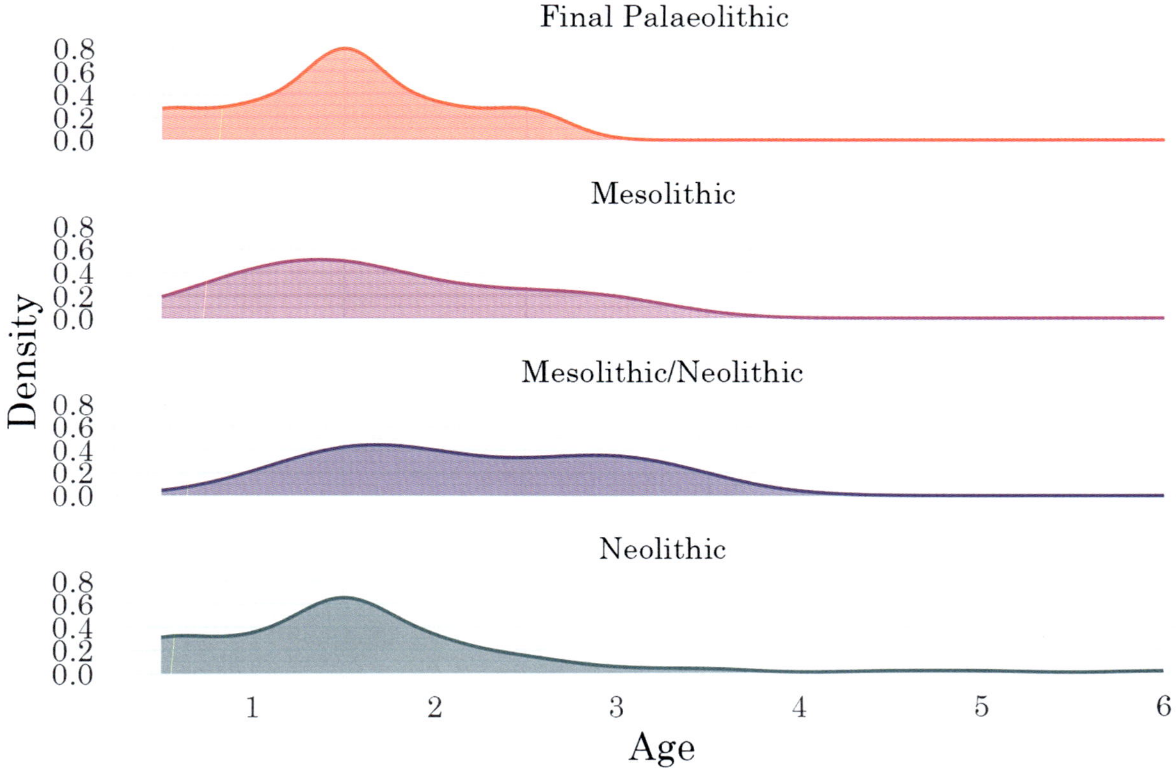

Fig. 155 Density plot comparing the biological age of the specimens across the different chronological periods (n = 125).

relation. For the length, this observation does not apply to the older shells (4.5 to 5 years old), but this group consists of only three specimens. Nonetheless, the width generally increases with age.

Furthermore, no statistically significant correlation was detected between the biological age of the archaeological samples and their chronological period (Spearman's rho = −0.099, p = 0.272).

The Neolithic samples, being the most numerous of the assemblage, cover the entire age spectrum, while specimens from the other chronological periods do not exhibit specific tendencies. Neolithic shells are generally younger, with 52 % of them being equal to or less than 1.5 years old. As indicated in **figure 155**, the Final Palaeolithic shells have a higher concentration of younger specimens (1 to 1.5 years old), which is broadly characteristic of most other periods.

The examination of the different growth rates (length/biological age) across time, their mean value, their range and their variability (standard error) led to the following comparisons (**tab. 15**). The Mesolithic period stands out with the highest mean growth rate (30.43 mm/years), suggesting faster growth rates of the specimens during this period, while the Final Palaeolithic period has the lowest mean growth rate (23.03 mm/years). Nonetheless, there are very small differences between the dif-

Grouped chronologies	Mean growth rate	Range of the growth rate	Standard error of the growth rate
Final Palaeolithic	23.03	48.27	7.56
Mesolithic	30.43	58.02	5.40
Mesolithic/Neolithic	24.32	46.91	8.37
Neolithic	24.77	70.83	1.31

Tab. 15 Summary statistics of the growth rate (in mm/years) by grouped chronology.

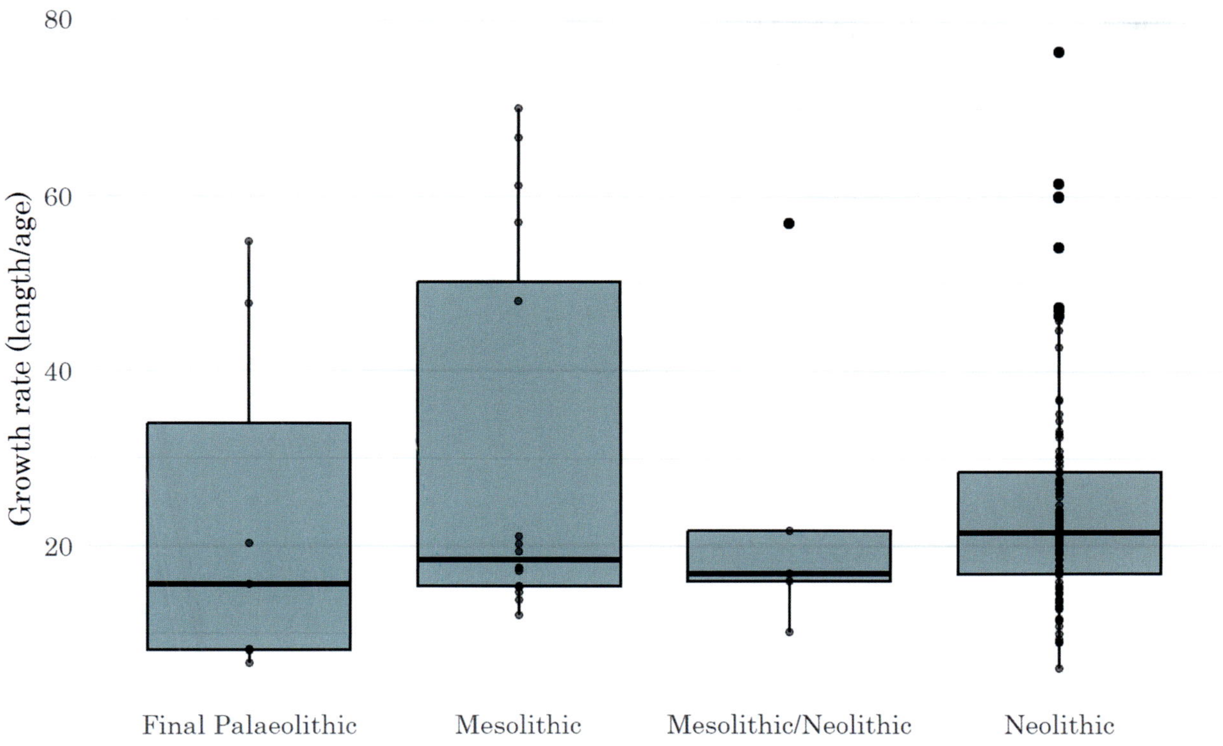

Fig. 156 Growth rate (length/age) across time (n = 125).

ferent means of the different periods. The greatest variability in growth rates is found during the Neolithic period, as indicated by the highest range (70.83 mm/years). The Mesolithic/Neolithic period shows the smallest range (46.91 mm/years) in growth rates, thus having more consistent results, while the high standard error of this period is due to the outlier. Lastly, the lowest standard error is detected during the Neolithic period, indicating that there is less variability of the estimated mean growth rate (**fig. 156**).

Based on the evidence reviewed, while differences between the periods may not reach statistical significance, several interpretations can be proposed for the observed patterns, likely influenced by biological, environmental, climatic and anthropogenic factors. Regarding growth rate distribution, modern studies have shown that small limpets have faster growth rates (Boaventura et al. 2003, 440). Second, various environmental factors can affect shell size and age (Claassen 1998, 134–137). For example, differences in the populations' habitat and substrate can also influence the specimens' growth rate and longevity (Vafidis et al. 2020, 13). Limpet growth rate is highly dependent on algae (Bosman/Hockey 1988), while differences in exposed and sheltered shores may affect the way the molluscs grow. Sheltered areas have shown to have slower growth rates in comparison with exposed areas, even though evidence on the matter is not conclusive (Jenkins/Hartnoll 2001, 134–136). Third, increasing sea surface temperature, which is attested by the isotope results of this study, can also be responsible for an increase in the limpet growth rate (Lima et al. 2006; Muñoz-Colmenero et al. 2012), a find that is present in the Neolithic group of samples.

5.2.4. Coeval Specimens

Elemental mapping offers a powerful means to assess the contemporaneity (coeval nature) and temporal scales of the shell specimens. In this section, I show how elemental mapping can contribute to the identification of coeval specimens and assemblages as high-resolution mapping can reveal unique patterns indicative of shared environmental conditions among specimens.

More specifically, the elemental maps revealed an interesting feature of multiple thin summer in-

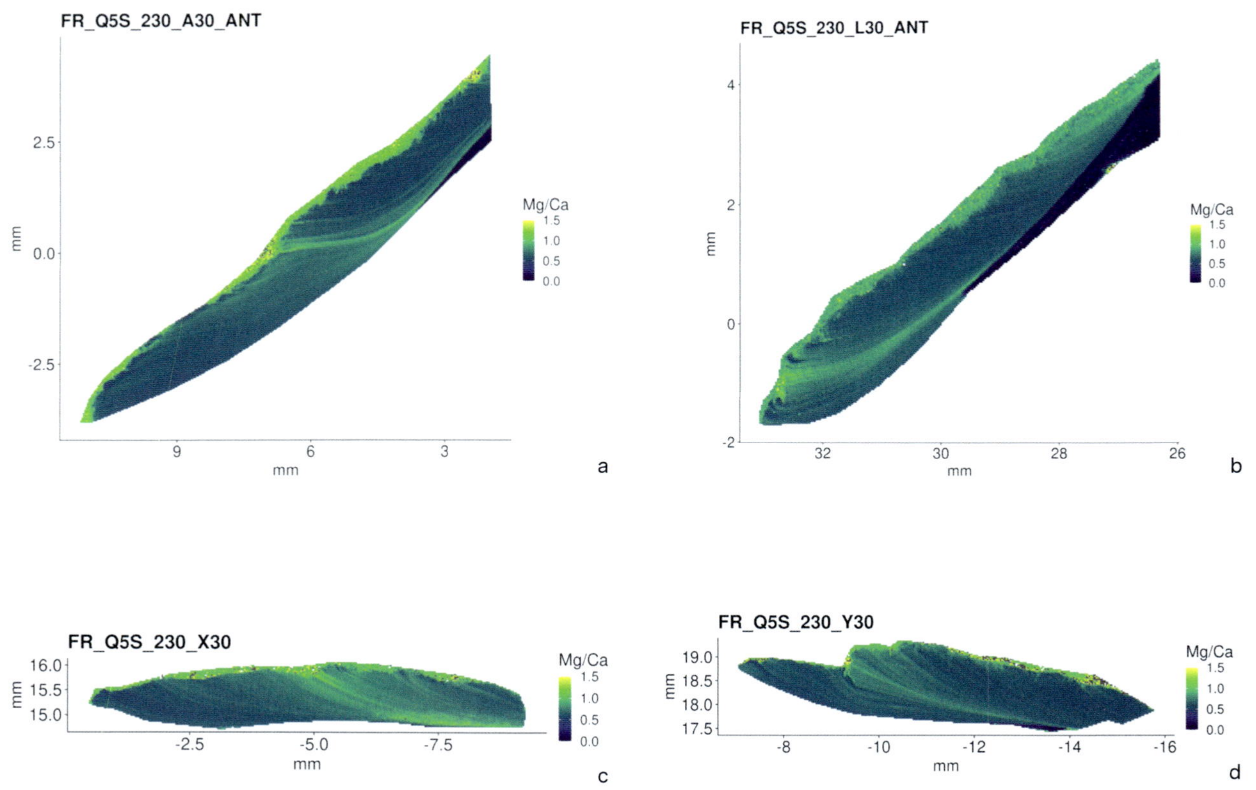

Fig. 157 Elemental maps of coeval specimens from excavation unit Q5S 230: **a** Q5S_230_A. – **b** Q5S_230_L. – **c** Q5S_230_X. – **d** Q5S_230_Y.

crements equally occurring on a number of specimens (9) that belonged to the same excavation unit (Q5S 230, n = 27, 63 % winter collection) (**fig. 157**). This recurrent feature led to the hypothesis that these same specimens could be coeval and that they may have been collected together forming part of the same collection episode. When comparing the $\delta^{18}O$ values of these samples six of them were very similar (Q5S_230_A2, E1, G1, L1, X2 and Y1) (**fig. 158**). In addition, this distinct elemental image of two winters and one thin and distinct summer is also found in a specimen from another Early Neolithic excavation unit (Q5S_189_E), which was also collected during winter.

A similar case was detected in the Middle Neolithic excavation unit FAN 127 (**fig. 159**). Even though this excavation unit contains far fewer shells (n = 7), two of the specimens FAN_127_A and B had strikingly identical elemental images and line scans. In addition, isotope analysis produced identical summer values (–1.32 ‰). A third specimen of this excavation unit FAN_127_E which was not sampled for isotopes, also produced almost identical elemental linear and 2D results with specimens A and B. Specimen FAN_127_F had also a very similar summer isotope value (–1.40 ‰).

These two excavation units are examples of how similarities between samples can be identified. Deith and Shackleton argue that »shells from a single meal could equally well be found in different excavation units, given the kind of blurring of resolution that results from trampling« (1988, 153). However, in the case of specific shells from Franchthi it appears that trampling and other disturbance activities do not completely mix the shells among the neighbouring layers. Specimens deposited at the same time may well belong to the same stratigraphic unit.

The concept of the contemporaneity among the specimens and the problem of the unclear amount of time accumulated in each unit has also been addressed by other authors. The growth rates of shells have for long been the subject of study in order to identify the seasonality of shellfish gathering (Koike 1973). Large deposits, such as in the case of shell middens, are approached via seasonality and stable isotope measurements combined with radiocarbon dating as well as the analysis of growth bands. While radiocarbon dating offers approaches at relatively

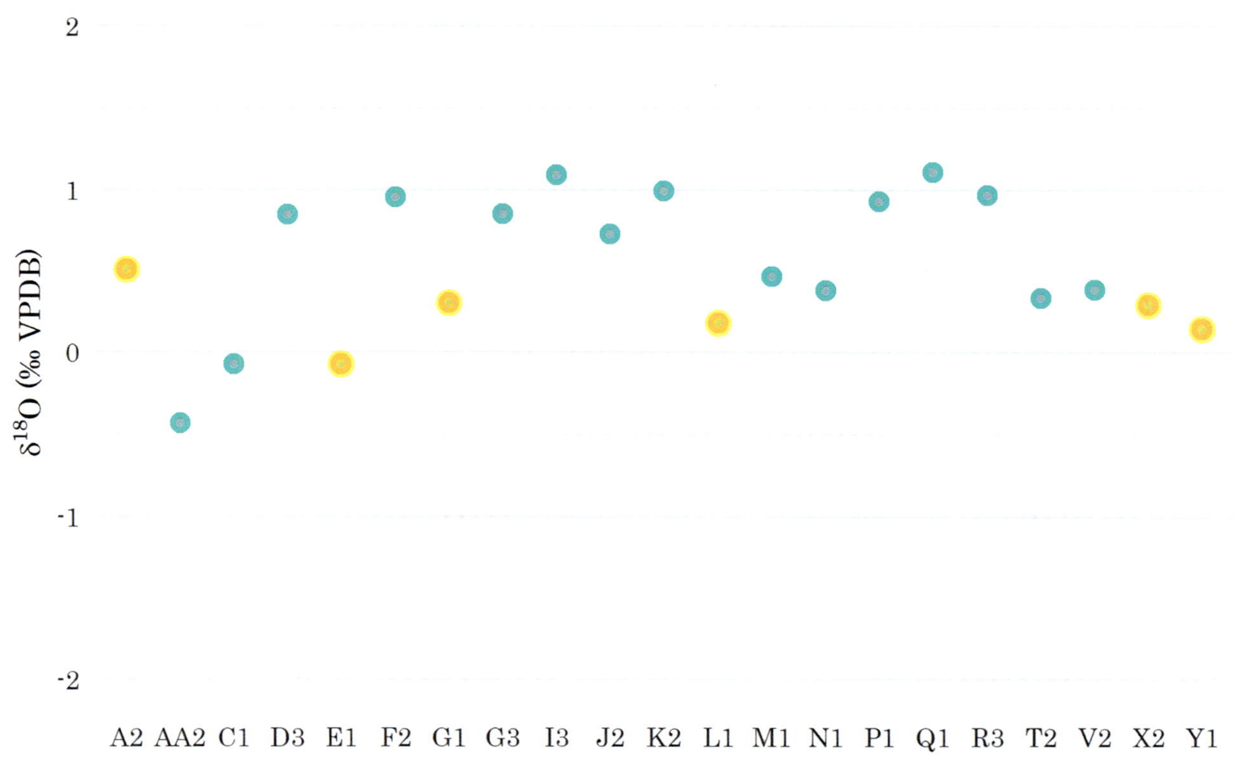

Fig. 158 All summer increment isotope values of excavation unit Q5S 230 (marked in light blue) with six thin summers isotope values (A2, E1, G1, L1, X2, Y1) marked in yellow.

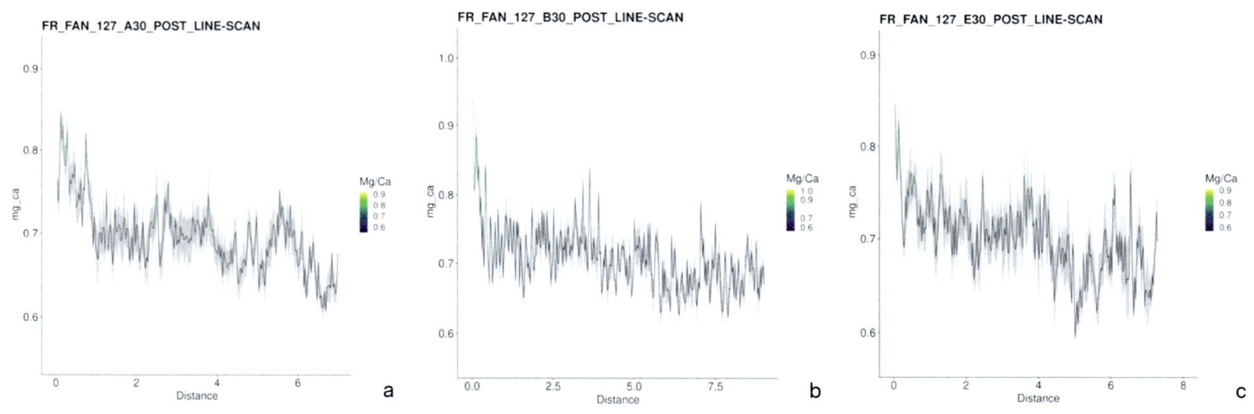

Fig. 159 Elemental line scans of coeval specimens from excavation unit FAN 127: **a** FAN_127_A. – **b** FAN_127_B. – **c** FAN_127_E.

broad time-scales (Stein et al. 2003), their combination with oxygen isotope analysis enables the reconstructions of individual or short-term events (Hausmann/Meredith-Williams 2017b, 777). The question of accumulation rates can provide insights into everyday consumption practices or feasting events related to the monumentality of the finds, as is the case of the Sapelo island shell ring complex (Thompson/Andrus 2011).

Nonetheless, in the case of Franchthi, we are not dealing with shell midden thick depositions, but shell specimens that do not belong to specific features or closed assemblages. The novelty of the method employed and the Mg/Ca variability patterns show that they can function as screening tools reflecting the same SST fluctuations across different specimens. In contrast with the studies mentioned above, the elemental maps can point towards potentially single events of collection and consumption and thus specific points in time, rather than seasons, i. e. spanning over a few months, or centuries, i. e. as is the case of radiocarbon dating. The examples of the Q5S 230 summers and the FAN 127 Mg/Ca patterns thus offer a window to specific events in time that took place on a human scale. Nonetheless, this requires a large amount of specimens per excavation unit to be preserved and analysed.

Thus, given that we do not know how many years are accumulated in each excavation unit and also that the shells do not belong to closed assemblages, elemental mapping may contribute to the identification of assemblages and specimens that are coeval in the sense of the exact year. The high-resolution mapping in these cases provided distinct patterns which indicated the affinity of the specimens which recorded the same environmental conditions. Isotope analyses can further corroborate this observation.

5.2.5. Seasonality Studies in the Eastern and Central Mediterranean

In this section, I discuss how the overall seasonality results from the site of Franchthi relate to other similar studies in the Eastern and Central Mediterranean. Several seasonality studies have been carried out in the area regarding the Mesolithic and the Neolithic period (Mannino et al. 2007; Colonese et al. 2009; Mannino et al. 2014; Prendergast et al. 2016; Colonese et al. 2018; Branscombe et al. 2020).

To start with, a seasonality study by Branscombe et al. (2020) constitutes an example in the Mediterranean region and offers an opportunity to explore and compare seasonality patterns with Franchthi. More specifically, the Branscombe study examined seasonal shellfishing practices during the Neolithic transition via oxygen isotope analysis using specimens (n = 28) of the gastropod *Phorcus turbinatus* from Vela Spila, Korčula (Croatia). They extracted seasonal information from Mesolithic, transitional Mesolithic to Neolithic and Neolithic samples and identified a preference in summer and autumn collection during the Mesolithic, which changes into a preference in autumn and winter collection during the Neolithic period (Branscombe et al. 2020, 8).

Their conclusion regarding the Mesolithic period derived from ten specimens and is in agreement with the conclusion by Deith and Shackleton for the Mesolithic Franchthi shells who make the hypothesis for the use of the cave as a summer/autumn camp (1988, 154). The Mesolithic shells from Franchthi (n = 67) also show a tendency for collection towards the warmer months of the year, i. e. autumn (n = 22) and summer (n = 18), while there are also winter-collected specimens in the record (n = 12). Results on the specimens from the transitional Mesolithic-Neolithic layers of Vela Spila were not conclusive (Branscombe et al. 2020, 8).

As for the Neolithic specimens, Branscombe et al. identify a preference for autumn and winter collection at Vela Spila (2020, 8). The winter element is also strong in the Early Neolithic shells (n = 56) that were analysed from Franchthi, while there is overall clear evidence for year-round collection of molluscs. The second most prominent season of collection is summer (n = 42), which is then followed by fewer autumn-collected specimens (n = 15) (see **fig. 151**).

As the study by Branscombe et al. does not refer to sub-divisions of the chronological periods to which the specimens they analysed belong, more direct comparisons on stricter chronological levels cannot be made. Overall, Branscombe et al. (2020) found a shift from summer and autumn collection during the Mesolithic to autumn and winter collection in the Neolithic. These Mesolithic findings align with Deith and Shackleton's hypothesis of summer/autumn usage of Franchthi Cave, supported by Mesolithic shell data from Franchthi. The Neolithic data from Franchthi suggest a preference for winter collection while showing evidence of year-round collection, whereas the Vela Spila shells show an autumn and winter collection preference. Despite differences in the number of specimens analysed, the general seasonality patterns are consistent between the two sites.

A seasonality study has also been carried out on *Phorcus turbinatus* shells from Grotta dell'Uzzo by Mannino et al. (2007). 50 specimens were analysed via $\delta^{18}O$ analysis of which, five were analysed sequentially (Mannino et al. 2007, tab. 3). According to this study, during the Early Mesolithic period molluscs were exploited during winter and some during early spring. During the later phases of the Mesolithic, collection takes place during all the seasons of the years or at least year-round visits take place at the cave (Mannino et al. 2007, 127). During this time, the site is interpreted as a home base camp that also acquires a ritual character (Di Salvo et al. 2012a; Borgognini Tarli 1993). Furthermore, during the transitional period, the shell isotope evidence indicates only autumn and winter collection, in contrast with the evidence from the avifauna and the fish that indicate site occupation during a broader seasonal range (Mannino et al. 2007, 129). During the Early Neolithic period, seasonal patterns point towards a year-round exploitation of the molluscs, with the summer months being under-represented. Mannino et al. suggest that the cave was used during autumn, winter and spring and that this occupation was associated with fishing, wildfowling and shellfish collection activities. Lastly, during the later phases of the Early Neolithic period shellfish are mostly collected during winter and autumn (2007, 130).

The results from Grotta dell'Uzzo resonate in some ways with those from Franchthi. Year-round collection is attested during the Upper Mesolithic period at Franchthi by the present study, a trend that is also evident during the Late Mesolithic at Grotta dell'Uzzo. The further broadening of the seasonal range is attested in both sites during the Early Neolithic. The strong winter element at Franchthi Paralia during the Early Neolithic period is also present during the later phases of the Early Neolithic at Grotta dell'Uzzo.

Similar seasonality studies have been conducted on *Phorcus turbinatus* specimens (n = 74) at Grotta d'Oriente on the Sicilian island Favignana (Colonese et al. 2009; Mannino et al. 2014; Colonese et al. 2018), where a recurrent trend is present. $\delta^{18}O$ analysis indicated that during the Upper Palaeolithic, the Early Mesolithic and the Late Mesolithic/Early Neolithic period the collection of molluscs took place at short-term episodes during the colder months of the year (Colonese et al. 2009, 1942; 2018, 120–122; Mannino et al. 2014, 141). The winter exploitation during the Mesolithic constitutes a trait that is common also in other sites in Sicily (e. g. Grotta della Serratura, Grotta delle Incisioni) (Colonese et al. 2009; Mannino et al. 2011; 2014). In addition, mollusc exploitation is more restricted during the Early Mesolithic, indicating a more sporadic occupation of the site, while during the later Mesolithic phases exploitation appears to occur over longer time periods (Mannino et al. 2014, 144). This sporadic evidence of exploitation at Grotta d'Oriente has led to the conclusion that mollusc consumption was not part of the subsistence practices and diet. It is assumed that molluscs may have been consumed because of their taste and not considered a primary source of food (Mannino et al. 2012; 2014, 145).

The winter exploitation of molluscs attested in the Mesolithic contexts of Grotta d'Oriente is a trend that is not present in the Mesolithic Franchthi exploitation patterns. The Mesolithic shells from Franchthi (n = 67) exhibit a tendency for collection during the warmer months, with 22 specimens collected in autumn and 18 in summer. Additionally, twelve specimens were collected in winter. The winter emphasis attested in the Late Mesolithic/Early Neolithic layers of Grotta d'Oriente is present at Franchthi in the Early Neolithic Paralia layers.

At Haua Fteah in Libya *Phorcus* and *Patella* shells are abundant in the Late Pleistocene and Early Holocene deposits, species that are indicative of the rocky shores that characterise the coastal environment (McBurney 1967; Barker et al. 2010; Hunt et al. 2011). Oxygen isotope analyses on *Phorcus turbinatus* shells (n = 46) indicated that mollusc exploitation was year-round during the Early Epipalaeolithic (or Oranian) period (approximately 15–13,000 cal BP). Nonetheless, it is not clear if this evidence indicates permanent occupation or episodic visits at the site. During the Late Epipalaeolithic (or Capsian) period and the Neolithic period seasonal exploitation patterns shift towards winter exploitation almost exclusively (Prendergast et al. 2016, 89–91).

The year-round evidence of the Early Epipalaeolithic exploitation patterns at Haua Fteah is similar to the Final Palaeolithic evidence from Franchthi even though the samples analysed are not numerous. In the present study, I identified two shells collected in winter, one in spring/summer, one in autumn, one in autumn/winter (see **fig. 141**), while Deith and Shackleton recognise an emphasis on autumnal collection (20 shells collected during autumn alongside one or two winter-collected specimens) (1988, 151). Furthermore, the Early Neolithic strong winter element at Franchthi is in agreement with the exclusive winter exploitation at Haua Fteah during the Neolithic as all the Neolithic specimens were collected during winter or autumn (Prendergast et al. 2016, fig. 5d).

Overall, the evaluation of the various studies revealed certain tendencies in the seasonal exploitation patterns among the different sites. Regarding the Late Palaeolithic period, seasonal evidence points towards year-round exploitation of molluscs as is the case for Haua Fteah (Prendergast et al. 2016, 89) and Franchthi. During the Mesolithic, similar exploitation patterns are attested between Franchthi and Vela Spila when there is an emphasis in summer and autumn collection (Branscombe et al. 2020, 8), while year-round exploitation evidence is also attested at Franchthi at a smaller scale. The Late Mesolithic phases of Grotta dell'Uzzo also indicate collection during all the seasons of the year, while the Early Mesolithic phases indicate a stronger winter element (Mannino et al. 2007, 127). At Grotta d'Oriente during the Mesolithic, collection happens predominantly during winter (Colonese et al. 2009, 1942; Mannino et al. 2014, 141; Colonese et al. 2018, 120–122). In addition, evidence from the Neolithic phases of Vela Spila, Grotta d'Oriente and Haua Fteah points towards mollusc exploitation during the colder months of the year (Colonese et al. 2009, 1942; Mannino et al. 2014, 141; Prendergast et al. 2016, 90–91; Colonese et al. 2018, 120–122; Branscombe et al. 2020, 8). Grotta dell'Uzzo has more similarities with Franchthi as seasonal patterns show year-round exploitation during the early phases of the Early Neolithic and during the later phases of the Early Neolithic period exploitation during winter and autumn (Mannino et al. 2007, 130).

Thus, it appears that there are three different exploitation patterns in the sites presented: Year-round exploitation, emphasis on autumn collection and emphasis on winter collection. The patterns of the sites can be grouped in the following way through time: First, the year-round exploitation patterns of the hunter-gatherer lifeways that are attested at Haua Fteah and Franchthi during the Palaeolithic period. Second, collection during the colder months of the years during the Mesolithic attested at Grotta d'Oriente, the early phases at Grotta dell'Uzzo and the Capsian phases at Haua Fteah. Third, emphasis on autumnal collection during the Mesolithic phases of Vela Spila and Franchthi. Fourth, collection during the colder months of the year during the Neolithic period at Vela Spila, Grotta d'Oriente and Haua Fteah. Fifth, year-round exploitation of molluscan resources during the Early Neolithic at Grotta dell'Uzzo and Franchthi, while the latter site also exhibits a strong winter element.

Year-round exploitation during the Palaeolithic and to some degree during the Mesolithic at Franchthi is indicative of hunter-gatherer subsistence practices relying on wild food resources (e. g. Waselkov 1987, 136–138; Erlandson 2001, 289; Mannino et al. 2003; 2007; Colonese et al. 2011). It also attests to the episodic character of occupation of the site by these groups. Regarding seasonal consumption preferences, winter consumption of molluscs can have multiple interpretations: The dietary contribution of molluscs has been often seen as merely supplemental as suggested by various ethnographic parallels (Claassen 1998, 178–179) and thus winter-collected specimens are interpreted as seasonal dietary supplements. In historical and ethnoarchaeological accounts, mollusc consumption has been identified as famine food in times of food scarcity and adversity (Brothwell/Brothwell 1969, 66; Jameson 1983, 9; Garnsey 1983, 56–60; 1988, 28–39; Theodoraki 2018, 39). Beyond famine, as shellfish are available year-round, focused seasonal exploitation may represent gathering preferences and patterns and may be related to people's perceptions regarding their interaction with the environment (Waselkov 1987, 111). In addition, year-round exploitation of molluscs may indicate the importance of molluscs as a dietary staple.

In the Central and Eastern Mediterranean sites discussed here, seasonal exploitation patterns are interpreted in various manners in terms of the dietary significance of the molluscs as well as the seasonality of occupation. At Grotta dell'Uzzo, the high number of winter-collected shells is interpreted as evidence for periodic visits to the site for ritual purposes (Mannino et al. 2007, 129; Di Salvo et al. 2012b, 346). At Grotta d'Oriente, the short-term cold month exploitation during the Mesolithic, is interpreted in relation to the seasonal migratory movements of hunting game which may have influenced hunter-gatherer mobility and coastal exploitation patterns (Colonese et al. 2009, 1943). During the Late Mesolithic/Early Neolithic, the winter exploitation is interpreted as a result of the intermittent use of the cave (Colonese et al. 2018, 122). At Vela Spila, the varying seasonal patterns are interpreted in relation to the changing season of occupation during the Mesolithic. During the Neolithic period, even though people inhabit the cave year-round, shellfishing predominantly takes place during the colder months of the year (Branscombe et al. 2020, 8–10). At Haua Fteah, winter/autumn collection during the Capsian and the Mesolithic period is interpreted as a means to add to the limited winter resources (Prendergast et al. 2016, 91).

At Franchthi, it appears that visits to the site happen year-round during the Mesolithic and the Neolithic period, and to a smaller extent during the Final Palaeolithic. During the latter period it may have served more as a summer/autumn base for the visiting groups – as suggested by Deith and Shackleton for the Mesolithic period – even though winter-collected specimens are present, as well as a few spring/summer and autumn/winter specimens. The Mesolithic and Neolithic periods at Franchthi contain a large number of specimens well distributed across the year, with a stronger winter component during the Early Neolithic and a stronger summer component during the Middle Neolithic. This more evident year-round collection aspect of the Franchthi shells in comparison to the other sites in the Central and Eastern Mediterranean may be due also to the larger sample sizes, increasing the chances of identifying specimens that belong to a broader seasonal spectrum.

5.3. Palaeoclimatic Record

5.3.1. Long-Term Sea Surface Temperature

As mentioned in Chapter 1 and in section 2.2, several hypotheses have been proposed to explain the westward movement of the pioneer farmers, which include demographic pressure, social stress, and environmental stress due to climatic deterioration (e. g. Demoule/Perlès 1993; Cauvin 2000, 1–3; Price 2000, 2–3; Guilaine 2000–2001, 272; Perlès 2001, 38; Guilaine 2015, 85). The climatic theories base their arguments on different datasets in each case of shorter or longer temporal and geographical scales to argue on the factors influencing the advent of the Neolithic, while the authors that advocate for »collapse« narratives tend to emphasise the intensity of the climatic phenomena implementing pressure on resources (e. g. Weiss et al. 1993; deMenocal 2001; Weninger et al. 2006; Roffet-Salque et al. 2018; Kaniewski et al. 2020). Evidently, the Early Neolithic across the Mediterranean is not a homogeneous entity and is largely varied across space (Guilaine 2015, 95), and site-specific or regional climatic and archaeological evidence is evaluated in most of the cases to support arguments of the climatic impact on past human communities. The palaeoclimatic evidence employed in such studies tends to be proxy-based climate reconstructions as direct records of past climate. These reconstructions often involve uncertainties regarding their temporal resolution or based on sparse palaeoclimate evidence, rendering them challenging to direct comparisons (Degroot et al. 2021, 540). Banded coral archives, ice core, cave deposits may be subject to age uncertainties while the correlation of the various climatic evidence may often be based on subjective comparisons (Crowley 1999; Comboul et al. 2014).

In the context of addressing the environmental stress hypothesis and the question of societal responses to environmental stress, the climatic conditions at Franchthi were examined on various temporal scales (long-term, by period, and lastly on a stratigraphic level [section 5.3.2]). To this end, the long-term Sea Surface Temperature (SST) at the site is reconstructed using oxygen isotope analyses, which allow for the detection of sea surface temperature changes on various, particularly subannual, levels. These analyses provide a high-resolution picture of climatic changes at the time of human occupation, which enables the assessment of the potential impact of these changes on newly established domestic economies.

In this section, I compare the Franchthi palaeoclimatic evidence with the larger scale climate variability picture of the Early Holocene in the Eastern Mediterranean and more specifically with the findings of the marine sediment core LC21 and core SL21 (e. g. Hayes et al. 1999; de Rijk et al. 1999; Mercone et al. 2000; Marino et al. 2007; 2009). Core LC21, recovered from the south-eastern Aegean Sea (26.58°E, 35.66°N), and core SL21, recovered from the north-eastern Aegean Sea (25.25°E, 39.01°N), constitute regional proxy records for the Early Holocene in the Aegean Sea (**fig. 160**). In addition, they have been associated with evidence of Rapid Climate Change (RCC) events in the Eastern Mediterranean (e. g. Rohling et al. 2002a; Casford et al. 2003; van der Meer et al. 2007; Rohling et al. 2009) and are thus discussed in the following section alongside the site-specific evidence from Franchthi.

The Aegean cores LC21 and SL21 contain distinct organic-rich dark intervals which represent the depositions of sapropel S1 (Mercone et al. 2000; Casford et al. 2003; Rohling et al. 2002). The cores were sampled in 1 and 0.5 cm intervals respectively and allowed for the investigation of climate variability at multi-decadal to centennial resolution (Marino et al. 2009, 3249). SST reconstruction was enabled via the study of planktonic foraminifera (*Globigerinoides ruber*) in the case of core LC21 (de Rijk et al. 1999; Hayes et al. 1999; Casford et al. 2002), which were converted into summer and winter SSTs using artificial neural networks (Hayes et al. 2005) (see **figs 161–162**). The chronology of the cores is quite robust as Accelerator Mass Spectrometry (AMS) radiocarbon

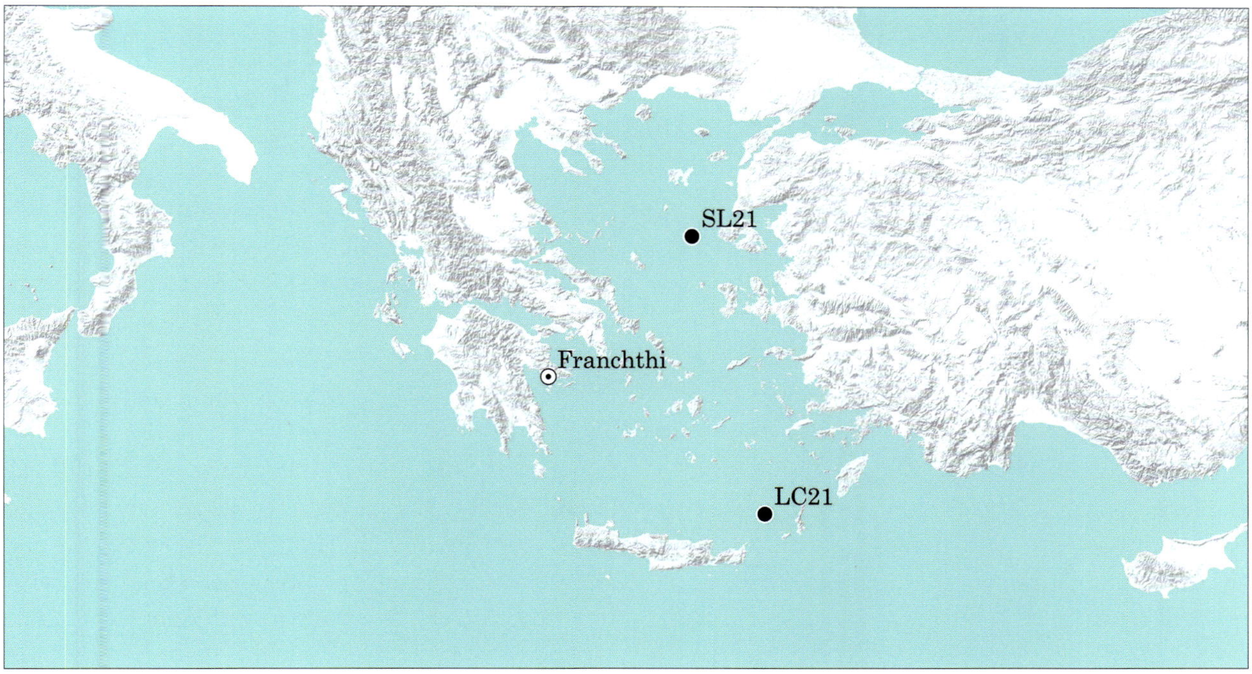

Fig. 160 Map of the Mediterranean basin with the positions of the core sampling sites (LC21, SL21) and Franchthi. – (Basemap source Esri and contributors).

datings have been applied (Casford et al. 2007). The records of the two cores present many similarities and in this section the LC21 core is firstly taken into account as it is geographically closer to Franchthi. Secondarily, the shell-derived climate data of Franchthi, are also compared with the $\delta^{18}O_{ice}$ record of the North Greenland Ice Core Project (NGRIP) ice core which bears evidence for the Early Holocene period (Andersen et al. 2004).

Within the chronological uncertainties, the NGRIP core, the Aegean and the Franchthi records exhibit resemblances in patterns during the Early Holocene. **Figures 161** and **162** provide a comparison of the local data with the regional proxies, as suggested by Finné and Holmgren (2010, 53–54). While the palaeoclimatic record of Franchthi is based on relative chronologies, whereas the regional and global scale proxies have been absolutely dated, there is an agreement in the general tendencies of the different datasets.

The results of the summer SSTs in LC21 indicate that summer SST was higher than the present-day value of 23.5 °C throughout the Early to Middle Holocene, which may reflect the orbital boreal summer solar radiation maximum (Marino et al. 2009, 3251 fig. 3b), periods when the Northern Hemisphere (boreal) received the maximum amount of solar radiation during summer (Fohlmeister et al. 2023). Interestingly, the average summer isotope value of Franchthi (−0.72 ‰) is translated into 26.0 ± 1.3 °C, when applying the equation by Hays and Grossman (1991) for calcite as well as the site-specific offset calculated (+0.85 ‰) described in section 4.1.3.2. This average temperature is very similar to the present-day average summer-months value of 25.5 °C at Franchthi based on satellite data, whereas the average summer SST values in the LC21 core have a mean value of 24.7 °C (Marino et al. 2009, 3251 fig. 3b). The higher SST values in the climatic data from Franchthi compared to the LC21 core data could be attributed to temperature differences between the near-shore and off-shore sampling locations, with limpets reflecting the warmer near-shore conditions and reduced water-exchange with deeper waters.

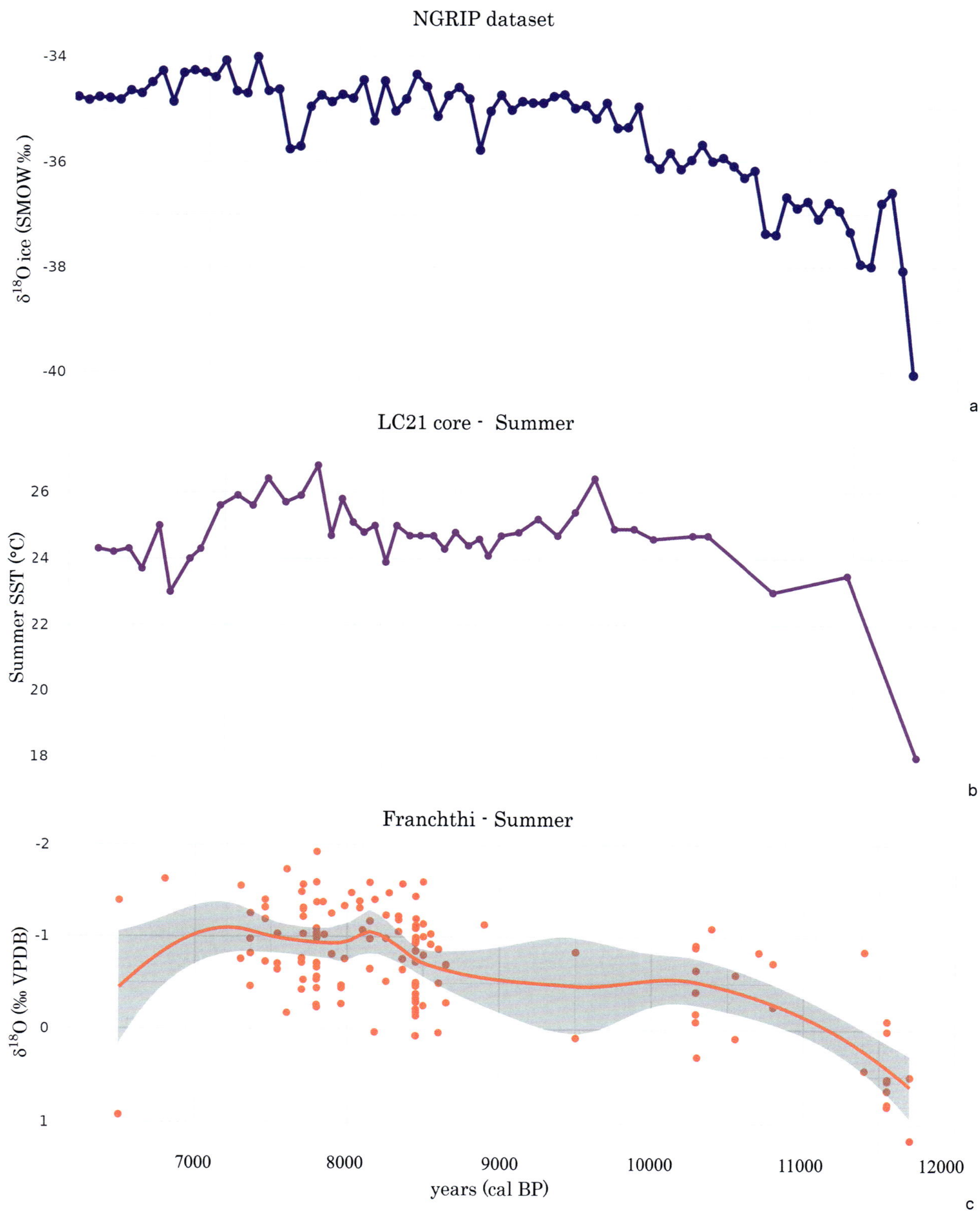

Fig. 161 **a** NGRIP ice core data (after Andersen et al. 2004, Suppl. data). – **b** Summer SST (°C) data from core LC21 (after Marino et al. 2009, fig. 3b). – **c** Summer shell-derived local data from Franchthi.

The comparison of winter climate variability from Franchthi also presents similarities with the winter values from core LC21. The regional winter SST data from LC21 and SL21 do not show any abrupt cooling, but a gradual increase following the Younger Dryas is attested in core LC21 culminating to a SST maximum of 17.2 °C at 9,400 BP. This gradual warming trend is also identified during the

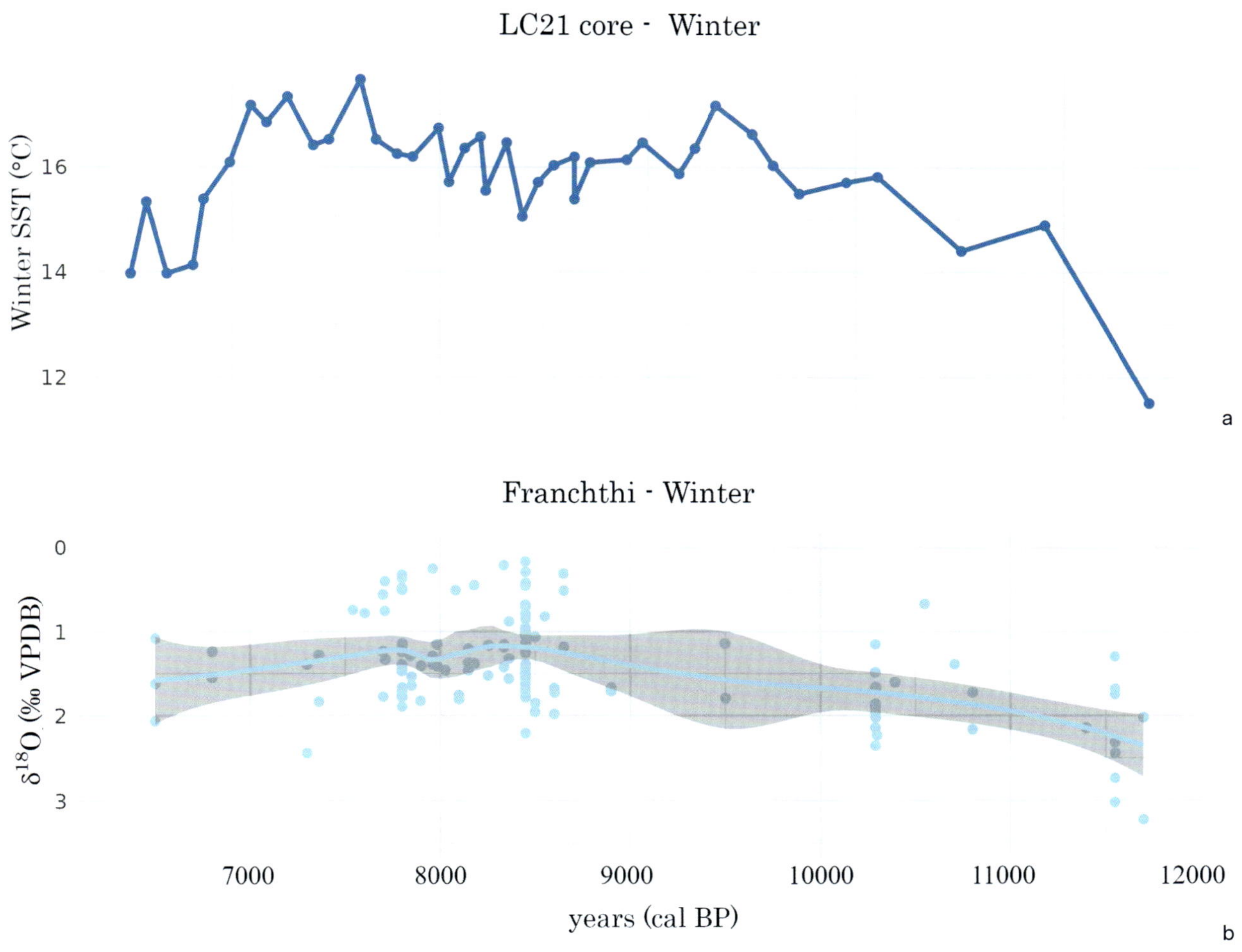

Fig. 162 **a** Winter SST (°C) data from core LC21 (after Marino et al. 2009, fig. 5f). – **b** Winter shell-derived local data from Franchthi.

same period in the summer values, but in a slightly steeper way (Marino et al. 2009, 3253 fig. 5f). A similar contemporaneous trend is also present in the $\delta^{18}O$ speleothem data from Soreq Cave in northern Israel during the Late Pleistocene and the Early Holocene (Bar-Matthews et al. 2000; 2003).

This gradual warming following the Younger Dryas is also reflected in the winter shell-derived data from Franchthi (**fig. 162b**). A peak at 9,400 BP is not present in this case, as the few data points from this period are not indicative of a temperature peak. In the case of winter $\delta^{18}O$ evidence from Franchthi, the culmination of the warming trend is evident at approximately around 8,500 BP, i.e. during the Early Neolithic period. Subsequently, there is a slight decrease at approximately 8,000 BP, while afterwards values increase again at 7,700–7,800 BP. This increase is also present in the LC21 core winter SST data at approximately 7,700 BP (**fig. 162a**). A slightly decreasing trend is observed in the winter $\delta^{18}O$ values of Franchthi after this period (**fig. 162b**).

According to Marino et al., three distinct sea surface cooling events are evident in the regional proxies of the Aegean Sea at 10,500, 9,500–9,030 and 8,800–7,800 BP (Marino et al. 2009, 3255; Rohling et al. 2002a). In the case of the local record of Franchthi, only the latter event could be discerned in the summer dataset as at around 8,000–7,500 a small drop in the $\delta^{18}O$ values is present. The cooling events at 10,500, 9,500–9,030 BP are not reflected, likely because shell samples from Franchthi dating to this period are not numerous.

Overall, the site-specific isotope data from Franchthi are in agreement with the regional and global scale temperature and isotope evidence. Following the cooler and drier Younger Dryas, the oxygen isotope records show changing climatic conditions as the eastern Mediterranean becomes warmer

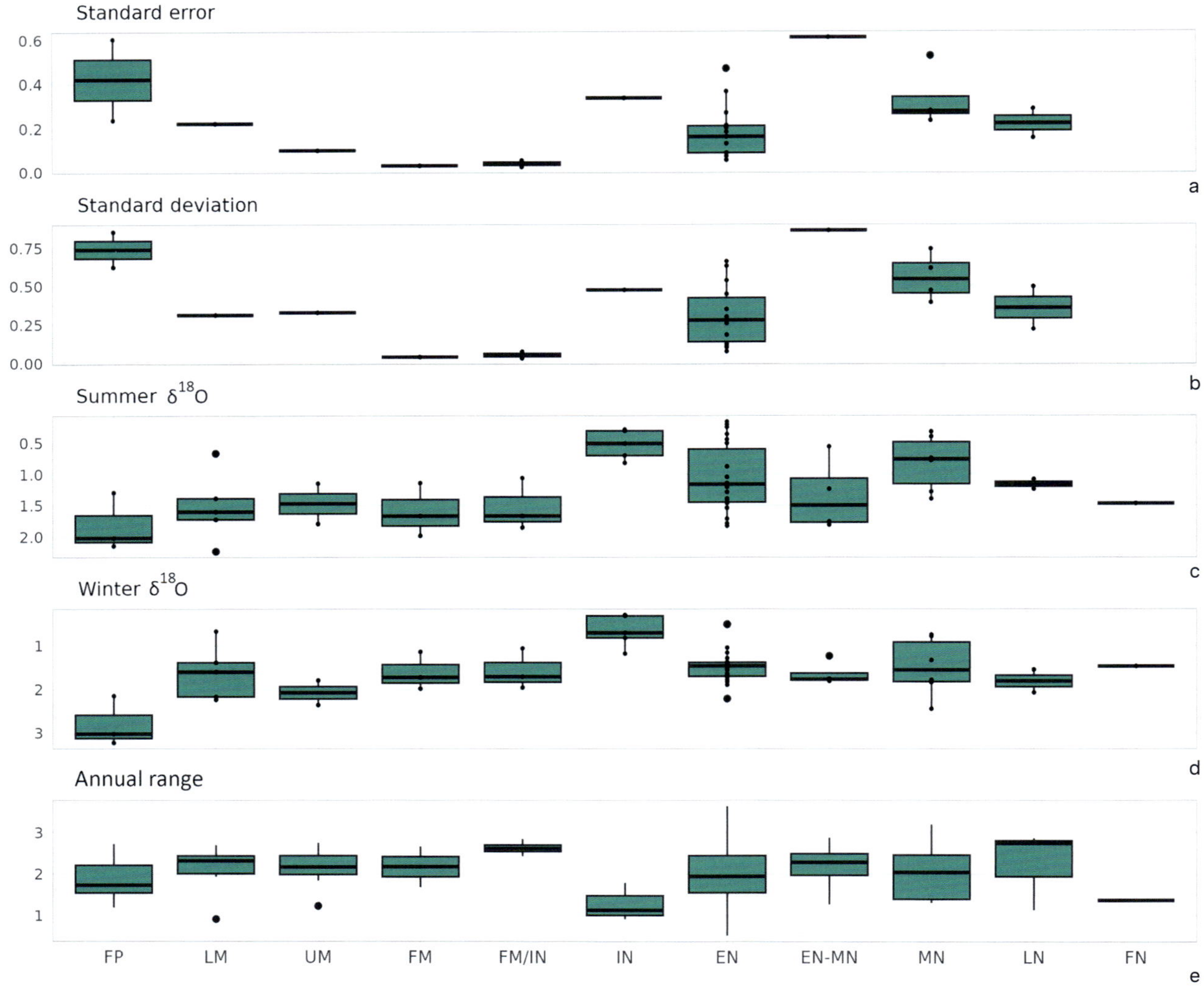

Fig. 163 Graph displaying summary statistics for the standard error (**a**), the standard deviation (**b**), the summer $\delta^{18}O$ (**c**), the winter $\delta^{18}O$ (**d**) and the annual range (**e**). – Chronology legend: FP: Final Palaeolithic; LM: Lower Mesolithic; UM: Upper Mesolithic; FM: Final Mesolithic; FM/IN: Final Mesolithic/Initial Neolithic; EN: Early Neolithic; MN: Middle Neolithic; LN: Late Neolithic; FN: Final Neolithic.

and more humid conditions into the Holocene Climatic Optimum (Rohling/de Rijk 1999; Casford et al. 2003, 109). The changing climatic conditions and SST temperature trends are well-reflected in the overall shell-derived climatic record of Franchthi from 279 season-specific isotope values.

Interestingly, summary statistics of the isotope values show that higher winter temperature values are present during the Initial Neolithic period (8,650–8,450 cal BP/6,700–6,500 BC) (**fig. 163d**). Furthermore, the calculated standard deviation and standard error of isotope values by chronological period have more variable results showing that the Final Palaeolithic (11,700–11,400 cal BP/9,750–9,450 BC), the Early Neolithic (8,450–7,800 cal BP/6,500–5,850 BC), the Early-Middle Neolithic and the Middle Neolithic period (7,800–7,300 cal BP/5,850–5,350 BC) were characterised by higher climate variability (**fig. 163a–b**).

The $\delta^{18}O$ value results point towards a more variable climate as the mean annual range of the Final Palaeolithic shells is relatively large (2.95 ‰), compared to that of the Initial Neolithic (1.35 ‰) for example (see **tab. 14**). The oxygen isotope values of almost all chronological periods (with the exception of the Initial Neolithic and the Final Neolithic) vary relatively widely and have high standard error (e. g. 0.23 for the Late Mesolithic) and high standard deviation values (e. g. 0.61 for the Final Palaeolithic) (see also **figs 128–129**). This could mean either that the values are indicative of the long temporal scales that are accumulated in the deposits or that the climate was highly variable, thus rendering the isotope values more variable.

Given the variable temporal resolutions of the different chronological periods, climate variability could be happening on longer time-scales (decadal or centennial). If the isotope variability reflects short-term climatic events, it could suggest that the climate experienced significant fluctuations over relatively brief periods, such as a few years or decades. On the other hand, if the isotope variability corresponds to longer-term trends spanning over a century or more, it would imply that the climate experienced gradual changes. This could suggest a more stable, albeit changing, environmental context in which human populations had time to adapt their practices and technologies to the evolving conditions. Such long-term variability might be linked to broader climatic phenomena, which influenced the regional climate over extended periods, as already described above. In addition to the above, due to the uncertainty of the temporal resolution of the periods, it is not clear whether within a certain period, there are hidden short-term events that increase the variability or whether the variability is homogeneous across the period.

Regarding the summary statistics, seasonal stable isotope values of the Mesolithic specimens appear to be well-clustered together, while they also have smaller standard deviation and standard error values (**fig. 163a–b**). This is especially the case for the Upper Mesolithic and the Final Mesolithic/Initial Neolithic period. The small standard error and standard deviation of these periods could be interpreted as indicators of a more stable climate during these periods, suggesting that environmental conditions may have been relatively constant without significant fluctuations. This stability in climate could have contributed to a predictable environment for the prehistoric inhabitants, potentially facilitating consistent subsistence strategies and resource availability. Alternatively, this could be due to shorter episodic visits to the site, similar to the interpretation by Colonese et al. (2009, 1942), given the non-sedentary character of the cave during this time.

$\delta^{18}O$-value ranges of the Neolithic samples present wider distributions in all parameters of **figure 163**, especially during the Early Neolithic period, and secondly during the Middle Neolithic, which could be attributed to increased climate variability. Particularly narrow annual ranges are present during the Initial Neolithic which may be due to the short span of this period, while during the Early Neolithic period the mean annual ranges were larger (see **fig. 163**).

When isolating the various isotope parameters of the Early (8,450–7,800 cal BP), Early-Middle and Middle Neolithic periods (7,800–7,300 cal BP) (**fig. 164**), periods during which the characteristics of the domestic economy were well-established at the site, the Early and the Middle Neolithic share many similarities. More specifically, standard error and standard deviation values are very similar with those of the Early Neolithic being more variable and the values of the Middle Neolithic being slightly more increased (see statistical tests below). In addition, the summer (mean values: Early Neolithic: −0.85 ‰, Middle Neolithic: −0.99 ‰) and winter $\delta^{18}O$ values (Early Neolithic: 1.21 ‰, Middle Neolithic: 1.24 ‰) are also quite similar between the two periods with the Middle Neolithic values being slightly more elevated. The isotope ranges seem very similar across all three periods. As for the Early to Middle Neolithic transitional period it exhibits slightly lower values than the other two periods in the case of the summer and winter $\delta^{18}O$ values. This period also has narrow standard error and standard deviation values which could be attributed to the fact that this period has very few samples. Overall, the climatic differences between the Early and the Middle Neolithic periods do not appear to be significant, with the Early Neolithic period being much more variable.

A number of t-tests were run in order to compare the $\delta^{18}O$ values between the Early and the Middle Neolithic. While there are differences in the mean values of the different sets of values (minimum, maximum, standard error, range), none produced any significant differences (see **tab. A.16**). More specifically, there is no statistically significant difference between the minimum $\delta^{18}O$ (i. e. summer) mean values of the Early Neolithic (−1.10 ‰) and the Middle Neolithic (−1.30 ‰) groups of values ($p = 0.31$). The same is the case for the maximum $\delta^{18}O$ (i. e. winter)

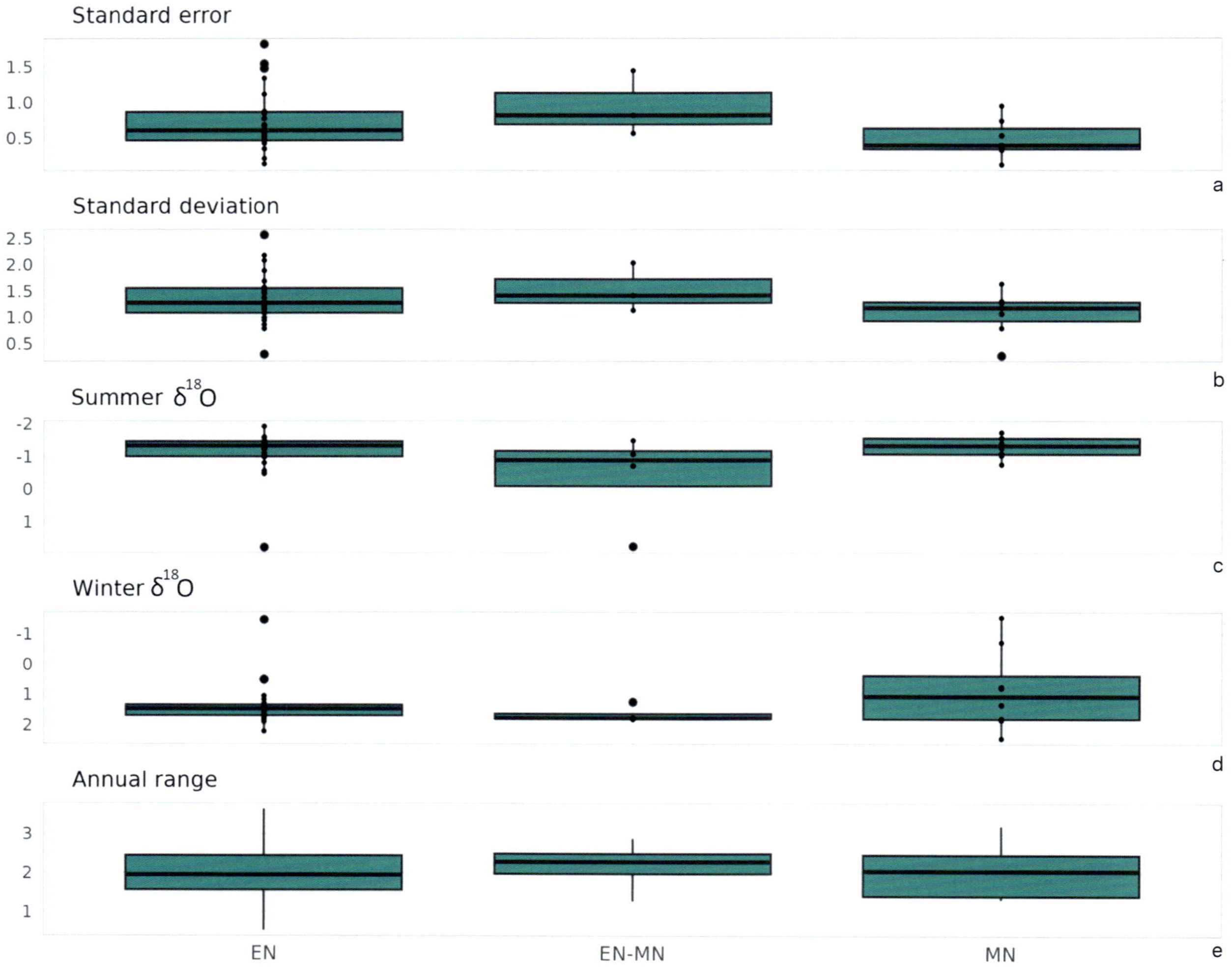

Fig. 164 Graph displaying summary statistics for the Early, Early-Middle and Middle Neolithic for the standard error (**a**), the standard deviation (**b**), the summer $\delta^{18}O$ (**c**), the winter $\delta^{18}O$ (**d**) and the annual range (**e**). – Chronology legend: EN: Early Neolithic; EN-MN: Early-Middle Neolithic; MN: Middle Neolithic.

values ($p = 0.35$), even though the mean values are more variable (Early Neolithic: 1.33 ‰, Middle Neolithic: 0.83 ‰). Similarly, the annual range showed that the variation in the two sets of values is random ($p = 0.53$; Early Neolithic: 2.43 ‰, Middle Neolithic: 2.13 ‰). The standard error was the only comparison that showed a slight trend towards significance ($p = 0.08$; Early Neolithic: 0.75 ‰, Middle Neolithic: 0.49 ‰).

To further contextualise the climatic changes with the environmental and cultural changes during the transition from the Early to the Middle Neolithic, it is essential to consider the palaeoenvironmental data and the archaeological findings. As presented in section 2.6.4, the pollen data of these time periods only offer very general climatic tendencies. The Late Pleistocene vegetation according to pollen archives from Northern and Central Greece – as the evidence for the Peloponnese is sparse – indicated a cold and dry climate which is in accordance with the climatic record described above. Pollen data also show a climatic amelioration at around 10,000 BP, which led to the rapid expansion of an open forest featuring junipers and deciduous oak, with a distinct presence of *Pistacia*, which suggests warmer conditions (Wijmstra et al. 1990). Nonetheless, as the local pollen data of a similar resolution are missing, more precise comparisons with the local shell-derived palaeoclimatic data are not possible.

The zooarchaeological and archaeobotanical data were analysed to identify potential shifts in subsistence strategies and resource exploitation. This includes detecting changes in the use of terrestrial resources, variations in the NISP of different species, and shifts in species composition across various periods. Archaeologically, such changes can be evidenced by variations in the types and quantities of animal remains, plant species, and artefacts, including lithic and ceramic materials, which may reflect broader socio-economic and environmental adaptations.

Regarding the zooarchaeological data and potential changes before and after 8,000 BP, the Early and Middle Neolithic assemblages largely resemble one another in their zooarchaeological characteristics (see section 2.6.8.2). Both periods are characterised by high taxonomic richness with increased use of pigs, red deer and fish compared to their preceding chronological period (Munro/Stiner 2020, 8–9). The same is the case with the archaeobotanical remains, presented in section 2.6.8.1, as there are no drastic species composition changes between the Early Neolithic (botanical zone VI) and the later Neolithic periods (botanical zone VII) which include the Middle to the Final Neolithic. Domesticated two-rowed barley (*Hordeum distichum*) and emmer wheat (*Triticum dicoccum*) first appear in stratum X2 dated to the Initial Neolithic together with weed taxa (Boraginaceae and *Teucrium* sp.). The difference with the subsequent Neolithic periods is that in botanical zone VII, the domesticated species (emmer wheat, barley, lentils) increase in numbers and einkorn wheat appears for the first time (Hansen 1991, 145. 163–164). Thus, no major changes in the zooarchaeological and the archaeobotanical archive are identified during the Early Neolithic and the subsequent phases which could indicate climatic influences on the fauna and the flora composition. In addition, the 8,200 event is not reflected in the animal and plant remains, as there are no significant changes in the zooarchaeological and archaeobotanical records that could indicate abrupt cooling conditions.

In addition, according to Perlès, the Early and the Middle Neolithic lithics and pottery show ample evidence of homogeneity across the two periods (Perlès 2004, 161; see sections 2.6.6 and 2.6.7). The continuity that characterises lithics and pottery during these periods shows that the subsistence economy does not change (Perlès 2004, 161). In the case of lithic tools there are some changes in the raw materials used, while the techniques and methods used remain the same (Perlès 2004, 69. 89). The same is the case with pottery even though there are certain changes in the composition of the ceramic assemblages, it appears that Early Neolithic traditions persist through the Middle Neolithic (Vitelli 1993, 39–41. 49–50. 208).

Overall, while the shell-derived climatic and palaeoenvironm ental data indicate a gradually changing climate, with slight but statistically insignificant differences in the climate variability between the Early and the Middle Neolithic, this transition at Franchthi shows elements of continuity in subsistence economy, as evidenced by homogeneous lithic and pottery assemblages. At the same time, the above evidence is based on broad chronological periods, which do not allow for the identification of short-term climate or subsistence shifts that could indicate instability or external influences. This evidence relies on broad chronological periods, which inherently obscure any short-term climate shifts or external influence that might suggest instability or rapid adaptation. The broad chronological resolution of the evidence constrains our interpretation. What currently appears to be the prehistoric inhabitants maintaining stable, unchanged lifeways could be due to the data's coarse scale, which masks any shorter-term adaptive responses or periods of instability. This inadequacy of resolution – our inability to capture dynamic short-term responses – is the central caveat that will be further addressed in section 5.4.

5.3.2. Direct Climate-Subsistence Comparisons

To examine human-environment interactions, it is essential to link different lines of evidence to understand how climate and human agency interconnect. Large-scale palaeoenvironmental proxies, such as pollen records, ice cores, and tree rings, are often not suitable for studying short-term, site-specific climate variability. This poses challenges for comparability due to different scales of analysis. To address this, synchronising archaeological data with climatic records is crucial. This approach, central to the current research question of the climatic record of Franchthi, involves directly comparing stratigraphically correlated shell-derived palaeoclimatic data with subsistence archaeological data to understand the interplay between palaeoclimate and human practices. Thus, the climatic shell-derived data discussed in the previous sections are used in order to reconstruct the human-environment interactions at the site.

To contextualise the climatic dynamics relevant to the archaeological samples from Franchthi, it is crucial to examine specific climatic events during the Holocene. One such event, occurring between 8,600–8,000 cal BP, is known as a Rapid Climate Change event, and is characterised by significant climatic anomalies on a global scale. Rapid Climate Change events constitute Holocene climate anomalies globally or widely spread. These events are considered »rapid« as they occur on decadal or centennial scales (Mayewski et al. 2004). In the Aegean record, they have been identified in core LC21 (already discussed in section 5.3.1) and are dated at about 8,600–8,000 cal BP, 6,300–5,500 cal BP, and 3,300–2,600 cal BP (Rohling et al. 2002a, 590; Mayewski et al. 2004). The Holocene cold events in core LC21, of multi-decadal scales, are attributed to the intensification and increased frequency of wintertime northerly (i. e. coming from the north) outbreaks of cold polar and continental air outbreaks across the Aegean and the eastern Mediterranean channelled through mountain ranges to the north of the Mediterranean basin. The expression of these Rapid Climate Change events would be cold and wet intervals, notable cold winters, of highly variable subannual conditions, leading to the reduction of the winter sea surface temperature by 2–4 °C in the Aegean (Rohling et al. 2002a, 590; Casford et al. 2007; Rohling et al. 2019, 39. 50; Mercone et al. 2000).

In addition, the climatic shift dating to 8,000 cal BP can roughly be matched in the archaeological record with the transition between Early Neolithic at Franchthi which is dated at 8,450–7,800 cal BP and the Middle Neolithic dated at 7,800–7,300 cal BP. Also, since there are a few layers (FAS 128, L5NE 43, L5NE 45) falling into a transitional period from the Early to the Middle Neolithic period, the examination of the evidence closer to this boundary can also be beneficial to understand potential changes and compare the climatic and the cultural record (Demoule/Perlès 1993, 366; Perlès et al. 2013).

As the focus in this work is on the onset of the Neolithic and the subsequent shifts in lifeways and subsistence patterns it introduced, changes in subsistence practices are examined to determine whether climatic changes had a significant or minor impact on the populations at the site. Thus, to understand the impact of the climate on the inhabitants of Franchthi, I explore the species-environment relationships by considering the archaeozoological and archaeobotanical data. The way to implement this is by comparing the shell-derived environmental results with the faunal and the botanical data.

To this end, a Redundancy Analysis (RDA) was carried out using the »vegan« package in R programming language (Oksanen et al. 2022; R Team 2023). The RDA is a constrained ordination method. It considers two different sets of variables, the explanatory variables and the response variables and aims to determine how the response variables are influenced by the explanatory variables. The RDA also visualises this relationship in a multidimensional space. The RDA identifies certain constraints which specify how the explanatory variables are related to the response variables and determines what percentage of the variance of the response variables is explained by this relationship (Legendre/Legendre 2012, 629–630). Thus, in this case the term »redundancy« describes the degree of variance that is shared between the climate and the fauna/flora datasets. This way, the analysis measures the percentage of the variance in one set of variables that can be explained or predicted by the other set and determines how much information is redundant or shared between them.

A similar analysis to the RDA described above was also carried out by Ivanova et al. (2018) who compared modern climatic data to archaeological subsistence data from a large number of Early Neolithic sites in the Balkans in order to understand the northward expansion of plant and animal husbandry, finding differences in animal husbandry and plant cultivation linked to certain climatic condi-

Tab. 16 Climatic parameters derived from $\delta^{18}O$ data. Note that reconstructed sea surface temperatures are based on $\delta^{18}O_{water}$ = 0 ‰.

Climate parameters	Explanation
Annual min	Temperature based on each unit's highest $\delta^{18}O$ value
Annual max	Each unit's lowest $\delta^{18}O$ value
Range summer	The difference of the highest and the lowest summer $\delta^{18}O$ value
Range winter	The difference of the highest and the lowest winter $\delta^{18}O$ value
Range	The difference of the highest and the lowest $\delta^{18}O$ value
Standard error	Standard error of all $\delta^{18}O$ values
Standard error summer	Summer standard error
Standard error winter	Winter standard error

tions. Similarly to the RDA, a Canonical Correspondence Analysis (CCA) was used to quantify and visualise distribution patterns among the total NISP of the archaeobotanical and zooarchaeological species of each site. In contrast to the present study where the faunal and botanical evidence is compared on a temporal scale, the authors of this study are comparing coeval evidence, in this case dating to the Early Neolithic, on a geographical scale.

In the present work, the layers that were selected to be included in the analysis were those that produced climatic, botanical and faunal data and date to the Neolithic period (**tab. A.17**). Thus, only a fraction of the total analysed layers could be used for the RDA. All three datasets are not compared simultaneously, because this would further decrease the number of available layers. Only the faunal data presented enough layers in both the Early and Middle Neolithic period, based on which assumptions could be made regarding the influence of the climate on subsistence before and after 8,000 BP.

Using the RDA I combined and compared the Neolithic layers with the specific climatic data from each excavation layer that also provides faunal or botanical information (**tab. 16**). The environmental variables are derived from the isotope data of each excavation unit (winter minimum and summer maximum temperature, range, standard error). The response variables are represented by the NISP of the zooarchaeological[11] and archaeobotanical data (**tab. A.21**). In addition to the RDA, a correlation analysis of the individual species and the climate parameters was done to further examine potential trends in the two datasets.

5.3.2.1. Integration with Faunal Data

Regarding the relationship between the environmental and the faunal data, the analysis showed that only 11.6 % of the variance of the datasets is explained by the environmental variables (**fig. 165**). In addition, three constraints were identified as response variables for this variance: winter temperature, the annual temperature range and the standard error, which reflects climate variability. The total variance (inertia) in the response data was 1.22. The analysis explained 11.8 % of the total variance through the constrained axes, with an inertia of 0.14. The remaining 88.2 % of the variance was captured by the unconstrained axes with an inertia of 1.08 (**tab. 17**).

In **figure 165**, the direction and length of the arrows (vectors) represent the strength and direction of the influence of these variables. Squares repre-

11 I am grateful to Prof. Natalie Munro for sharing the zooarchaeological NISP numbers per excavation unit with me.

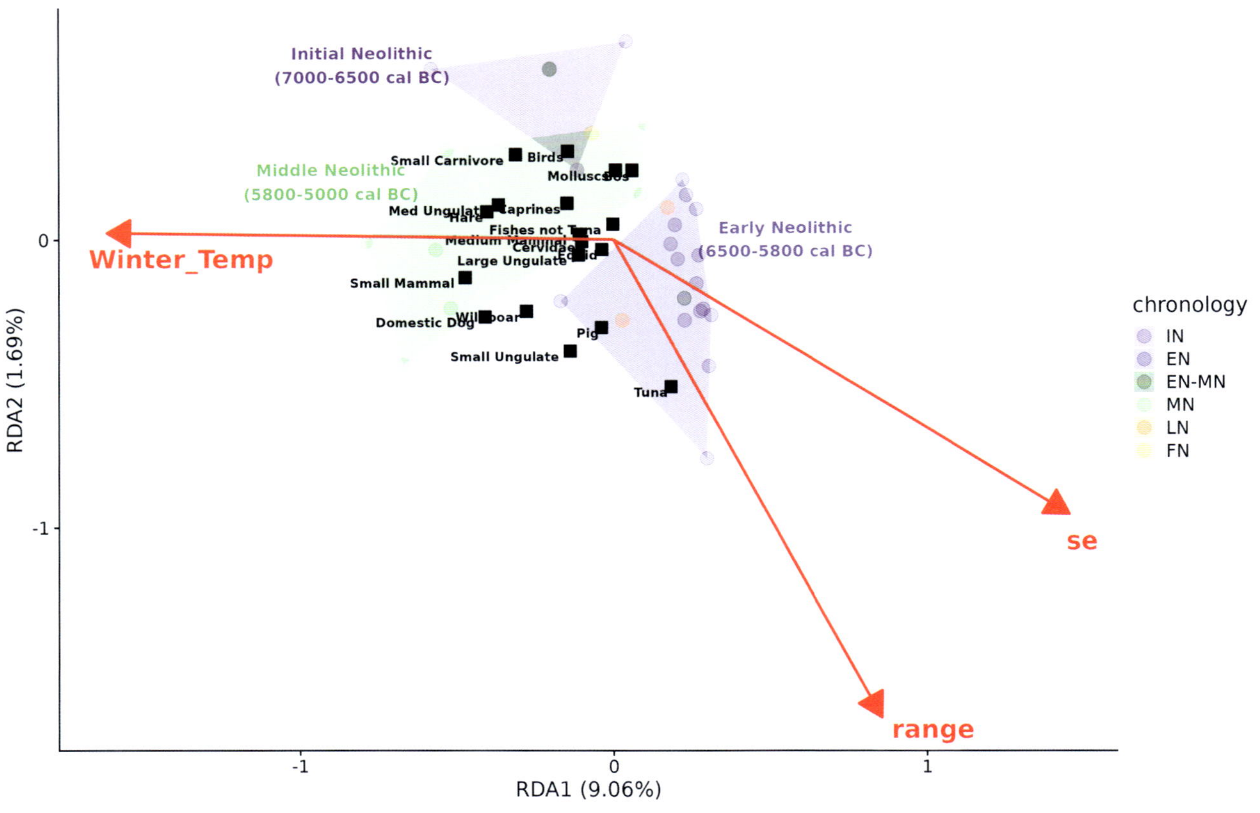

Fig. 165 RDA biplot comparing the faunal species remains to the environmental variables. Black squares represent animal species and points represent excavation units. Here the animal species are annotated. – Chronology legend: IN: Initial Neolithic; EN: Early Neolithic; EN-MN: Early-Middle Neolithic; MN: Middle Neolithic; LN: Late Neolithic; FN: Final Neolithic.

Component	Inertia	Proportion	Rank
Total	1.2222	1.00000	
Constrained	0.1441	0.1179	3
Unconstrained	1.0781	0.8821	18

Tab. 17 Summary of Redundancy Analysis (RDA) fauna results.

sent animal species and points represent excavation units. The proximity of the points/squares to these vectors indicates their correlation. Each arrow is labelled (winter temperature, standard error and range) and the position of the points of the excavation layers and the faunal species indicates their correlation with the environmental parameters.

The results show that certain associations are detected among the compared datasets and the clustering of the values are shown with polygons, each of which represents a chronological period. When evaluating the temporal distribution of the RDA results, there is a clustering of certain Early Neolithic layers in an area which is characterised by colder winter temperatures, higher annual range and higher standard error (**fig. 165**). In the same area, among the faunal remains the pigs (*Sus scrofa*) and tuna (*Thunnus thynnus*) appear to be more important, and their distinct grouping shows that these species within this cluster are more abundant when similar environmental conditions are present. The Early Neolithic cluster suggests that these faunal species during that time are closely associated with specific winter-related environmental conditions of higher variability. The strong association with the winter temperature may indicate that these species were consumed in higher numbers in years where winters were colder. The significant association with the standard error indicates variability and fluctuating winter conditions during the Early Neolithic.

Most of the rest of the faunal species share common characteristics with Middle Neolithic layers (birds, cattle, caprines, Cervidae, medium ungulates, molluscs, small carnivores and small mammals). Their wider spread indicates that they are influenced by a broader range of environmental factors. Also, a number of species (fishes-non tuna, medium mammals, Cervidae, large ungulates and equids) are located at the centre of the RDA plot and exhibit a neutral response to the environmental variables included in the analysis, which means that they do not have strong environmental preferences and can thrive across a range of conditions. In addition, their central positioning may suggest that these species are found consistently across different periods without a marked preference for the environmental conditions of a particular period and that their distribution remains are relatively stable through time.

This clustering of values shows that there seems to be a distinction among the Early and the Middle Neolithic layers and the climatic characteristics that they share. The Early Neolithic layers appear to be colder and more variable, while the Middle Neolithic ones warmer and more stable. Interestingly, the transitional period (Early-Middle Neolithic) also has one point (in dark green in **fig. 165**) in the Early Neolithic and one in the Initial Neolithic cluster. Two Late Neolithic layers are also part of the Early Neolithic cluster (in orange in **fig. 165**). The highly variable Early Neolithic conditions are also in agreement with the large distributions of the standard error and standard deviation values (**fig. 165**). This agrees with the observations by Rohling et al. (2019) during the Rapid Climate Change event, when the climatic conditions were colder and more variable. In addition, during this unstable and colder period of the Early Neolithic, one may suggest that people rely more on the pigs and the seasonally available tuna. This may indicate that during periods of heightened climate variability and unpredictability, the Early Neolithic inhabitants of Franchthi relied more heavily on these resilient resources to cope with the changing environmental conditions. These observations are further discussed in section 5.3.2.3.

Interestingly, Ivanova et al. identify the predominance of pigs in comparison to other species during the Early Neolithic in the sites of Thessaly and Aegean Macedonia, while they consider pig herding to be associated with the Mediterranean environments of the Central and Northern Greece sites (2018, 12–13). This is also identified in the present faunal RDA analysis during the Early Neolithic period. Tuna, a highly seasonally exploited fish, is also highlighted during this period by the analysis, however Ivanova et al. do not take into consideration any marine resources.

Separately to the RDA, a correlation analysis of the individual species and environmental parameters was run in order to detect the presence of more detailed trends (**tabs A.17–A.18**). The correlation analysis used the environmental parameters derived by the isotope analyses (range, winter standard error, summer standard error, winter temperature and summer temperature) and the NISP data of the fauna species in this case. Unlike the RDA, correlation analysis can handle missing values (NAs) in the data. This allowed for the inclusion of a greater number of specimens per excavation unit thus facilitating the calculation of the seasonal ranges and standard errors by excavation unit for these additional units.

Among the 144 combinations of the two datasets, in only 13 of the cases significant statistical correlations ($p < 0.05$) were found (**tab. 18**). After assessing the linear regression graphs that were produced, certain misleading correlations were excluded due to outlying values. Significant correlations in most of the cases (8 out of 10) are related to the winter temperatures (i. e. winter standard error, winter range, winter minimum temperature), standard error (4 out of 10) and range (3 out of 10) environmental variables. Species that stand out are cattle (*Bos primigenius*), Cervidae, domestic dog (*Canis familiaris*), hare (*Lepus europaeus*), small, medium and large ungulates[12] and small mammals. Nonetheless, the over-

12 Munro and Stiner (2020) do not specifically define which species are attributed to the small, medium, large ungulate categories.

	Species	Environmental variable	P-value
1	*Bos* sp.	Winter standard error	0.03
2	Cervidae	Winter range	0.03
3	Domestic dog	Winter minimum temperature	0.02
4	Equid	Winter range	0.05 (outlier)
5	Hare	Winter minimum temperature	0.01
6	Hare	Winter standard error	0.04
7	Hare	Standard error	0.05
8	Large ungulate	Winter range	0.04
9	Medium ungulate	Winter minimum temperature	0.00 (outlier) [0.52]
10	Medium ungulate	Standard error	0.02 [0.03]
11	Medium ungulate	Range	0.03 (outlier) [0.28]
12	Small mammal	Winter minimum temperature	0.03
13	Small ungulate	Winter range	0.01

Tab. 18 Selected fauna RDA results presenting significant statistical correlations ($p < 0.05$). Note that the test for Equid could not be carried out, because NISP was always 0. Values in brackets are derived when excluding outlying excavation units.

all significant correlation results represent a very small percentage (6.9 %, 10 out of 144 correlations), which does not allow for reliable considerations on the matter.

Given the large number of tests that were run, a Bonferroni correction was used to account for multiple comparisons in the faunal and the botanical data. Without the adjustment, there is a chance of falsely identifying certain results as significant. The Bonferroni correction is calculated by dividing the original threshold of 0.05 by the number of tests (Abdi 2007). The new Bonferroni-corrected p-value thresholds found are extremely small (0.000347 for the faunal data and 0.0004167 for the botanical data) which would mean that none of the tests is significant. These extremely small thresholds would mean that only very strong evidence (p-values below these thresholds) would be considered statistically significant, thus posing very strict criteria for significance. This suggests that the present data cannot show the significant correlations either due to insufficient sample size, the strength of the correlations, the specific distribution of the data points, or the categorical grouping of the points.

5.3.2.2. Integration with Botanical Data

As for the archaeobotanical remains, the comparison of the botanical data and the $\delta^{18}O$ datasets indicates that 22.3 % of the variance is explained by environmental variables. The annual temperature range and the standard error were identified also in this case as responsible variables for this variance alongside the summer maximum temperature. The results showed that the total variance (inertia) in the response data was 1.34. The model explained 22.3 % of the total variance through the constrained axes, with an inertia of 0.30. The remaining 77.7 % of the variance was captured by the unconstrained axes, with an inertia of 1.05 (**tab. 19**).

Component	Inertia	Proportion	Rank
Total	1.3446	1.00000	
Constrained	0.2997	0.2229	3
Unconstrained	1.0449	0.7771	10

Tab. 19 Summary of Redundancy Analysis (RDA) botanical results.

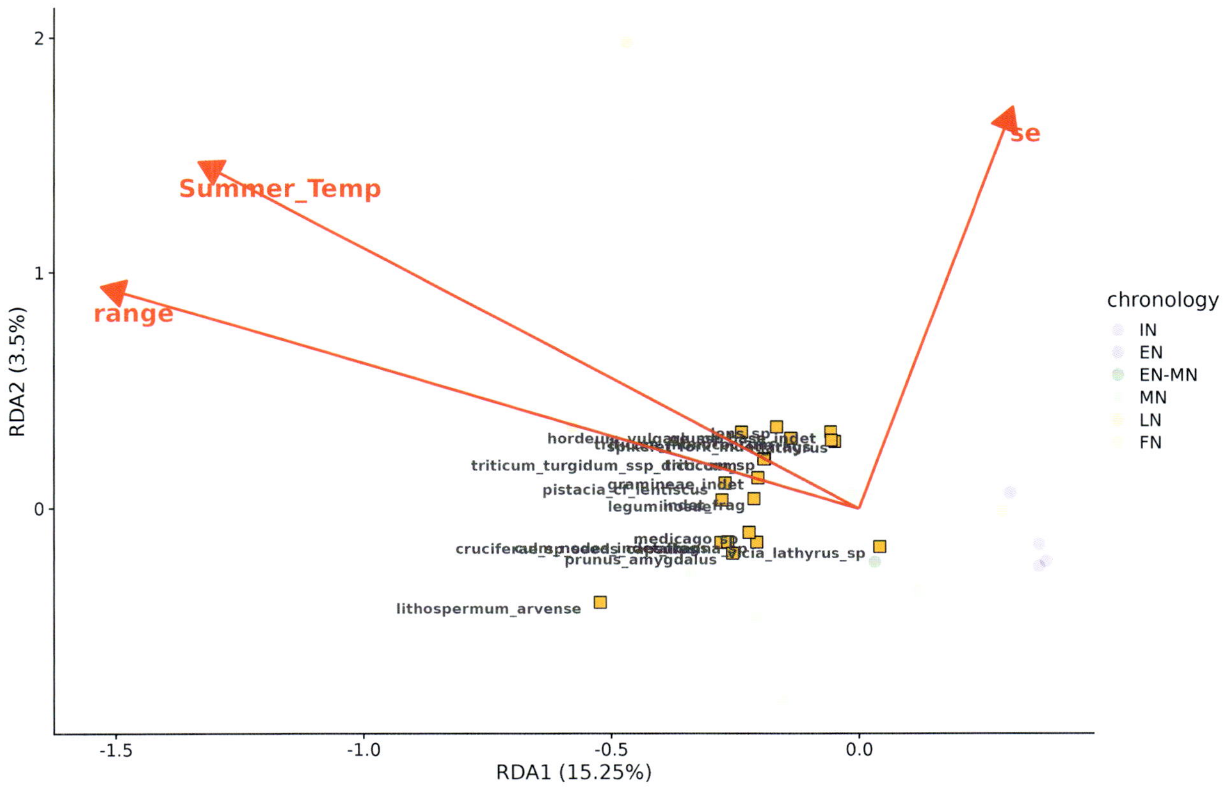

Fig. 166 RDA biplot comparing the botanical species remains to the environmental variables. Orange squares represent botanical species and points represent excavation units. Here the botanical species are annotated. – Chronology legend: IN: Initial Neolithic; EN: Early Neolithic; EN-MN: Early-Middle Neolithic; MN: Middle Neolithic; LN: Late Neolithic; FN: Final Neolithic.

The vector of the summer temperature points towards the top left of the plot, indicating that species aligned with this direction are more influenced by higher summer temperatures (**fig. 166**). The range vector is positioned slightly upwards and to the left, suggesting that species in this direction show greater variability in the climatic conditions, the standard error vector points towards the top right, indicating that species along this vector have higher uncertainty or variability. The positioning of agricultural crops like barley (*Hordeum vulgare*) and emmer wheat (*Triticum turgidum* ssp.) close to the summer temperature vector implies a significant reliance on these crops during warmer periods. The wider distribution of other species, particularly those further from the climatic vectors, could indicate a diverse subsistence strategy that included a variety of plant species.

As for the temporal distribution of the botanical RDA results, no clustering of chronologically similar layers was observed. This may be due to the fact that the botanical data are composed by different Early and Middle Neolithic layers in comparison with the faunal data, as different layers were studied in each case. For this reason, the two subsistence datasets are compared separately. The botanical RDA results show that the majority of the species concentrate in an area in the centre of the plot which is characterised by colder summer temperatures, lower annual range (**fig. 166**). No obvious pattern can be discerned regarding the species distribution or the temporal groupings. In addition, a few Initial Neolithic layers (FAS 139, FAS 145, FAN 156) tend towards colder summer temperatures, lower annual range while the only parameter where they seem centrally distributed is the standard error.

Among the findings of the Ivanova et al. study is that botanical analysis detected an association of the free-threshing wheat (*T. aestivum/durum*) towards warmer climate (2018, 5), as its Latin name (»of the summer«) suggests, while in the present study wheat is found to be associated with higher summer temperatures. In addition, due to a lack of structuring of the crop samples evident in the CCA plots, as most of the crops are located near the point of origin,

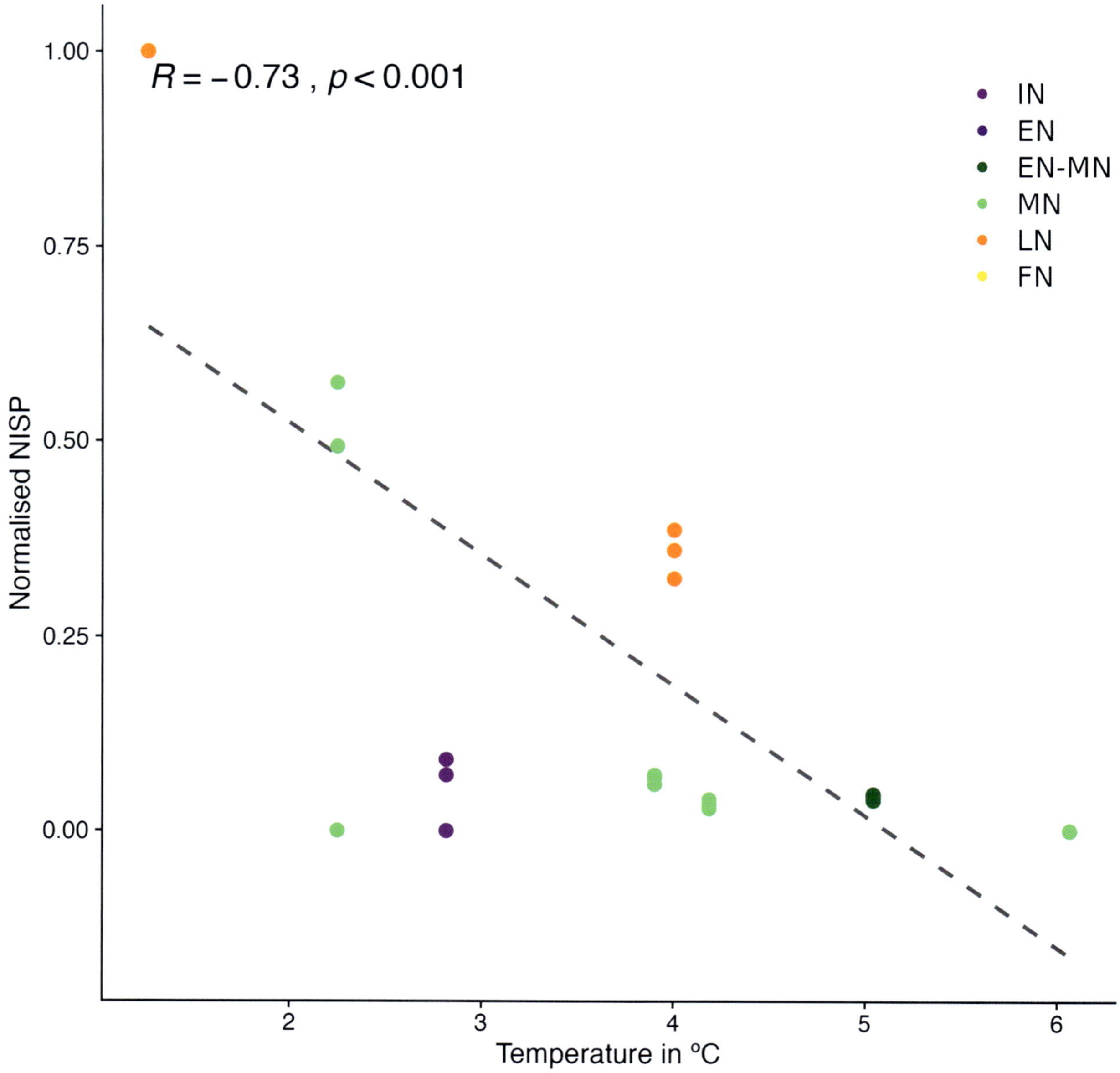

Fig. 167 Scatterplot showing the negative relationship of the *Triticum* sp. NISP abundance with the winter range. – Chronology legend: IN: Initial Neolithic; EN: Early Neolithic; EN-MN: Early-Middle Neolithic; MN: Middle Neolithic; LN: Late Neolithic; FN: Final Neolithic.

the authors contend that the various crop distributions do not correspond to climatic gradients (Ivanova et al. 2018, 15). The RDA analysis of the present study identified summer maximum temperature as a variable that influences the abundance of wheat. In addition, as the scatterplots revealed, wheat (see below) is also positively correlated with summer temperature (**fig. 167**).

Similarly to the faunal data, the NISP values of each species were compared separately to the environmental parameters and were evaluated (**tab. A.22**). Among the 152 combinations of the two datasets, in only 8 of the cases significant statistical correlations ($p < 0.05$) were found (**tab. 20**). False correlations were excluded due to outlying values and significant correlations are present among the standard error, summer range and winter range environmental variables. As was the case with the faunal remains, the overall significant correlation results of the botanical remains represent a very small percentage (3.8 %, 4 out of 104 correlations). However, when applying the Bonferroni correction, none of these remain significant, apart from the combined *Triticum* sp. vs. winter range.

	Species	Environmental variable	P-value
1	Gramineae indet.	Summer standard error	0.03 (outlier) [0.71]
2	Gramineae indet.	Summer range	0.04 (outlier) [0.25]
3	*Medicago* sp.	Summer standard error	0.03 (outlier) [0.53]
4	*Medicago* sp.	Summer range	0.04 (outlier) [0.72]
5	*Pistacia* cf. *lentiscus* sp.	Summer standard error	0.02 (outlier) [0.44]
6	*Pistacia* cf. *lentiscus* sp.	Summer range	0.02 (outlier) [0.41]
7	*Triticum* sp.	Winter range	0.02
8	*Triticum turgidum* ssp. *dicoccum*	Winter range	0.01
	Triticum sp. (combined)	Range	0.03
	Triticum sp. (combined)	Winter range	< 0.001
	Triticum sp. (combined)	Winter standard error	0.01
	Triticum sp. (combined)	Summer maximum temperature	0.02

Tab. 20 Selected botanical RDA results presenting significant statistical correlations ($p < 0.05$). Values in brackets are derived when excluding outlying excavation units.

Among the few species that are strongly correlated are Gramineae (grasses) indeterminate, *Medicago* sp. (medick), *Pistacia* cf. *lentiscus* sp. (pistachio) and *Triticum* sp. (wheat). However, the correlations of Gramineae, *Medicago* sp., and *Pistacia* cf. *lentiscus* sp. were ruled out when the outlying excavation units were excluded and the correlations ceased to appear. Thus, the only significant correlations were those of *Triticum* sp. ($R = -0.79$, $p = 0.02$) and *Triticum turgidum* ssp. *dicoccum* ($R = -0.82$, $p = 0.01$), which represent cultivated edible plants and both appear to be associated with the winter range variable. In addition, when combining all the different *Triticum* sp. categories together, stronger correlations were produced: The winter range ($R = -0.73$, $p < 0.001$) (**fig. 167**) and the winter standard error ($R = -0.56$, $p = 0.005$) present strong negative correlations with the wheat species and strong positive correlations with the summer temperature variable ($R = 0.36$, $p = 0.02$). In addition, when applying the Bonferroni correction (0.0004167 for the botanical data), none of these correlations are significant, except from the combined *Triticum* sp. vs. winter range correlation. The new p-value for this correlation is 0.000048 and thus remains highly significant. The combined *Triticum* sp. category benefits from a larger sample number than other groupings, suggesting that with more excavation units being part of the study, more significant correlations will become apparent.

5.3.2.3. Overall Integration of Datasets

Overall, the RDA revealed that annual temperature range and climate variability (standard error) were partially responsible for the explained variance of both the faunal and the botanical data and the climatic conditions. In the case of the fauna, the winter temperature was also identified as a significant parameter, while in the case of the botanical remains the summer maximum temperature was identified. The RDA of the botanical remains did not reveal any specific temporal patterns.

In the case of the faunal data, there is a clear difference between the Early and the Middle Neolithic period. The Early Neolithic layers appear to be characterised by colder and more variable conditions and the Middle Neolithic layers by warmer and more stable conditions. In addition, the RDA showed that there are associations among the Early Neolithic layers and the abundance of pigs (*Sus scrofa*) and tuna (*Thunnus thynnus*). This may suggest that at times of more variable and unpredictable conditions, the Early Neolithic inhabitants of Franchthi turned towards these resources that may have been more resilient towards climate variability.

Accordingly, the Ivanova et al. study (2018) concluded that animal exploitation shows evidence of climatic-related differentiation, while crop taxonomic proportions were not particularly influenced by climatic parameters. The study illustrates how changes in the plant and animal exploitation followed a geographic and climatic gradient that characterised the northward movement of the Neolithic farmers in south-east Europe. Moreover, the authors of the study detected the importance of animal adaptations for the establishment of agriculture in new environments (Ivanova et al. 2018, 16–18). Lastly, a similar finding with the current study when comparing evidence through time rather than spatially, is that of the importance of the fauna and specifically the pigs and tuna.

The correlations that were run among the botanical species abundance and the environmental parameters indicated that there are strong associations regarding the *Triticum* sp. data. More specifically, there is a negative correlation between the winter range and the winter standard error and a positive correlation with the summer maximum temperature. This means that when the range of winter temperatures and the climate variability (standard error) increase, the abundance of wheat decreases. Thus, wheat tends to be less abundant when winter temperatures vary widely and when they are less predictable. In addition, the correlation tests suggest that the summer temperature increases alongside the abundance of *Triticum* sp. This means that wheat tends to be more abundant when there are higher summer temperatures.

The linear regression plots depicting the relationships between the NISP of the faunal and the botanical species and the environmental values show that there are very few associations among them. The very small percentage of significant statistical correlations indicates that the current data does not provide sufficient evidence to confirm that species composition was influenced by environmental factors. However, this does not conclusively prove that there is no relationship, but may mean that with the current dataset no significant correlation can be established.

The rest of the faunal species share common traits with Middle Neolithic layers, where climatic conditions are more stable and warm. Nonetheless, there is little statistical influence of the climatic conditions on subsistence data. Thus, the observed differences in the climatic conditions do not appear to be substantial enough to affect the use of the specific resources. The distinct conditions of the Early and the Middle Neolithic period and their climatic conditions, may be characteristic of the climatic conditions within and outside the large-scale Rapid Climate Change event (8,600–8,000 cal BP), described above.

Overall, this context-specific analysis allows us to specifically assess the degree to which the local climate influenced and shaped past subsistence practices. The RDA results, as well as the long-term isotope results, indicate different levels of variability across the Early and Middle Neolithic periods, which are not however reflected on the subsistence practices of these periods. Early Neolithic societies may have been influenced by long-term climatic trends that characterise this period, while short-scale evidence shows signs of resilience and adaptation. The more stable climatic conditions of the Middle Neolithic period could potentially be associated with the further northward expansion of the pioneer farmers into the Balkans, as Ivanova et al. suggest (2018, 4). These observations are further discussed below in section 5.4.

5.4. Resilience in the Face of Climatic Change

As presented in section 2.3, three different directions exist in current literature on the climatic impact on the prehistoric communities during the Holocene in the Eastern Mediterranean. Basing their arguments on different case-studies of shorter or longer temporal and geographical scales, they argue on the factors that influenced the advent of the Neolithic and the aspects of the Neolithic dispersal (Weiss et al. 1993; deMenocal 2001; Weninger et al. 2006; Clare et al. 2008; Flohr et al. 2016; Krauß et al. 2017; Ivanova et al. 2018; Roffet-Salque et al. 2018; Kaniewski et al. 2020; Glais et al. 2023). This topic is ascribed in the larger discussion of »collapse« and »resilience« in archaeology that has been increasingly investigated in the recent decades (Heitz et al. 2021, 130–134).

First, some studies suggest that cultural adaptations followed regional and climatic patterns (Krauß et al. 2017; Ivanova et al. 2018), while second, other research links climatic events to societal disruptions, migrations, and warfare (Weiss et al. 1993; deMenocal 2001; Weninger et al. 2006; Clare et al. 2008; Roffet-Salque et al. 2018; Kaniewski et al. 2020). Third, some literature argues that communities adapted effectively to climatic changes without significant disruptions, highlighting resilience and adaptation strategies such as resource diversification and storage (Flohr et al. 2016; Glais et al. 2023).

In regard to this larger discussion, the following observations about how climate affected the prehistoric people of Franchthi can be made, based on the results produced. The long-term and seasonally-resolved climatic record of Franchthi indicates that the climate does gradually change from the end of the Pleistocene into the Early Holocene. The changing climatic patterns are in general agreement with the regional palaeoclimatic proxy records from the Aegean (see section 5.3.1). Interestingly, evidence of the 8,200 event is absent from this long-term archive, nonetheless none of the shell specimens have been absolutely dated to that period.

Moreover, while the climate changes, evidence for changes in the subsistence is not detected (see section 5.3.1). This conclusion was drawn via the integration of the climate with the archaeological data and the Redundancy Analysis that was carried out in order to understand the influences of the climate on past human populations, as proposed by Berger (2021). The RDA demonstrated that the Early and the Middle Neolithic period differ in certain climatic characteristics, with the first being characterised by colder and more variable conditions and the second by warmer and more stable conditions. The RDA also demonstrated that human subsistence was not substantially influenced by climatic factors. The comparison of the faunal data with the climatic data showed that only 11.8 % of the variance of the faunal composition could be explained by the environmental variables.

During the Early Neolithic, which is characterised by colder and more variable conditions, the pigs and tuna appear to become more important. Pigs have been found to be of importance also in other studies (Ivanova et al. 2018), while the fishing of tuna, a highly seasonal activity, could be due to resource diversification strategies due to more unstable climatic conditions. One interpretation could be that during periods of greater climate variability and unpredictability, the Early Neolithic inhabitants of Franchthi relied more on resilient resources such as pigs and the seasonally available tuna. These species likely provided a food resource amid the unstable climatic conditions of the Early Neolithic.

The reasons why the Early Neolithic coastal community of Franchthi appears to have been resilient towards the changing climate can be numerous and similar to those suggested by Flohr et al. (2016, 35). First, the effects of these events may not have been severe enough to significantly impact vegetation and harvest. Nonetheless, it is difficult to provide evidence for such an argument. Given that it is difficult to compare the intensity of specific climatic events identified in the climatic record, only general trends can be commented upon. In the case of the 9,200 and the 8,200 BP events, these were superimposed on generally wet and warm Early Holocene trends (Arz et al. 2003; Robinson et al. 2006, 1530; Finné et al. 2011, 3162; Göktürk et al. 2011, 2425; Flohr et al. 2016,

35), rendering the identification of the influence of short-term events on the archaeological record difficult. In the case of the climatic record of Franchthi (see **fig. 134**), the warm Early Holocene trend is clearly visible and compares well with the regional proxy record, while no abrupt events are identified (see section 5.3.1).

A second possible reason for resilience, is that plants, people, and animals might have already adapted to adverse conditions, with coping mechanisms in place. Storage and resource diversification could well have played an important role in this aspect (Flohr et al. 2016, 35). Regarding storage facilities, evidence on the use of the Early and Middle Neolithic pottery from Franchthi for storage is inconclusive, while Vitelli is reluctant to attribute specific functions to the different wares while placing an emphasis on the symbolic importance of pottery (1989, 26–27; 1993, 213–214). Storage pits are also not reported at Franchthi, although these may have been manufactured by perishable materials or could have existed in areas that were not excavated (Hansen 1991, 160; Perlès 2010, 122). Regarding the zooarchaeological remains, pig (*Sus scrofa*) and caprine (*Capra/Ovis*) significantly increased in numbers during the Middle Neolithic compared to the Early Neolithic, while pigs peak at 18 % of the ungulates during the Middle Neolithic (Munro/Stiner 2020, 5; Appendix A). This shift in the use of pigs at the site, alongside the results of the RDA (see section 5.3.2), pointed towards the significance of this species during the Early Neolithic and could be indicative of subsistence practices that focused on this food resource. Furthermore, no significant changes are observed in the archaeobotanical records during the Early Neolithic and subsequent phases that would suggest potential adaptations to climatic influences.

A third possible reason for resilience could be that these communities were inherently resilient and capable of withstanding extreme climatic events. However, social and political aspects are largely elusive in archaeology, making it difficult to determine to what extent changes in resource exploitation may mirror social transformations or climatic fluctuations. Also, archaeologists often underestimate internal factors – such as demographic, social, cultural, economic and political changes – while focusing more on observable external phenomena (Perlès 2010, 124). Additionally, attempts to link climate to societal changes often result in simplistic causation approaches that struggle to connect social dynamics to broader climatic trends and thus more nuanced approaches are deemed necessary (Degroot et al. 2021, 541).

In the case of Franchthi, any of the three reasons proposed could apply or a multi-causal combination of factors could be at play. The impact of the climatic expressions may not have been severe enough to significantly impact subsistence practices. Regarding mitigation mechanisms, evidence of at least large-scale storage is absent from the site, nonetheless, pigs may have been part of adaptive strategies to mitigate the effects of potentially adverse conditions. In addition, social and cultural factors could have contributed to the resilience of the community and may have reduced the risk of famine, adversity or collapse under environmental stress conditions. Finally, different environmental changes may have occurred but also the societal response to these and the society's vulnerability to the environmental influences may have differed at different times and phases (Heitz et al. 2021, 188).

Chapter 6: Conclusions

In this work, I investigated aspects of human-environment interactions at the coastal prehistoric settlement of Franchthi during the Neolithic period. The seasonal and climatic record of the site was reconstructed in order to identify potential climatic influences on its inhabitants. The contextual data on subsistence were examined alongside the palaeoclimatic shell-derived data to understand potential influences and consider the extent to which the climate posed an agricultural risk. The shell-derived data constitute local climatic information that refer to time-scales that would have been perceived by the prehistoric populations.

To this end, a modern pilot study was initially conducted on reference specimens to validate the methodology developed for the research. A successful reference study was carried out based on 131 modern shell specimens from locations in Greece, Italy and Libya. The results showed good agreement between estimated and actual time of collection (in 79 % of the cases) as well as sea surface temperature (± 1.5 °C), whereas the estimate for absolute temperature was based on the analysis of 20 shell specimens. Site-specific offsets in $\delta^{18}O$ values occurred in multiple cases but the resulting error margin at Franchthi was small and consistent (+0.85 ‰). The oxygen isotope results further demonstrated that the elemental data and the isotope data are highly correlated between them (1st objective). The correlation calculated between the oxygen isotope values and the Mg/Ca elemental results from the 17 modern samples analysed sequentially produced consistently good R^2 results (average $R^2 = 0.89$).

The combined analysis of modern shells with the use of LIBS and stable isotopes showed that the elemental mapping with LIBS can provide interpretable, fast and efficient high-resolution results on seasonal variability, while validating the efficacy of the employed methodology. With the use of LIBS the season of death of the molluscs is identified and the annual minima and maxima values were targeted for isotope analysis from the targeted spots. Via this combined approach, the sequential vs. edge sampling dilemma was overcome, while seasonal information became available. This way, an cost-effective analysis of large sample assemblages at a high-resolution became possible.

Having successfully shown the efficacy of the method, I proceeded to the analysis of the archaeological samples. The 180 shell samples from Franchthi that were analysed with the use of LIBS and oxygen isotope analysis produced multivariate seasonal and palaeoclimatic results, which revealed seasonal patterns of exploitation as well as sea surface temperature changes through time.

Seasonal collection patterns showed that there are certain changes in the exploitation patterns as well as the seasonality of occupation at the site. Most of the seasons of the year are represented during the Final Palaeolithic and the Mesolithic, while during the Early Neolithic there is a winter-time collection preference at the Paralia and during the Middle Neolithic a summer-time collection preference at the cave. Also, LIBS mapping enabled various considerations regarding size and the molluscs' biological age as well as the identification of coeval specimens. Among these considerations was the estimation of the biological age of the specimens which showed that the largest age group (44 %) was that of 1 to 1.5 years old and that there is a pattern of collecting younger specimens during winter time. Elemental mapping also enabled the identification of coeval specimens and assemblages as distinct patterns became visible which can be indicative of similar environmental conditions.

The diachronic changes of annual and seasonal temperature changes and climate variability were successfully reconstructed via the isotope analyses. 279 $\delta^{18}O$ seasonal values were extracted from 147 shells across 60 excavation units. This large number of shell specimens per excavation unit allowed for the reconstruction not only of the seasonal temperatures, but also of the annual isotope ranges and the seasonal annual ranges, as well as the calculation of indicators of climate variability, such as the standard error, per excavation unit or season. The long-term climate data results from Franchthi show that after the colder period of the Younger Dryas, there is a gradual shift towards warmer conditions during the Early Holocene. These climatic changes closely match the variability observed during the Late Pleistocene and the Early Holocene on the regional scale, as evidenced by the palaeoclimatic data from the Eastern Mediterranean, and particularly the marine sediment cores LC21 and SL21 from the Aegean Sea (e. g. Hayes et al. 1999; de Rijk et al. 1999; Mercone et al. 2000; Marino et al. 2007; 2009) (2nd objective).

The climatic parameters produced enabled the direct comparison of the oxygen isotope data with the faunal and botanical data of the site. This comparison was accomplished via a Redundancy Analysis that was carried out on the excavation unit level. The layer-by-layer comparison of the faunal and the botanical data showed that only a small percentage of the variance (11.8 %) in the subsistence remains was explained by the environmental variables and it appears that there is little statistical influence of climatic conditions on people's subsistence strategies. However, the multivariate analysis of the climatic parameters did point towards distinct characteristics among the layers of the different periods and revealed that the Early Neolithic period was climatically more variable, its seasonal temperatures more extreme, and it was characterised by colder winters while the Middle Neolithic was warmer and more stable. The RDA gave the opportunity to overcome the chronological constraints by enabling the strict contextual comparison at a stratigraphic level. This way, climatic data are layer-specific and are not assessed on an abstract site-level allowing for chronologically more reliable results. Specific excavation layers alongside their subsistence data, can be associated with specific environmental parameters. Thus, a more direct integration of climatic and archaeological data becomes possible (3rd objective). This enables a data and evidence driven bottom-up approach that combines archaeological information with palaeoclimatic proxies as proposed by Heitz et al. (2021, 154).

The combined archaeometric approach of this thesis together with the subsistence and palaeoenvironmental previously published data offered a glimpse into the short and long-term fluctuations of the climate and their potential impact on the prehistoric inhabitants of the site. Having successfully overcome the limitations of the disconnect between the archaeological and the climatic data by linking the different lines of evidence, it appears that even though there is evidence of climate variability locally, this did not affect human subsistence nor did it impose a significant agricultural risk. The coastal community of Franchthi appears to have been resilient towards the changing climate with subsistence practices being largely unaffected by these climatic factors. Only 11.8 % of the faunal composition variance was explained by the environmental variables as demonstrated by the RDA. Possible reasons for resilience include mild climate impacts, pre-existing adaptations, and inherent community resilience that may have contributed to the community's ability to withstand climatic changes.

The application of this combined method to other sites in the Eastern and the Central Mediterranean in the future should lead to more concrete seasonal and palaeoclimatic considerations of larger geographical scales.

6.1. Future Research

The potential of the methodology applied in this study is multi-faceted and can be further applied on archaeological materials in numerous ways. First, future work should aim towards the investigation of the palaeoclimate on a larger geographical scale. More archaeological shells should be analysed, from coastal sites that document the Neolithisation process and depending on site preservation, more detailed analyses of the faunal and botanical remains could be undertaken, should the preservation of these find categories overlap with the preservation of limpets. Such research is already ongoing and will be carried out on specimens from Cyclops's Cave, Haua Fteah, Grotta dell'Uzzo, and Grotta d'Oriente. This work can contribute to the further investigation of the seasonally-resolved climatic record in the Central and the Eastern Mediterranean.

Further future research can employ LIBS combined with isotope samples on limpet specimens of other chronological periods. Limpets as a palaeoclimatic archive, as explained in Chapter 4, constitute a potent tool that can be used to investigate the climatic impact on past coastal communities. In addition, the present approach can also be used on other shell species or other accretionary hard tissues, such as fish otoliths, in order to extract seasonal and environmental information (e. g. Coulson et al. 2014; Vignon 2015; Lea et al. 2000; Warburton et al. 2017; Watanabe et al. 2019; Perhada et al. 2021). The application of this work can further extend to ecological, marine science or historical studies (e. g. Nigra et al. 2015; Cobo García et al. 2017; Hausmann et al. 2023).

As it has been demonstrated in this thesis, the applied methodology on limpet shells can contribute to the investigation of past seasonal exploitation patterns as well as to the reconstruction of the climate of the past and its impact on the prehistoric communities. Thus, future research can concentrate on any of these topics or expand on larger temporal and spatial scales.

The resilience of the prehistoric inhabitants of Franchthi in the face of climate change highlights the adaptive capacity of human societies. Even though the scale and pace of the present-day unprecedented warming conditions are vastly different from those of the Early Holocene, the comparison of historical resilience to contemporary challenges posed by climate change can offer certain parallels and insights that oppose accounts of crisis and collapse. Societies today can also adapt and innovate in response to environmental pressure and the example of prehistoric Franchthi may offer a hopeful example of resilience which may contribute to present-day efforts towards adaptation.

Bibliography

Abdi 2007: H. Abdi, The Bonferroni and Šidák Corrections for Multiple Comparisons. In: N. J. Salkind (ed.), Encyclopedia of Measurement and Statistics (Thousand Oaks CA 2007) 103–107.

Abrantes et al. 2012: F. Abrantes / A. Voelker / F. J. Sierro / F. Naughton / T. Rodrigues / I. Cacho / D. Ariztegui / D. Brayshaw / M.-A. Sicre / L. Batista, Paleoclimate Variability in the Mediterranean Region. In: P. Lionello (ed.), The Climate of the Mediterranean Region: From the Past to the Future (Amsterdam 2012) 1–86. DOI: 10.1016/B978-0-12-416042-2.00001-X.

Adam 1989: E. Adam, A Technological and Typological Analysis of Upper Palaeolithic Stone Industries of Epirus, Northwestern Greece. BAR Internat. Ser. 512 (Oxford 1989).

1999a: E. Adam, The Upper Palaeolithic stone industries of Epirus in their regional setting. In: Bailey et al. 1999, 137–147. www.jstor.org/stable/40960222.

1999b: E. Adam, Preliminary presentation of the Upper Palaeolithic and Mesolithic stone industries of Theopetra Cave, Western Thessaly. In: Bailey et al. 1999, 266–270. www.jstor.org/stable/40960235.

2007: E. Adam, Looking out for the Gravettian in Greece. Paleo 19, 2007, 145–158. DOI: 10.4000/paleo.411.

2016: E. Adam, Methodological issues underlying the analysis of Upper Palaeolithic stone assemblages in Greece. In: P. Elefanti / N. Andreasen / N. Kardulias / G. Marshall (eds), Lithics Past and Present. Perspectives on Chipped Stone Studies in Greece. Stud. Mediterranean Arch. 144 (Uppsala 2016) 33–46.

Allaby 2013: M. Allaby, A Dictionary of Geology and Earth Sciences (Oxford [4]2013). DOI: 10.1093/acref/9780199653065.001.0001.

Alley et al. 1997: R. B. Alley / P. A. Mayewski / T. Sowers / M. Stuiver / K. C. Taylor / P. U. Clark, Holocene climatic instability: A prominent, widespread event 8200 yr ago. Geology 25(6), 1997, 483–486. DOI: 10.1130/0091-7613(1997)025%3C0483:HCIAPW%3E2.3.CO;2.

Ammerman/Cavalli-Sforza 1984: A. J. Ammerman / L. L. Cavalli-Sforza, The Neolithic Transition and the Genetics of Populations in Europe (Princeton 1984). www.jstor.org/stable/j.ctt7zvqz7.

van Andel 1987a: T. H. van Andel, The modern landscape. In: T. H. van Andel / S. B. Sutton (eds), Landscape and People of the Franchthi Region. Excav. Franchthi Cave, Greece 2 (Bloomington IN 1987) 3–11.

1987b: T. H. van Andel, The adjacent sea. In: T. H. van Andel / S. B. Sutton (eds), Landscape and People of the Franchthi Region. Excav. Franchthi Cave, Greece 2 (Bloomington IN 1987) 31–54.

van Andel/Lianos 1983: T. H. van Andel / N. Lianos, Prehistoric and historic shorelines of the southern Argolid Peninsula: A subbottom profiler study. Internat. Journal Nautical Arch. and Underwater Exploration 12(4), 1983, 303–324. DOI: 10.1111/J.1095-9270.1983.TB01165.X.

1984: T. H. van Andel / N. Lianos, High-resolution seismic reflection profiles for the reconstruction of postglacial transgressive shorelines: An example from Greece. Quaternary Research 22(1), 1984, 31–45. DOI: 10.1016/0033-5894(84)90004-8.

van Andel et al. 1980: T. H. van Andel / T. W. Jacobsen / J. B. Jolly / N. Lianos, Late Quaternary History of the Coastal Zone near Franchthi Cave, Southern Argolid, Greece. Journal Field Arch. 7(4), 1980, 389–402. DOI: 10.2307/529518.

Andersen et al. 2004: K. K. Andersen / N. Azuma / J.-M. Barnola / M. Bigler / P. Biscaye / N. Caillon / J. Chappellaz / H. B. Clausen / D. Dahl-Jensen / H. Fischer / J. Flückiger / D. Fritzsche / Y. Fujii / K. Goto-Azuma / K. Grønvold / N. S. Gundestrup / M. Hansson / C. Huber / C. S. Hvidberg / S. J. Johnsen / U. Jonsell / J. Jouzel / S. Kipfstuhl / A. Landais / M. Leuenberger / R. Lorrain / V. Masson-Delmotte / H. Miller / H. Motoyama / H. Narita / T. Popp / S. O. Rasmussen / D. Raynaud / R. Rothlisberger / U. Ruth / D. Samyn / J. Schwander / H. Shoji / M.-L. Siggaard-Andersen / J. P. Steffensen / T. Stocker / A. E. Sveinbjörnsdóttir / A. Svensson / M. Takata / J.-L. Tison / T. Thorsteinsson / O. Watanabe / F. Wilhelms / J. W. C. White / North Greenland Ice Core Project members, High-resolution record of Northern Hemisphere climate extending into the last interglacial period. Nature 431(7005), 2004, 147–151. DOI: 10.1038/nature02805.

Andrus 2011: C. F. T. Andrus, Shell midden sclerochronology. Quaternary Scien. Reviews 30(19–20), 2011, 2892–2905. DOI: 10.1016/j.quascirev.2011.06.026.

Anglos 2001: D. Anglos, Laser-induced breakdown spectroscopy in art and archaeology. Applied Spectroscopy 55(6), 2001, 186A–205A. DOI: 10.1366/0003702011952398.

Anglos/Detalle 2014: D. Anglos / V. Detalle, Cultural heritage applications of LIBS. In: S. Musazzi / U. Perini (eds), Laser-Induced Breakdown Spectroscopy. Theory and Applications. Springer Ser. Optical Scien. 182 (Heidelberg et al. 2014) 531–554. DOI: 10.1007/978-3-642-45085-3-20?urlappend=%3Futm_source%3Dresearchgate.net%26utm_medium%3Darticle.

Anglos/Miller 2006: D. Anglos / J. C. Miller, Cultural heritage applications of LIBS. In: A. Miziolek / V. Palleschi / I. Schechter (eds), Laser-Induced Breakdown Spectroscopy. Fundamentals and Applications (Cambridge 2006) 332–367. DOI: 10.1017/CBO9780511541261.010.

Arz et al. 2003: H. W. Arz / F. Lamy / J. Pätzold / P. J. Müller / M. Prins, Mediterranean moisture source for an early-Holocene humid period in the northern Red Sea. Science 300(5616), 2003, 118–121. DOI: 10.1007/978-3-642-45085-3.

Bada 1972: J. L. Bada, The dating of fossil bones using the racemization of isoleucine. Earth and Planetary Scien. Letters 15(3), 1972, 223–231. DOI: 10.1016/0012-821X(72)90167-7.

Bailey 1981: G. N. Bailey, Concepts, time-scales and explanations in economic prehistory. In: A. Sheridan / G. Bailey (eds), Economic Archaeology. Towards an Integration of Ecological and Social Approaches. BAR Internat. Ser. 96 (Oxford 1981) 97–117.

1997: G. N. Bailey (ed.), Klithi: Palaeolithic Settlement and Quaternary Landscapes in Northwest Greece. 2: Klithi in its Local and Regional Setting (Cambridge 1997).

2007: G. N. Bailey, Time perspectives, palimpsests and the archaeology of time. Journal Anthr. Arch. 26(2), 2007, 198–223. DOI: 10.1016/j.jaa.2006.08.002.

2008: G. N. Bailey, Time Perspectivism: Origins and Consequences. In: S. Holdaway / L. Wandsnider (eds), Time in Archaeology: Time Perspectivism Revisited (Salt Lake City 2008) 13–30.

Bailey et al. 1983a: G. N. Bailey / M. R. Deith / N. J. Shackleton, Oxygen isotope analysis and seasonality determinations: Limits and potential of a new technique. Am. Ant. 48(2), 1983, 390–398. DOI: 10.1017/S0002731600065033.

1983b: G. N. Bailey / P. L. Carter / C. S. Gamble / H. P. Higgs, Asprochaliko and Kastritsa: Further investigations of Palaeolithic settlement and economy in Epirus (northwest Greece). Proc. Prehist. Soc. 49(1), 1983, 15–42. DOI: 10.1017/S0079497X00007945.

1999: G. N. Bailey / E. Adam / E. Panagopoulou / C. Perlès / K. Zachos (eds), The Palaeolithic Archaeology of Greece and Adjacent Areas. Proceedings of the ICOPAG Conference, Ioannina. Brit. School Athens Stud. 3 (London 1999). www.jstor.org/stable/i40043001.

Bar-Matthews et al. 2000: M. Bar-Matthews / A. Ayalon / A. Kaufman, Timing and hydrological conditions of Sapropel events in the Eastern Mediterranean, as evident from speleothems, Soreq cave, Israel. Chemical Geol. 169(1–2), 2000, 145–156. DOI: 10.1016/S0009-2541(99)00232-6.

2003: M. Bar-Matthews / A. Ayalon / M. Gilmour / A. Matthews / C. J. Hawkesworth, Sea–land oxygen isotopic relationships from planktonic foraminifera and speleothems in the Eastern Mediterranean region and their implication for paleorainfall during interglacial intervals. Geochimica et Cosmochimica Acta 67(17), 2003, 3181–3199. DOI: 10.1016/S0016-7037(02)01031-1.

Barber et al. 1999: D. C. Barber / A. Dyke / C. Hillaire-Marcel / A. E. Jennings / J. T. Andrews / M. W. Kerwin / G. Bilodeau / R. McNeely / J. Southon / M. D. Morehead / J. M. Gagnon, Forcing of the cold event of 8,200 years ago by catastrophic drainage of Laurentide lakes. Nature 400(6742), 1999, 344–348. DOI: 10.1038/22504.

Barker 1958: H. Barker, Radio Carbon Dating: Its Scope and Limitations. Antiquity 32(128), 1958, 253–263. DOI: 10.1017/S0003598X00027289.

Barker et al. 2010: G. Barker / A. Antoniadou / S. J. Armitage / I. Brooks / I. Candy / K. Connell / K. Douka / L. Farr / E. Hill / C. O. Hunt / R. Inglis / S. Jones / C. Lane / G. Lucarini / J. Meneely / J. Morales / G. Mutri / A. L. Prendergast / R. Rabett / H. Reade / T. Reynolds / N. Russell / D. Simpson / B. Smith / C. Stimpson / M. Twati / K. White, The Cyrenaican Prehistory Project 2010: The fourth season of investigations of the Haua Fteah cave and its landscape, and further results from the 2007–2009 fieldwork. Libyan Stud. 41, 2010, 63–88. DOI: 10.1017/S0263718900000273.

2012: G. Barker / P. Bennett / L. Farr / E. Hill / C. O. Hunt / G. Lucarini / J. Morales / G. Mutri / A. L. Prendergast / A. Pryor / R. Rabett / T. Reynolds / P. Spry-Marques, The Cyrenaican Prehistory Project 2012: The fifth season of investigations of the Haua Fteah cave. Libyan Stud. 43, 2012, 115–136. DOI: 10.1017/S026371890000008X.

Batenburg et al. 2011: S. J. Batenburg / G. J. Reichart / T. Jilbert / M. Janse / F. P. Wesselingh / W. Renema, Interannual climate variability in the Miocene: High resolution trace element and stable isotope ratios in giant clams. Palaeogeogr., Palaeoclimatol., Palaeoecol. 306(1–2), 2011, 75–81. DOI: 10.1016/j.palaeo.2011.03.031.

Beck/Koutsoumba 2014: J. Beck / D. Koutsoumba, Baie de Kiladha 2013. Ant. Kunst 57, 2014, 162–165. www.jstor.org/stable/45137293.

2015: J. Beck / D. Koutsoumba, Baie de Kiladha 2014. Expédition Terra Submersa. Ant. Kunst 58, 2015, 187–190. www.jstor.org/stable/26663623.

2016: J. Beck / D. Koutsoumba, Baie de Kiladha 2015. Ant. Kunst 59, 2016, 153–156. www.jstor.org/stable/26663645.

2017: J. Beck / D. Koutsoumba, Baie de Kiladha 2016. Ant. Kunst 60, 2017, 164–167. www.jstor.org/stable/26663667.

Beck/Sotiriou 2020: J. Beck / A. Sotiriou, Baie de Kiladha 2019. Ant. Kunst 63, 2020, 120–124. www.jstor.org/stable/26974639.

Beck et al. 2021: J. Beck / D. Koutsoumba / D. Sakellariou / M. Surdez / F. Anselmetti / N. Papadopoulos / I. Morfis / I. Panagiotopoulos / G. Rousakis / D. Oikonomou / K. Simyrdanis / G. Cantoro / A. Argyriou / P. Birchler Emery / A. Krijnen / K. Tsampouraki-Kraounaki, Searching for Neolithic sites in the Bay of Kiladha, Greece. Quaternary Internat. 584, 2021, 129–140. DOI: 10.1016/j.quaint.2020.12.025.

Berger 2021: J.-F. Berger, Geoarchaeological and Paleo-Hydrological Overview of the Central-Western Mediterranean Early Neolithic Human–Environment Interactions. Open Arch. 7(1), 2021, 1371–1397. DOI: 10.1515/opar-2020-0199.

Berger/Guilaine 2009: J.-F. Berger / J. Guilaine, The 8200 calBP abrupt environmental change and the Neolithic transition: A Mediterranean perspective. Quaternary Internat. 200(1–2), 2009, 31–49. DOI: 10.1016/j.quaint.2008.05.013.

Bernabeu Aubán et al. 2015: J. Bernabeu Aubán / C . M. Barton / S. Pardo Gordó / S. M. Bergin, Modeling initial Neolithic dispersal. The first agricultural groups in West Mediterranean. Ecol. Modelling 307, 2015, 22–31. DOI: 10.1016/j.ecolmodel.2015.03.015.

Berry/Barker 1968: W. B. N. Berry / R. M. Barker, Fossil Bivalve Shells indicate Longer Month and Year in Cretaceous than Present. Nature 217(5132), 1968, 938–939. DOI: 10.1038/217938b0.

Binford 1968: L. R. Binford, Post Pleistocene adaptations. In: S. Binford / L. R. Binford (eds), New Perspectives in Archaeology (Chicago 1968) 313–341.

Bintliff 1977: J. L. Bintliff, Natural environment and human settlement in prehistoric Greece based on original fieldwork (Parts I and II). BAR Suppl. Ser. 28 (Oxford 1977).

Bird-David 1990: N. Bird-David, The giving environment: Another perspective on the economic system of gatherer-hunters. Current Anthr. 31(2), 1990, 189–196. www.jstor.org/stable/2743592.

Boaventura et al. 2003: D. Boaventura / L. C. Da Fonseca / S. J. Hawkins, Size matters: Competition within populations of the limpet *Patella depressa*. Journal Animal Ecol. 72(3), 2003, 435–446. DOI: 10.1046/j.1365-2656.2003.00713.x.

Borgognini Tarli et al. 1993: S. M. Borgognini Tarli / A. Canci / M. Piperno / E. Repetto, Dati archeologici e antropologici sulle sepolture mesolitiche della Grotta dell'Uzzo (Trapani). Bull. Paletnol. Italiana 84, 1993, 85–179.

Bosman/Hockey 1988: A. L. Bosman / P. A. R. Hockey, Life-history patterns of populations of the limpet *Patella granularis*: The dominant roles of food supply and mortality rate. Oecologia 75(3), 1988, 412–419. DOI: 10.1007/BF00376945.

Bottema 1979: S. Bottema, Pollen analytical investigations in Thessaly (Greece). Palaeohistoria 21, 1979, 19–40. https://ugp.rug.nl/Palaeohistoria/article/view/24996.

1982: S. Bottema, Palynological investigations in Greece with special reference to pollen as an indicator of human activity. Palaeohistoria 24, 1982, 257–288. https://ugp.rug.nl/Palaeohistoria/article/view/24837.

1990: S. Bottema, Holocene environment of the Southern Argolid: A pollen core from Kiladha Bay. In: T. J. Wilkinson / S. T. Duhon (eds), Franchthi Paralia: The Sediments, Stratigraphy, and Offshore Investigations. Excav. Franchthi Cave, Greece 6 (Bloomington IN 1990) 117–138.

1995: S. Bottema, The Younger Dryas in the eastern Mediterranean. Quaternary Scien. Reviews 14(9), 1995, 883–891. DOI: 10.1016/0277-3791(95)00069-0.

2003: S. Bottema, The vegetation history of the Greek Mesolithic. In: Galanidou/Perlès 2003, 33–49. www.jstor.org/stable/40960304.

Brami 2023: M. N. Brami, Theory versus data: Dealing with the interpretive dilemma in the biomolecular era. In: C. Heitz / M. Wunderlich / M. Hinz / M. Furholt (eds), Rethinking Neolithic Societies. New Perspectives on Social Relations, Political Organization and Cohabitation. Open Ser. Prehist. Arch. 5 (Leiden 2023) 283–290. DOI: 10.59641/ndr028gp.

Branscombe et al. 2020: T. L. Branscombe / M. D. Bosch / P. T. Miracle, Seasonal Shellfishing across the East Adriatic Mesolithic–Neolithic Transition: Oxygen Isotope Analysis of *Phorcus turbinatus* from Vela Spila (Croatia). Environmental Arch. 25(1), 2020, 1–14. DOI: 10.1080/14614103.2020.1721695.

Brooks et al. 1990: A. S. Brooks / P. E. Hare / J. E. Kokis / G. H. Miller / R. D. Ernst / F. Wendorf, Dating Pleistocene archaeological sites by protein diagenesis in ostrich eggshell. Science 248(4951), 1990, 60–64. DOI: 10.1126/science.248.4951.60.

Brothwell/Brothwell 1969: D. Brothwell / P. Brothwell, Food in Antiquity: A Survey of the Diet of Early Peoples (London 1969).

Burchell et al. 2013: M. Burchell / A. Cannon / N. Hallmann / H. P. Schwarcz / B. R. Schöne, Inter-site variability in the season of shellfish collection on the central coast of British Columbia. Journal Arch. Scien. 40(1), 2013, 626–636. DOI: 10.1016/j.jas.2012.07.002.

Cáceres et al. 2017: J. O. Cáceres / F. Pelascini / V. Motto-Ros / S. Moncayo / F. Trichard / G. Panczer / A. Marín-Roldán / J. A. Cruz / I. Coronado / J. Martín-Chivelet, Megapixel multi-elemental imaging by Laser-Induced Breakdown Spectroscopy, a technology with considerable potential for paleoclimate studies. Scien. Reports 7(5080), 2017, 1–11. DOI: 10.1038/s41598-017-05437-3.

Carapiperis 1952: L. N. Carapiperis, On the surface temperature of Greek waters. Geofisica Pura e Applicata 23(1), 1952, 153–161. DOI: 10.1007/BF01992467.

1957: L. N. Carapiperis, Halcyon days. Weather 12(10), 1957, 311–313. DOI: 10.1002/j.1477-8696.1957.tb00391.x.

Casford et al. 2002: J. S. L. Casford / E. J. Rohling / R. H. Abu-Zied / S. Cooke / C. Fontanier / M. Leng / V. Lykousis, Circulation changes and nutrient concentrations in the late Quaternary Aegean Sea: A nonsteady state concept for sapropel formation. Paleoceanogr. and Paleoclimatol. 17(2), 2002, 1024. DOI: 10.1029/2000pa000601.

2003: J. S. L. Casford / E. J. Rohling / R. H. Abu-Zied / C. Fontanier / F. J. Jorissen / M. J. Leng / G. Schmiedl / J. Thomson, A dynamic concept for eastern Mediterranean circulation and oxygenation during sapropel formation. Palaeogeogr., Palaeoclimatol., Palaeoecol. 190(1–2), 2003, 103–119. DOI: 10.1016/S0031-0182(02)00601-6.

2007: J. S. L. Casford / R. H. Abu-Zied / E. J. Rohling / S. Cooke / C. Fontanier / M. Leng / A. Millard / J. Thomson, A stratigraphically controlled multi-proxy chronostratigraphy for the eastern Mediterranean. Paleoceanogr. and Paleoclimatol. 22(4), 2007, PA4215. DOI: 10.1029/2006PA001401.

Cassoli et al. 1987: P. F. Cassoli / M. Piperno / A. Tagliacozzo, Dati paleoeconomici relativi al processo di neolitizzazione alla Grotta dell'Uzzo (Trapani). In: Atti della XXVI Riunione Scientifica: Il Neolitico in Italia 2 (Firenze 1987) 809–817.

Cauvin 2000: J. Cauvin, The Birth of the Gods and the Origins of Agriculture (Cambridge 2000).

Childe 1936: V. G. Childe, Man Makes Himself (London 1936).

1958: V. G. Childe, The Dawn of European Civilisation (New York [6]1958).

Chourmouziadis 1979: G. C. Chourmouziadis, Το Νεολιθικό Διμήνι [Neolithic Dimini] (Volos 1979).

Çilingiroğlu 2005: Ç. Çilingiroğlu, The concept of »Neolithic package«: Considering its meaning and applicability. Doc. Praehist. 32, 2005, 1–13. DOI: 10.4312/dp.32.1.

Claassen 1998: C. Claassen, Shells (Cambridge 1998).

Clare et al. 2008: L. Clare / E. J. Rohling / B. Weninger / J. Hilpert, Warfare in Late Neolithic/Early Chalcolithic Pisidia, southwestern Turkey. Climate-induced social unrest in the late 7th millennium calBC. Doc. Praehist. 35, 2008, 65–92. DOI: 10.4312/dp.35.6.

Clark 1968: G. R. Clark II, Mollusk Shell: Daily Growth Lines. Science 161(3843), 1968, 800–802. DOI: 10.1126/science.161.3843.800.

Clark 1952: J. G. D. Clark, Prehistoric Europe: The Economic Basis (New York 1952).

Cobo García et al. 2017: A. Cobo García / A. García Escárzaga / F. I. Gutiérrez Zugasti / J. Setién Marquínez / M. R. González Morales / J. M. López Higuera, Automated Measurement of Magnesium/Calcium Ratios in Gastropod Shells Using Laser-Induced Breakdown Spectroscopy for Paleoclimatic Applications. Applied Spectroscopy 71(4), 2017, 591–599. DOI: 10.1177/0003702816687570.

Colonese et al. 2009: A. C. Colonese / S. Troelstra / P. Ziveri / F. Martini / D. Lo Vetro / S. Tommasini, Mesolithic shellfish exploitation in SW Italy: Seasonal evidence from the oxygen isotopic composition of *Osilinus turbinatus* shells. Journal Arch. Scien. 36(9), 2009, 1935–1944. DOI: 10.1016/j.jas.2009.04.021.

2011: A. C. Colonese / M. A. Mannino / D. E. Bar-Yosef Mayer / D. Fa / J. C. Finlayson / D. Lubell / M. C. Stiner, Marine mollusc exploitation in Mediterranean prehistory: An overview. Quaternary Internat. 239(1–2), 2011, 86–103. DOI: 10.1016/j.quaint.2010.09.001.

2012: A. C. Colonese / E. Verdun-Castello / M. Alvarez / I. B. Godino / D. Zurro / R. Salvatelli, Oxygen isotopic composition of limpet shells from the Beagle Channel: Implications for seasonal studies in shell middens of Tierra del Fuego. Journal Arch. Scien. 39(6), 2012, 1738–1748. DOI: 10.1016/j.jas.2012.01.012.

2013: A. C. Colonese / G. Zanchetta / C. Perlès / R. N. Drysdale / G. Manganelli / I. Baneschi / E. Dotsika / H. Valladas, Deciphering late Quaternary land snail shell $\delta^{18}O$ and $\delta^{13}C$ from Franchthi Cave (Argolid, Greece). Quaternary Research 80(1), 2013, 66–75. DOI: 10.1016/j.yqres.2013.03.006.

2014: A. C. Colonese / D. Lo Vetro / F. Martini, Holocene coastal change and intertidal mollusc exploitation in the central Mediterranean: Variations in shell size and morphology at Grotta d'Oriente (Sicily). Archaeofauna 23(23), 2014, 181–192. DOI: 10.15366/archaeofauna2014.23.012.

2018: A. C. Colonese / D. Lo Vetro / W. Landini / Z. Di Giuseppe / N. Hausmann / B. Demarchi / C. d'Angelo / M. J. Leng / A. Incarbona / A. C. Whitwood / F. Martini, Late Pleistocene-Holocene coastal adaptation in central Mediterranean: Snapshots from Grotta d'Oriente (NW Sicily). Quaternary Internat. 493, 2018, 114–126. DOI: 10.1016/j.quaint.2018.06.018.

Comboul et al. 2014: M. Comboul / J. Emile-Geay / M. N. Evans / N. Mirnateghi / K. M. Cobb / D. M. Thompson, A probabilistic model of chronological errors in layer-counted climate proxies: Applications to annually banded coral archives. Climate of the Past 10(2), 2014, 825–841. DOI: 10.5194/cp-10-825-2014.

Coulson et al. 2014: P. G. Coulson / B. A. Black / I. C. Potter / N. G. Hall, Sclerochronological studies reveal that patterns of otolith growth of adults of two co-occurring species of Platycephalidae are synchronised by water temperature variations. Marine Biol. 161(2), 2014, 383–393. DOI: 10.1007/s00227-013-2343-0.

Craig 1957: H. Craig, Isotopic standards for carbon and oxygen and correction factors for mass-spectrometric analysis of carbon dioxide. Geochimica et Cosmochimica Acta 12(1–2), 1957, 133–149. DOI: 10.1016/0016-7037(57)90024-8.

Cremers/Radziemski 2006a: D. A. Cremers / L. J. Radziemski, Handbook of Laser-Induced Breakdown Spectroscopy (Chichester 2006). DOI: 10.1002/9781118567371.

2006b: D. A. Cremers / L. J. Radziemski, History and fundamentals of LIBS. In: A. Miziolek / V. Palleschi / I. Schechter (eds), Laser-Induced Breakdown Spectroscopy. Fundamentals and Applications (Cambridge 2006) 1–39. DOI: 10.1017/CBO9780511541261.002.

Crisp 2013: M. K. Crisp, Amino acid racemization dating: Method development using African ostrich (*Struthio camelus*) eggshell [PhD thesis, Univ. York 2013]. https://etheses.whiterose.ac.uk/id/eprint/4770/7/FINAL%20v6.pdf (23/02/2026).

Crowley 1999: T. J. Crowley, Correlating high-frequency climate variations. Paleoceanogr. and Paleoclimatol. 14(3), 1999, 271–272. DOI: 10.1029/1999PA900003.

Cruz-Silva et al. 2023: E. Cruz-Silva / S. P. Harrison / I. Colin Prentice / E. Marinova / P. J. Bartlein / H. Renssen / Y. Zhang, Pollen-based reconstructions of Holocene climate trends in the eastern Mediterranean region. EGUsphere Preprint Repository 2023. DOI: 10.5194/egusphere-2023-864.

Cubizolle et al. 2007: H. Cubizolle / P. Bonnel / C. Oberlin / A. Tourman / J. Porteret, Advantages and limits of radiocarbon dating applied to peat inception during the end of the Late Glacial and the Holocene: The example of mires in the eastern Massif Central (France). Quaternaire 18(2), 2007, 187–208. DOI: 10.4000/quaternaire.1049.

Cullen 1995: T. Cullen, Mesolithic mortuary ritual at Franchthi Cave, Greece. Antiquity 69(263), 1995, 270–289. DOI: 10.1017/S0003598X00064681.

Darlas/Psathi 2016: A. Darlas / E. Psathi, The Middle and Upper Paleolithic on the Western Coast of the Mani Peninsula (Southern Greece). In: K. Harvati / M. Roksandic (eds), Paleoanthropology of the Balkans and Anatolia: Human Evolution and Its Context (Dordrecht 2016) 95–118. DOI: 10.1007/978-94-024-0874-4_7.

Davenport 1938: C. B. Davenport, Growth Lines in Fossil Pectens as Indicators of Past Climates. Journal Paleont. 12(5), 1938, 514–515. www.jstor.org/stable/1298671.

De la Pierre et al. 2014: M. De la Pierre / C. Carteret / L. Maschio / E. André / R. Orlando / R. Dovesi, The Raman spectrum of $CaCO_3$ polymorphs calcite and aragonite: A combined experimental and computational study. Journal Chemical Physics 140(16), 2014, 164509. DOI: 10.1063/1.4871900.

De Vivo et al. 2001: B. De Vivo / G. Rolandi / P. B. Gans / A. Calvert / W. A. Bohrson / F. J. Spera / H. E. Belkin, New constraints on the pyroclastic eruptive history of the Campanian volcanic Plain (Italy). Mineral. and Petrol. 73, 2001, 47–65. DOI: 10.1007/s007100170010.

Degroot et al. 2021: D. Degroot / K. Anchukaitis / M. Bauch / J. Burnham / F. Carnegy / J. Cui / K. de Luna / P. Guzowski / G. Hambrecht / H. Huhtamaa / A. Izdebski / K. Kleemann / E. Moesswilde / N. Neupane / T. Newfield / Q. Pei / E. Xoplaki / N. Zappia, Towards a rigorous understanding of societal responses to climate change. Nature 591(7851), 2021, 539–550. DOI: 10.1038/s41586-021-03190-2.

Deith 1983: M. R. Deith, Seasonality of shell collecting determined by oxygen isotope analysis of marine shells from Asturian sites in Cantabria. In: C. Grigson / J. Clutton-Brock (eds), Animals and Archaeology. 2: Shell Middens, Fishes and Birds. BAR Internat. Ser. 183 (Oxford 1983) 67–76.

Deith/Shackleton 1988: M. R. Deith / N. J. Shackleton, Oxygen isotope analyses of marine molluscs from Franchthi Cave. In: J. Shackleton (ed.), Marine Molluscan Remains from Franchthi Cave. Excav. Franchthi Cave, Greece 4 (Bloomington IN 1988) 133–156.

Demarchi/Collins 2014: B. Demarchi / M. J. Collins, Amino Acid Racemization Dating. In: W. J. Rink / J. W. Thompson (eds), Encyclopedia of Scientific Dating Methods (Dordrecht 2014) 1–22. DOI: 10.1007/978-94-007-6326-5_73-1.

Demarchi et al. 2011: B. Demarchi / M. G. Williams / N. Milner / N. Russell / G. Bailey / K. Penkman, Amino acid racemization dating of marine shells: A mound of possibilities. Quaternary Internat. 239(1–2), 2011, 114–124. DOI: 10.1016/j.quaint.2010.05.029.

2013: B. Demarchi / M. J. Collins / P. J. Tomiak / B. J. Davies / K. E. H. Penkman, Intra-crystalline protein diagenesis (IcPD) in *Patella vulgata*. Part II: Breakdown and temperature sensitivity. Quaternary Geochronol. 16, 2013, 158–172. DOI: 10.1016/j.quageo.2012.08.001.

Demoule/Perlès 1993: J.-P. Demoule / C. Perlès, The Greek Neolithic: A new review. Journal World Prehist. 7(4), 1993, 355–416. DOI: 10.1007/BF00997801.

Desse/Desse-Berset 2002: J. Desse / N. Desse-Berset, Le cortège de Neptune: Les poissons de la Méditerranée durant l'Holocène. In: A. Gardeisen (ed.), Mouvements ou déplacements de populations animales en Méditerranée au cours de l'Holocène. Séminaire de recherche du thème 15 »Archéologie de l'Animal« (UMR 154-CNRS), Lattes-Montpellier (France), 29 septembre 2000. BAR Internat. Ser. 1017 (Oxford 2002) 83–96.

Dettman et al. 1999: D. L. Dettman / A. K. Reische / K. C. Lohmann, Controls on the stable isotope composition of seasonal growth bands in aragonitic fresh-water bivalves (unionidae). Geochimica et Cosmochimica Acta 63, 1999, 1049–1057. DOI: 10.1016/S0016-7037(99)00020-4.

Di Salvo et al. 2012a: R. Di Salvo / M. A. Mannino / V. Schimmenti / K. D. Thomas, Nuovi dati sulle sepolture della Grotta dell'Uzzo. In: Atti della XLI Riunione scientifica: Dai ciclopi agli ecisti: Società e territorio nella Sicilia preistorica e protostorica, San Cipirello (PA), 16–19 novembre 2006 (Firenze 2012) 461–470.

2012b: R. Di Salvo / G. Mannino / M. A. Mannino / V. Schimmenti / L. Sineo / K. D. Thomas, Le sepolture della Grotta d'Oriente (Favignana). In: Atti della XLI Riunione scientifica: Dai ciclopi agli ecisti: Società e territorio nella Sicilia preistorica e protostorica, San Cipirello (PA), 16–19 novembre 2006 (Firenze 2012) 341–351.

Diamant 1979: S. Diamant, A short history of archaeological sieving at Franchthi Cave, Greece. Journal Field Arch. 6(2), 1979, 203–217. www.jstor.org/stable/529364.

Dickinson 2018: M. R. Dickinson, Enamel Amino Acid Racemization Dating and its Application to Building Proboscidean Geochronologies [PhD thesis, Univ. York 2018]. https://etheses.whiterose.ac.uk/id/eprint/22261/1/Enamel%20amino%20acid%20racemisation%20dating%20and%20its%20application%20to%20building%20Proboscidean%20geochronologies.pdf (23/02/2026).

Dickinson et al. 2019: M. R. Dickinson / A. M. Lister / K. E. Penkman, A new method for enamel amino acid racemization dating: A closed system approach. Quaternary Geochronol. 50, 2019, 29–46. DOI: 10.1016/j.quageo.2018.11.005.

Dornan 2002: J. Dornan, Agency and Archaeology: Past, Present and Future Directions. Journal Arch. Method and Theory 9(4), 2002, 303–329. www.jstor.org/stable/20177466.

Douka et al. 2011: K. Douka / C. Perlès / H. Valladas / M. Vanhaeren / R. E. M. Hedges, Franchthi Cave revisited: The age of the Aurignacian in southeastern Europe. Antiquity 85, 2011, 1131–1150. DOI: 10.1017/S0003598X00061962.

2014: K. Douka / R. Grün / Z. Jacobs / C. Lane / L. Farr / C. O. Hunt / R. Inglis / T. Reynolds / P. Albert / M. Aubert / V. Cullen / E. Hill / L. Kinsley / R. G. Roberts / E. L. Tomlinson / S. Wulf / G. Barker, The chronostratigraphy of the Haua Fteah Cave (Cyrenaica, northeast Libya). Journal Human Evol. 66, 2014, 39–63. DOI: 10.1016/j.jhevol.2013.10.001.

Elefanti/Marshall 2018: P. Elefanti / G. Marshall, Mobility during the Upper Palaeolithic in Greece: Some suggestions for the Argolid Peninsula. Ex Novo Journal Arch. 3, 2018, 7–22. DOI: 10.32028/exnovo.v3i0.378.

Elliot et al. 2009: M. Elliot / K. Welsh / C. Chilcott / M. McCulloch / J. Chappell / B. Ayling, Profiles of trace elements and stable isotopes derived from giant long-lived *Tridacna gigas* bivalves: Potential applications in paleoclimate studies. Palaeogeogr., Palaeoclimatol., Palaeoecol. 280(1–2), 2009, 132–142. DOI: 10.1016/j.palaeo.2009.06.007.

Emiliani et al. 1964: C. Emiliani / L. Cardini / T. Mayeda / C. B. M. McBurney / E. Tongiorgi, Palaeotemperature analysis of marine molluscs (food refuse) from the site of Arene Candide Cave, Italy and the Haua Fteah Cave, Cyrenaica. In: H. Craig / S. L. Miller / G. J. Wasserburg (eds), Isotopic and Cosmic Chemistry. Dedicated to Harold C. Urey on his Seventieth Birthday April 29, 1963 (Amsterdam 1964) 133–156.

Epstein et al. 1951: S. Epstein / R. Buchsbaum / H. Lowenstam / H. C. Urey, Carbonate-water isotopic temperature scale. Geol. Soc. Am. Bull. 62(4), 1951, 417–426. DOI: 10.1130/0016-7606(1951)62[417:CITS]2.0.CO;2.

1953: S. Epstein / R. Buchsbaum / H. A. Lowenstam / H. C. Urey, Revised carbonate-water isotopic temperature scale. Geol. Soc. Am. Bull. 64(11), 1953, 1315–1325. DOI: 10.1130/0016-7606(1953)64[1315:RCITS]2.0.CO;2.

Erlandson 2001: J. M. Erlandson, The Archaeology of Aquatic Adaptations: Paradigms for a New Millennium. Journal Arch. Research 9, 2001, 287–350. DOI: 10.1023/A:1013062712695.

Evans 2003: J. G. Evans, Environmental Archaeology and the Social Order (London 2003). DOI: 10.4324/9780203711767.

Facorellis 2003: Y. Facorellis, Radiocarbon dating the Greek Mesolithic. In: Galanidou/Perlès 2003, 51–67. www.jstor.org/stable/40960305.

2011: Y. Facorellis, Sequential radiocarbon dating and calculation of the marine reservoir effect. In: A. Sampson (ed.), The Cave of the Cyclops. Mesolithic and Neolithic Networks in the Northern Aegean Basin. 2: Bone Tool Industries, Dietary Resources and the Paleoenvironment, and Archaeometrical Studies. Prehist. Monogr. 31 (Philadelphia 2011) 31–44.

Farr et al. 2014: L. Farr / R. Lane / F. Abdulazeez / P. Bennett / J. Holman / A. Marasi / A. Prendergast / M. Al-Zweyi / G. Barker, The Cyrenaican Prehistory Project 2013: The seventh season of excavations in the Haua Fteah Cave. Libyan Stud. 45, 2014, 163–173. DOI: 10.1017/lis.2014.2.

Farrand 2003: W. R. Farrand, Depositional environments and site formation during the Mesolithic occupations of Franchthi Cave, Peloponnesos, Greece. In: Galanidou/Perlès 2003, 69–78. www.jstor.org/stable/40960306.

Fenger et al. 2007: T. Fenger / D. Surge / B. Schöne / N. Milner, Sclerochronology and geochemical variation in limpet shells (*Patella vulgata*): A new archive to reconstruct coastal sea surface temperature. Geochemistry, Geophysics, Geosystems 8(7), 2007, 1–17. DOI: 10.1029/2006GC001488.

Ferguson et al. 2011: J. E. Ferguson / G. M. Henderson / D. A. Fa / J. C. Finlayson / N. R. Charnley, Increased seasonality in the Western Mediterranean during the last glacial from limpet shell geochemistry. Earth and Planetary Scien. Letters 308, 2011, 325–333. DOI: 10.1016/j.epsl.2011.05.054.

Finné/Holmgren 2010: M. Finné / K. Holmgren, Climate variability in the Eastern Mediterranean and the Middle East during the Holocene. In: P. J. J. Sinclair / G. Nordquist / F. Herschend / C. Isendahl (eds), The Urban Mind. Cultural and Environmental Dynamics. Stud. Global Arch. 15 (Uppsala 2010) 29–60.

Finné et al. 2011: M. Finné / K. Holmgren / H. S. Sundqvist / E. Weiberg / M. Lindblom, Climate in the eastern Mediterranean, and adjacent regions, during the past 6000 years – A review. Journal Arch. Scien. 38(12), 2011, 3153–3173. DOI: 10.1016/j.jas.2011.05.007.

Flannery 1967: K. V. Flannery, Culture History v. Cultural Process: A Debate in American Archaeology. Scien. Am. 217(2), 1967, 119–122.

Flessa 2001: K. W. Flessa, Time-averaging. In: D. E. G. Briggs / P. Crowther (eds), Palaeobiology II (Oxford 2001) 292–296.

Flohr et al. 2016: P. Flohr / D. Fleitmann / R. Matthews / W. Matthews / S. Black, Evidence of resilience to past climate change in Southwest Asia: Early farming communities and the 9.2 and 8.2 ka events. Quaternary Scien. Reviews 136, 2016, 23–39. DOI: 10.1016/j.quascirev.2015.06.022.

Fowler et al. 2015: C. Fowler / J. Harding / D. Hofmann, The Oxford Handbook of Neolithic Europe: An Introduction. In: C. Fowler / J. Harding / D. Hofmann (eds), The Oxford Handbook of Neolithic Europe (Oxford 2015) 3–26. DOI: 10.1093/oxfordhb/9780199545841.013.061.

Frech/Wang 1980: R. Frech / E. C. Wang, The i. r. and Raman spectra of $CaCO_3$ (aragonite). Spectrochimica Acta A: Molecular Spectroscopy 36(10), 1980, 915–919. DOI: 10.1016/0584-8539(80)80044-4.

Freitas et al. 2005: P. Freitas / L. J. Clarke / H. Kennedy / C. Richardson / F. Abrantes, Mg/Ca, Sr/Ca, and stable-isotope ($\delta^{18}O$ and $\delta^{13}C$) ratio profiles from the fan mussel *Pinna nobilis*: Seasonal records and temperature relationships. Geochemistry, Geophysics, Geosystems 6(4), 2005, 1–16. DOI: 10.1029/2004GC000872.

2006: P. Freitas / L. J. Clarke / H. Kennedy / C. Richardson / F. Abrantes, Environmental and biological controls on elemental (Mg/Ca, Sr/Ca and Mn/Ca) ratios in shells of the king scallop *Pecten maximus*. Geochimica et Cosmochimica Acta 70(20), 2006, 5119–5133. DOI: 10.1016/j.gca.2006.07.029.

Gaitanakis/Fotiadis 1991: P. Gaitanakis / A. D. Fotiadis, Γεωλογική δομή της ΝΔ Αργολίδας [Geological structure of SW Argolis (Peloponnesus, Greece)]. Bull. Geol. Soc. Greece 25(1), 1991, 319–338.

Galanidou 2011: N. Galanidou, Mesolithic cave use in Greece and the mosaic of human communities. Journal Mediterranean Arch. 24(2), 2011, 219–242. DOI: 10.1558/jma.v24i2.219.

Galanidou/Perlès 2003: N. Galanidou / C. Perlès (eds), The Greek Mesolithic: Problems and Perspectives. Brit. School Athens Stud. 10 (London 2003). www.jstor.org/stable/i40043007.

García-Escárzaga et al. 2015: A. García-Escárzaga / S. Moncayo / I. Gutierrez-Zugasti / M. R. Gonzalez-Morales / J. Martin-Chivelet / J. O. Caceres, Mg/Ca ratios measured by laser induced breakdown spectroscopy (LIBS): A new approach to decipher environmental conditions. Journal Analytical Atomic Spectrometry 30, 2015, 1913–1919. DOI: 10.1039/C5JA00168D.

Garnsey 1988: P. Garnsey, Famine and food supply in the Graeco-Roman world: Responses to risk and crisis (Cambridge 1988). DOI: 10.1017/CBO9780511583827.

Gassiot 2005: E. B. Gassiot, Shell middens on the Caribbean coast of Nicaragua: Prehistoric patterns of mollusc collection and consumption. In: D. E. Bar-Yosef Mayer (ed.), Archaeomalacology: Molluscs in former environments of human behaviour (Oxford 2005) 40–53. DOI: 10.2307/j.ctt1kw2959.8.

Gavrielides 1976: N. Gavrielides, The impact of olive growing on the landscape in the Fourni Valley. Ann. New York Acad. Scien. 268(1), 1976, 143–157. DOI: 10.1111/j.1749-6632.1976.tb47640.x.

Giakoumaki et al. 2007: A. Giakoumaki / K. Melessanaki / D. Anglos, Laser-induced breakdown spectroscopy (LIBS) in archaeological science – applications and prospects. Analytical and Bioanalytical Chemistry 387, 2007, 749–760. DOI: 10.1007/s00216-006-0908-1.

Gifford 1990: J. A. Gifford, Analysis of submarine sediments off Paralia. In: T. J. Wilkinson / S. T. Duhon (eds), Franchthi Paralia: The Sediments, Stratigraphy, and Offshore Investigations. Excav. Franchthi Cave, Greece 6 (Bloomington IN 1990) 85–116.

Glais et al. 2023: A. Glais / L. Lespez / J. A. Lopez-Saez / Z. Tsirtsoni / C. Virmoux / M. Ghilardi / R. Davidson / D. Malamidou / K. Pavlopoulos, Rapid climate changes and human dynamics during the holocene in the eastern Mediterranean (Lower Strymon Valley, northern Greece). Quaternary Scien. Reviews 313, 2023, 108130. DOI: 10.1016/j.quascirev.2023.108130.

Godfrey 1988: M. C. S. Godfrey, Oxygen isotope analysis: a means for determining the seasonal gathering of the pipi (*Donax deltoides*) by Aborigines in prehistoric Australia. Arch. Oceania 23, 1988, 17–21. DOI: 10.1002/j.1834-4453.1988.tb00179.x.

Göktürk et al. 2011: O. M. Göktürk / D. Fleitmann / S. Badertscher / H. Cheng / R. L. Edwards / M. Leuenberger / A. Fankhauser / O. Tüysüz / J. Kramers, Climate on the southern Black Sea coast during the Holocene: implications from the Sofular Cave record. Quaternary Scien. Reviews 30(19–20), 2011, 2433–2445. DOI: 10.1016/j.quascirev.2011.05.007.

Gonfiantini et al. 1995: R. Gonfiantini / W. Stichler / K. Rozanski, Standards and intercomparison materials distributed by the International Atomic Energy Agency for stable isotope measurements. In: International Atomic Energy Agency (ed.), Reference and Intercomparison Materials for Stable Isotopes of Light Elements. Proceedings of a Consultants Meeting held in Vienna, 1–3 December 1993. Internat. Atomic Energy Agency: Technical Doc. 825 (Vienna 1995) 13–29.

Goodwin et al. 2004: D. H. Goodwin / K. W. Flessa / M. A. Téllez-Duarte / D. L. Dettmann / B. R. Schöne / G. A. Avila-Serrano, Detecting time-averaging and spatial mixing using oxygen isotope variation: A case study. Palaeogeogr., Palaeoclimatol., Palaeoecol. 205(1–2), 2004, 1–21. DOI: 10.1016/j.palaeo.2003.10.020.

de Graaf et al. 2021: S. de Graaf / H. B. Vonhof / E. J. Levy / M. Markowska / G. H. Haug, Isotope ratio infrared spectroscopy analysis of water samples without memory effects. Rapid Commun. Mass Spectrometry 35(8), 2021, e9055. DOI: 10.1002/rcm.9055.

Grimal 1986: P. Grimal, A concise dictionary of classical mythology (Oxford 1986).

Gröcke/Gillikin 2008: D. R. Gröcke / D. P. Gillikin, Advances in mollusc sclerochronology and sclerochemistry: tools for understanding climate and environment. Geo-Marine Letters 28, 2008, 265–268. DOI: 10.1007/s00367-008-0108-4.

Gronenborn 2016: D. Gronenborn, Climate Fluctuations, Human Migrations, and the Spread of Farming in Western Eurasia: Refining the Argument. In: P. F. Biehl / O. P. Nieuwenhuyse (eds), Climate and Cultural Change in Prehistoric Europe and the Near East. IEMA Proc. 6 (New York 2016) 127–200. DOI: 10.1515/9781438461847-013?urlappend=%3Futm_source%3Dresearchgate.net%26utm_medium%3Darticle.

Grossman/Ku 1986: E. L. Grossman / T. Ku, Oxygen and carbon isotope fractionation in biogenic aragonite: Temperature effects. Chemical Geol. 59, 1986, 59–74. DOI: 10.1016/0168-9622(86)90057-6.

Guilaine 2000-2001: J. Guilaine, La diffusion de l'agriculture en Europe: une hypothèse arythmique. Zephyrus 53–54, 2000–2001, 267–272.

2015: J. Guilaine, The Neolithization of Mediterranean Europe: Mobility and Interactions from the Near East to the Iberian Peninsula. In: C. Fowler / J. Harding / D. Hofmann (eds), The Oxford Handbook of Neolithic Europe (Oxford 2015) 81–98. DOI: 10.1093/oxfordhb/9780199545841.013.064.

Gutiérrez-Zugasti et al. 2017: I. Gutiérrez-Zugasti / R. Suárez-Revilla / L. J. Clark / B. R. Schöne / G. N. Bailey / M. R. González-Morales, Shell oxygen isotope values and sclerochronology of the limpet *Patella vulgata* Linnaeus 1758 from Northern Iberia: Implications for the reconstruction of past seawater temperatures. Palaeogeogr., Palaeoclimatol., Palaeoecol. 475, 2017, 162–175. DOI: 10.1016/j.palaeo.2017.03.018.

Hallmann et al. 2009: N. Hallmann / M. Burchell / B. R. Schöne / G. V. Irvine / D. Maxwell, High-resolution sclerochronological analysis of the bivalve mollusk *Saxidomus gigantea* from Alaska and British Columbia: techniques for revealing environmental archives and archaeological seasonality. Journal Arch. Scien. 36(10), 2009, 2353–2364. DOI: 10.1016/j.jas.2009.06.018.

Halstead 1989: P. Halstead, Like rising damp? An ecological approach to the spread of farming in southeast and central Europe. In: A. Milles / D. Williams / N. Gardner (eds), The Beginnings of Agriculture. BAR Internat. Ser. 496 (Oxford 1989) 23–53.

1999: P. Halstead, Neighbours from Hell? The Household in Neolithic Greece. In: P. Halstead, Neolithic Society in Greece. Sheffield Stud. Aegean Arch. 2 (Sheffield 1999) 77–95.

Hansen 1991: J. M. Hansen, The Palaeoethnobotany of Franchthi Cave. Excav. Franchthi Cave, Greece 7 (Bloomington IN 1991).

2001: J. Hansen, Macroscopic Plant Remains from Mediterranean Caves and Rockshelters: Avenues of Interpretation. Geoarchaeology 16(4), 2001, 401–432. DOI: 10.1002/gea.1010.

Hausmann/Meredith-Williams 2017a: N. Hausmann / M. Meredith-Williams, Seasonal patterns of coastal exploitation on the Farasan Islands, Saudi Arabia. Journal Island and Coastal Arch. 12(3), 2017, 360–379. DOI: 10.1080/15564894.2016.1216478.

2017b: N. Hausmann / M. Meredith-Williams, Exploring Accumulation Rates of Shell Deposits Through Seasonality Data. Journal Arch. Method and Theory 24, 2017, 776–795. DOI: 10.1007/s10816-016-9287-x.

Hausmann et al. 2017: N. Hausmann / P. Siozos / A. Lemonis / A. C. Colonese / H. K. Robson / D. Anglos, Elemental mapping of Mg/Ca intensity ratios in marine mollusc shells using laser-induced breakdown spectroscopy. Journal Analytical Atomic Spectrometry 32(8), 2017, 1467–1472. DOI: 10.1039/C7JA00131B.

2019a: N. Hausmann / A. L. Prendergast / A. Lemonis / J. Zech / P. Siozos / D. Anglos, Extensive elemental mapping unlocks Mg/Ca ratios as climate proxy in seasonal records of Mediterranean limpets. Scien. Reports 9, 2019, 3698. DOI: 10.1038/s41598-019-39959-9.

2019b: N. Hausmann / H. K. Robson / C. Hunt, Annual Growth Patterns and Interspecimen Variability in Mg/Ca Records of Archaeological *Ostrea edulis* (European Oyster) from the Late Mesolithic Site of Conors Island. Open Quaternary 5(9), 2019, 1–18. DOI: 10.5334/oq.59.

2023: N. Hausmann / D. Theodoraki / V. Piñon / P. Siozos / A. Lemonis / D. Anglos, Applying laser induced breakdown spectroscopy (LIBS) and elemental imaging on marine shells for archaeological and environmental research. Scien. Reports 13, 2023, 19812. DOI: 10.1038/s41598-023-46453-w.

Hayes et al. 1999: A. Hayes / E. J. Rohling / S. de Rijk / D. Kroon / W. J. Zachariasse, Mediterranean planktonic foraminiferal faunas during the last glacial cycle. Marine Geol. 153(1–4), 1999, 239–252. DOI: 10.1016/S0025-3227(98)00019-X.

2005: A. Hayes / M. Kucera / N. Kallel / L. Sbaffi / E. J. Rohling, Glacial Mediterranean sea surface temperatures based on planktonic foraminiferal assemblages. Quaternary Scien. Reviews 24(7–9), 2005, 999–1016. DOI: 10.1016/j.quascirev.2004.02.018.

Hays/Grossman 1991: P. Hays / E. Grossman, Oxygen isotopes in meteoric calcite cements as indicators of continental paleoclimate. Geology 19(5), 1991, 441–444. DOI: 10.1130/0091-7613(1991)019%3C0441:OIIMCC%3E2.3.CO;2.

Heitz et al. 2021: C. Heitz / J. Laabs / M. Hinz / A. Hafner, Collapse and resilience in prehistoric archaeology: Questioning concepts and causalities in models of climate-induced societal transformations. In: P. Erdkamp / J. G. Manning / K. Verboven (eds), Climate Change and Ancient Societies in Europe and the Near East: Diversity in Collapse and Resilience (London 2021) 127–200. DOI: 10.1007/978-3-030-81103-7_5.

Hendy et al. 2012: E. J. Hendy / P. J. Tomiak / M. J. Collins / J. Hellstrom / A. W. Tudhope / J. M. Lough / K. Penkman, Assessing amino acid racemization variability in coral intra-crystalline protein for geochronological applications. Geochimica et Cosmochimica Acta 86(9–2), 2012, 338–353. DOI: 10.1016/j.gca.2012.02.020.

Heurtley 1939: W. A. Heurtley, Prehistoric Macedonia: An archaeological reconnaissance of Greek Macedonia (West of the Struma) in the Neolithic, Bronze, and Iron Ages (Cambridge 1939).

Hodder 1982: I. Hodder, Symbols in Action: Ethnoarchaeological Studies of Material Culture (Cambridge 1982).

1990: I. Hodder, The Domestication of Europe: Structure and Contingency in Neolithic Societies (Oxford 1990).

Hodder/Hutson 2010: I. Hodder / S. Hutson, Διαβάζοντας το παρελθόν: Τρέχουσες ερμηνευτικές προσεγγίσεις στην αρχαιολογία [Reading the Past: Current Approaches to Interpretation in Archaeology], trans. N. Kourkoulos (Athens 2010).

Hofmanová et al. 2016: Z. Hofmanová / S. Kreutzer / G. Hellenthal / C. Sell / Y. Diekmann / D. Díez-del-Molino / L. van Dorp / S. López / A. Kousathanas / V. Link / K. Kirsanow / L. M. Cassidy / R. Martiniano / M. Strobel / A. Scheu / K. Kotsakis / P. Halstead / S. Triantaphyllou / N. Kyparissi-Apostolika / D. Urem-Kotsou / C. Ziota / F. Adaktylou / S. Gopalan / D. M. Bobo / L. Winkelbach / J. Blöcher / M. Unterländer / C. Leuenberger / Ç. Çilingiroğlu / B. Horejs / F. Gerritsen / S. J. Shennan / D. G. Bradley / M. Currat / K. R. Veeramah / D. Wegmann / M. G. Thomas / C. Papageorgopoulou / J. Burger, Early farmers from across Europe directly descended from Neolithic Aegeans. PNAS 113(25), 2016, 6886–6891. DOI: 10.1073/pnas.1523951113.

Honea 1975: K. Honea, Prehistoric Remains on the Island of Kythnos. Am. Journal Arch. 79(3), 1975, 277–279. DOI: 10.2307/503483.

Horejs et al. 2015: B. Horejs / B. Milic / F. Ostmann / U. Thanheiser / B. Weninger / A. Galik, The Aegean in the early 7th millennium BC: Maritime networks and colonization. Journal World Prehist. 28, 2015, 289–330. DOI: 10.1007/s10963-015-9090-8.

Hunt et al. 2011: C. O. Hunt / T. G. Reynolds / H. A. El-Rishi / A. Buzaian / E. Hill / G. W. Barker, Resource pressure and environmental change on the North African littoral: Epipalaeolithic to Roman gastropods from Cyrenaica, Libya. Quaternary Internat. 244(1), 2011, 15–26. DOI: 10.1016/j.quaint.2011.04.045.

Ingold 2000: T. Ingold, The Perception of the Environment: Essays on Livelihood, Dwelling and Skill (London, New York 2000).

Ivanova et al. 2018: M. Ivanova / B. De Cupere / J. Ethier / E. Marinova, Pioneer farming in southeast Europe during the early sixth millennium BC: Climate-related adaptations in the exploitation of plants and animals. PLoS ONE 13(8), 2018, e0202668. DOI: 10.1371/journal.pone.0202668.

Jacob et al. 2011: D. E. Jacob / R. Wirth / A. L. Soldati / U. Wehrmeister / A. Schreiber, Amorphous calcium carbonate in the shells of adult Unionoida. Journal Structural Biol. 173(2), 2011, 241–249. DOI: 10.1016/j.jsb.2010.09.011.

Jacobs et al. 2017: Z. Jacobs / B. Li / L. Farr / E. Hill / C. Hunt / S. Jones / R. Rabett / T. Reynolds / R. G. Roberts / D. Simpson / G. Barker, The chronostratigraphy of the Haua Fteah cave (Cyrenaica, northeast Libya) – Optical dating of early human occupation during Marine Isotope Stages 4, 5 and 6. Journal Human Evol. 105, 2017, 69–88. DOI: 10.1016/j.jhevol.2017.01.008.

Jacobsen 1973a: T. W. Jacobsen, Excavation in the Franchthi Cave, 1969–1971, Part I. Hesperia 42(1), 1973, 45–88. DOI: 10.2307/147470.

1973b: T. W. Jacobsen, Excavation in the Franchthi Cave, 1969–1971, Part II. Hesperia 42(3), 1973, 253–283. DOI: 10.2307/147518.

1976: T. W. Jacobsen, 17,000 years of Greek prehistory. Scien. Am. 234(6), 1976, 76–87. www.jstor.org/stable/10.2307/24950373.

1981: T. W. Jacobsen, Franchthi Cave and the Beginning of Settled Village Life in Greece. Hesperia 50(4), 1981, 303–319. DOI: 10.2307/147874.

Jacobsen/Farrand 1987: T. W. Jacobsen / W. R. Farrand, Introduction. In: T. W. Jacobsen / W. R. Farrand, Franchthi Cave and Paralia. Maps, Plans, and Sections. Excav. Franchthi Cave, Greece 1 (Bloomington IN 1987) 1–10.

Jahns 1993: S. Jahns, On the Holocene vegetation history of the Argive plain (Peloponnese, southern Greece). Vegetation Hist. and Archaeobot. 2, 1993, 187–203. DOI: 10.1007/BF00198161.

Jameson 1983: M. Jameson, Famine in the Greek world. In: P. Garnsey / C. R. Whittaker (eds), Trade and Famine in Classical Antiquity. Proc. Cambridge Philol. Soc. 8 (Cambridge 1983) 6–16. DOI: 10.2307/j.ctv1r07g2v.

Jameson et al. 1994: M. H. Jameson / C. N. Runnels / T. H. van Andel, A Greek Countryside. The Southern Argolid from Prehistory to the Present Day (Stanford CA 1994).

Jenkins/Hartnoll 2001: S. R. Jenkins / R. G. Hartnoll, Food supply, grazing activity and growth rate in the limpet *Patella vulgata* L.: A comparison between exposed and sheltered shores. Journal Experimental Marine Biol. and Ecol. 258(1), 2001, 123–139. DOI: 10.1016/S0022-0981(01)00211-8.

Jerardino 1997: A. Jerardino, Changes in Shellfish Species Composition and Mean Shell Size from a Late-Holocene Record of the West Coast of Southern Africa. Journal Arch. Scien. 24(11), 1997, 1031–1044. DOI: 10.1006/jasc.1997.0182.

Jew et al. 2013: N. P. Jew / J. M. Erlandson / J. Watts / F. J. White, Shellfish, Seasonality, and Stable Isotope Sampling: $\delta^{18}O$ Analysis of Mussel Shells from an 8,800-Year-Old Shell Midden on California's Channel Islands. Journal Island and Coastal Arch. 8(2), 2013, 170–189. DOI: 10.1080/15564894.2012.736917.

Kaczanowska et al. 2010: M. Kaczanowska / J. K. Kozłowski / K. Sobczyk, Upper Palaeolithic human occupations and material culture at Klissoura Cave 1. Eurasian Prehist. 7(2), 2010, 133–285.

Kalogiropoulou/Sarris 2018: E. Kalogiropoulou / A. Sarris, Communities, landscapes, and interaction: An introduction. In: A. Sarris / E. Kalogiropoulou / T. Kalayci / E. Karimali (eds), Communities, Landscapes, and Interaction in Neolithic Greece. Internat. Monogr. Prehist. Arch. Ser. 20 (New York NY 2018) 1–6. DOI: 10.2307/j.ctvw049k3.7.

Kaniewski et al. 2020: D. Kaniewski / N. Marriner / R. Cheddadi / P. M. Fischer / T. Otto / F. Luce / E. Van Campo, Climate Change and Social Unrest: A 6,000-Year Chronicle From the Eastern Mediterranean. Geophysical Research Letters 47(7), 2020, 1–10. DOI: 10.1029/2020GL087496.

Kaufman/Manley 1998: D. S. Kaufman / W. F. Manley, A new procedure for determining DL amino acid ratios in fossils using reverse phase liquid chromatography. Quaternary Scien. Reviews 17(11), 1998, 987–1000. DOI: 10.1016/S0277-3791(97)00086-3.

Kennett/Voorhies 1996: D. J. Kennett / B. Voorhies, Oxygen Isotopic Analysis of Archaeological Shells to Detect Seasonal Use of Wetlands on the Southern Pacific Coast of Mexico. Journal Arch. Scien. 23(5), 1996, 689–704. DOI: 10.1006/jasc.1996.0065.

Kim et al. 1998: T. Kim / C. T. Lin / Y. Yoon, Compositional Mapping by Laser-Induced Breakdown Spectroscopy. Journal Physical Chemistry B, 102(22), 1998, 4284–4287. DOI: 10.1021/jp980245m.

Krauß et al. 2018: R. Krauß / E. Marinova / H. De Brue / B. Weninger, The rapid spread of early farming from the Aegean into the Balkans via the Sub-Mediterranean–Aegean vegetation zone. Quaternary Internat. 496, 2018, 24–41. DOI: 10.1016/j.quaint.2017.01.019.

Koike 1973: H. Koike, Daily growth lines of the clam Meretrix lusoria. A basic study of prehistoric seasonal gathering. Anthr. Soc. Nippon 81(2), 1973, 122–138. www.jstage.jst.go.jp/article/ase1911/81/2/81_2_122/_pdf.

Kosnik et al. 2017: M. A. Kosnik / Q. Hua / D. S. Kaufman / M. Kowalewski / K. Whitacre, Radiocarbon-calibrated amino acid racemization ages from Holocene sand dollars (Peronella peronii). Quaternary Geochronol. 39, 2017, 174–188. DOI: 10.1016/j.quageo.2016.12.001.

Kotjabopoulou et al. 1999: E. Kotjabopoulou / E. Panagopoulou / E. Adam, The Boila rockshelter: Further evidence of human activity in the Voidomatis Gorge. In: Bailey et al. 1999, 197–210.

Kotsakis 1999: K. Kotsakis, What tells can tell: Social space and settlement in the Greek Neolithic. In: P. Halstead, Neolithic Society in Greece. Sheffield Stud. Aegean Arch. 2 (Sheffield 1999) 66–76.

2001: K. Kotsakis, Mesolithic to Neolithic in Greece. Continuity, discontinuity or change of course? Doc. Praehist. 28, 2001, 63–73. DOI: 10.4312/dp.28.4.

2004: K. Kotsakis, Ο νεολιθικός οικισμός. Χώρος παραγωγής και ιδεολογίας. In: A. F. Lagopoulos (ed.), Η ιστορία της ελληνικής πόλης [History of the Greek polis] (Athens 2004) 55–68.

2014: K. Kotsakis, Domesticating the periphery: New research into the Neolithic of Greece. Pharos 20(1), 2014, 41–73. DOI: 10.2143/PHA.20.1.3064536.

Koumouzelis et al. 2001: M. Koumouzelis / B. Ginter / J. Kozłowski / M. Pawlikowski / O. Bar-Yosef / R. M. Albert / M. Litynska-Zajac / E. Stworzewicz / P. Wojtal / G. Lipecki / T. Tomek / Z. Bochenski / A. Pazdur, The Early Upper Palaeolithic in Greece: The Excavations in Klisoura Cave. Journal Arch. Scien. 28(5), 2001, 515–539. DOI: 10.1006/jasc.2000.0599.

Kowalewski et al. 2014: M. Kowalewski / R. Domènech / J. Martinell, Vanishing Clams on an Iberian Beach: Local Consequences and Global Implications of Accelerating Loss of Shells to Tourism. PLoS ONE 9(1), 2014, e83615. DOI: 10.1371/journal.pone.0083615.

Kyparissi-Apostolika 2003: N. Kyparissi-Apostolika, The Mesolithic in Theopetra Cave: New data on a debated period of Greek prehistory. In: Galanidou/Perlès 2003, 189–198. www.jstor.org/stable/40960316.

Lafuente et al. 2015: B. Lafuente / R. T. Downs / H. Yang / N. Stone, The power of databases: The RRUFF project. In: T. Armbruster / R. M. Danisi (eds), Highlights in Mineralogical Crystallography (Berlin 2015) 1–30. DOI: 10.1515/9783110417104-003.

Laidler 1984: K. J. Laidler, The development of the Arrhenius equation. Journal Chemical Education 61(6), 1984, 494. DOI: 10.1021/ed061p494.

Larkin 2011: P. Larkin, Infrared and Raman Spectroscopy: Principles and Spectral Interpretation (Amsterdam 2011). DOI: 10.1016/C2015-0-00806-1.

Lea et al. 2000: D. W. Lea / D. K. Pak / H. J. Spero, Climate Impact of Late Quaternary Equatorial Pacific Sea Surface Temperature Variations. Science 289(5485), 2000, 1719–1724. DOI: 10.1126/science.289.5485.1719.

Legendre/Legendre 2012: P. Legendre / L. Legendre, Numerical Ecology. Developments in Environmental Modelling 3 (Amsterdam ³2012).

Leng/Lewis 2016: M. J. Leng / J. P. Lewis, Oxygen isotopes in Molluscan shell: Applications in environmental archaeology. Environmental Arch. 21(3), 2016, 295–306. DOI: 10.1179/1749631414Y.0000000048.

Lima et al. 2006: F. P. Lima / N. Queiroz / P. A. Ribeiro / S. J. Hawkins / A. M. Santos, Recent changes in the distribution of a marine gastropod, *Patella rustica* Linnaeus, 1758, and their relationship to unusual climatic events. Journal Biogeogr. 33(5), 2006, 812–822. DOI: 10.1111/j.1365-2699.2006.01457.x.

Lionello et al. 2006: P. Lionello / P. Malanotte-Rizzoli / R. Boscolo / P. Alpert / V. Artale / L. Li / J. Luterbacher / W. May / R. Trigo / M. Tsimplis / U. Ulbrich / E. Xoplaki, The Mediterranean climate: An overview of the main characteristics and issues. In: P. Lionello / P. Malanotte-Rizzoli / R. Boscolo (eds), Mediterranean Climate Variability. Developments in Earth and Environmental Scien. 4 (Amsterdam 2006) 1–27. DOI: 10.1016/S1571-9197(06)80003-0.

Lipson et al. 2017: M. Lipson / A. Szécsényi-Nagy / S. Mallick / A. Pósa / B. Stégmár / V. Keerl / N. Rohland / K. Stewardson / M. Ferry / M. Michel / J. Oppenheimer / N. Broomandkhoshbacht / E. Harney / S. Nordenfelt / B. Llamas / B. G. Mende / K. Köhler / K. Oross / M. Bondár / T. Marton / A. Osztás / J. Jakucs / T. Paluch / F. Horváth / P. Csengeri / J. Koós / K. Sebők / A. Anders / P. Raczky / J. Regenye / J. P. Barna / S. Fábián / G. Serlegi / Z. Toldi / E. G. Nagy / J. Dani / E. Molnár / Gy. Pálfi / L. Márk / B. Melegh / Z. Bánfai / L. Domboróczki / J. Fernández-Eraso / J. A. Mujika-Alustiza / C. Alonso Fernández / J. Jiménez Echevarría / R. Bollongino / J. Orschiedt / K. Schierhold / H. Meller / A. Cooper / J. Burger / E. Bánffy / K. W. Alt / C. Lalueza-Fox / W. Haak / D. Reich, Parallel palaeogenomic transects reveal complex genetic history of early European farmers. Nature 551, 2017, 368–372. DOI: 10.1038/nature24476.

Lu et al. 2015: Yu. Lu / Yu. Li / Yi. Li / Ya. Wang / S. Wang / Z. Bao / R. Zheng, Micro spatial analysis of seashell surface using laser-induced breakdown spectroscopy and Raman spectroscopy. Spectrochimica Acta B: Atomic Spectroscopy 110, 2015, 63–69. DOI: 10.1016/j.sab.2015.05.012.

Maheras/Anagnostopoulou 2003: P. Maheras / C. Anagnostopoulou, Circulation types and their influence on the interannual variability and precipitation changes in Greece. In: H.-J. Bolle (ed.), Mediterranean Climate: Variability and Trends (Berlin, Heidelberg 2003) 215–239. DOI: 10.1007/978-3-642-55657-9_12.

Maheras et al. 2006: P. Maheras / H. Flocas / K. Tolika / C. Anagnostopoulou / M. Vafiadis, Circulation types and extreme temperature changes in Greece. Climate Research 30(2), 2006, 161–174. www.jstor.org/stable/24869244.

Mannino 1972: G. Mannino, Grotta d'Oriente. Riv. Scien. Preist. 27, 1972, 470.

1973: G. Mannino, Il Riparo dell'Uzzo (Trapani). Sicilia Arch. 23, 1973, 21–39.

Mannino/Thomas 2004: M. A. Mannino / K. D. Thomas, Studio archeozoologico dei reperti faunistici dalla Grotta d'Oriente a Favignana (Trapani). Quad. Mus. Arch. Regionale »Antonino Salinas« 8, 2004, 23–54.

Mannino et al. 2003: M. A. Mannino / B. F. Spiro / K. D. Thomas, Sampling shells for seasonality: Oxygen isotope analysis on shell carbonates of the inter-tidal gastropod *Monodonta lineata* (da Costa) from populations across its modern range and from a Mesolithic site in southern Britain. Journal Arch. Scien. 30(6), 2003, 667–679. DOI: 10.1016/S0305-4403(02)00238-8.

2006: M. A. Mannino / K. D. Thomas / M. Piperno / S. Tusa / A. Tagliacozzo, Fine-tuning the radiocarbon chronology of the Grotta dell'Uzzo (Trapani). Atti Soc. Preist. e Protost. Regione Friuli-Venezia Giulia 15, 2006, 17–31.

2007: M. A. Mannino / K. D. Thomas / M. J. Leng / M. Piperno / S. Tusa / A. Tagliacozzo, Marine resources in the Mesolithic and Neolithic at the Grotta dell'Uzzo (Sicily): Evidence from isotope analyses of marine shells. Archaeometry 49(1), 2007, 117–133. DOI: 10.1111/j.1475-4754.2007.00291.x.

2011: M. A. Mannino / K. D. Thomas / M. J. Leng / R. Di Salvo / M. P. Richards, Stuck to the shore? Investigating prehistoric hunter-gatherer subsistence, mobility and territoriality in a Mediterranean coastal landscape through isotope analyses on marine mollusc shell carbonates and human bone collagen. Quaternary Internat. 244(1), 2011, 88–104. DOI: 10.1016/j.quaint.2011.05.044.

2012: M. A. Mannino / G. Catalano / S. Talamo / G. Mannino / R. Di Salvo / V. Schimmenti / C. Lalueza-Fox / A. Messina / D. Petruso / D. Caramelli / M. P. Richards / L. Sineo, Origin and Diet of the Prehistoric Hunter-Gatherers on the Mediterranean Island of Favignana (Ègadi Islands, Sicily). PLoS ONE 7(11), 2012, e49802. DOI: 10.1371/journal.pone.0049802.

2014: M. A. Mannino / K. D. Thomas / E. R. Crema / M. J. Leng, A matter of taste? Mode and periodicity of marine mollusc exploitation on the Mediterranean island of Favignana (Ègadi Islands, Italy) during its isolation in the early Holocene. Archaeofauna 23(23), 2014, 133–147. DOI: 10.15366/archaeofauna2014.23.009.

Marali et al. 2017: S. Marali / B. R. Schöne / R. Mertz-Kraus / S. M. Griffin / A. D. Wanamaker Jr. / P. G. Butler / H. A. Holland / K. P. Jochum, Reproducibility of trace element time-series (Na/Ca, Mg/Ca, Mn/Ca, Sr/Ca, and Ba/Ca) within and between specimens of the bivalve Arctica islandica – A LA-ICP-MS line scan study. Palaeogeogr., Palaeoclimatol., Palaeoecol. 484, 2017, 109–128. DOI: 10.1016/j.palaeo.2016.11.024.

Marchi et al. 2022: N. Marchi / L. Winkelbach / I. Schulz / M. Brami / Z. Hofmanová / J. Blöcher / C. S. Reyna-Blanco / Y. Diekmann / A. Thiéry / A. Kapopoulou / V. Link / V. Piuz / S. Kreutzer / S. M. Figarska / E. Ganiatsou / A. Pukaj / T. J. Struck / R. N. Gutenkunst / N. Karul / F. Gerritsen / J. Pechtl / J. Peters / A. Zeeb-Lanz / E. Lenneis / M. Teschler-Nicola / S. Triantaphyllou / S. Stefanović / C. Papageorgopoulou / D. Wegmann / J. Burger / L. Excoffier, The genomic origins of the world's first farmers. Cell 185(11), 2022, 1842–1859. e18. DOI: 10.1016/j.cell.2022.04.008.

Marino et al. 2007: G. Marino / E. J. Rohling / W. I. C. Rijpstra / F. Sangiorgi / S. Schouten / J. S. Sininghe Damste, Aegean Sea as driver of hydrographic and ecological changes in the eastern Mediterranean. Geology 35(8), 2007, 675–678. DOI: 10.1130/G23831A.1.

2009: G. Marino / E. J. Rohling / F. Sangiorgi / A. Hayes / J. L. Casford / A. F. Lotter / M. Kucera / H. Brinkhuis, Early and middle Holocene in the Aegean Sea: Interplay between high and low latitude climate variability. Quaternary Scien. Reviews 28(27–28), 2009, 3246–3262. DOI: 10.1016/j.quascirev.2009.08.011.

Masters/Bada 1977: P. M. Masters / J. L. Bada, Racemization of isoleucine in fossil molluscs from Indian middens and interglacial terraces in southern California. Earth and Planetary Scien. Letters 37(2), 1977, 173–183. DOI: 10.1016/0012-821X(77)90162-5.

Mathieson et al. 2018: I. Mathieson / S. Alpaslan-Roodenberg / C. Posth / A. Szécsényi-Nagy / N. Rohland / S. Mallick / I. Olalde / N. Broomandkhoshbacht / F. Candilio / O. Cheronet / D. Fernandes / M. Ferry / B. Gamarra / G. González Fortes / W. Haak / E. Harney / E. Jones / D. Keating / B. Krause-Kyora / I. Kucukkalipci / M. Michel / A. Mittnik / K. Nägele / M. Novak / J. Oppenheimer / N. Patterson / S. Pfrengle / K. Sirak / K. Stewardson / S. Vai / S. Alexandrov / K. W. Alt / R. Andreescu / D. Antonović / A. Ash / N. Atanassova / K. Bacvarov / B. G. Mende / H. Bocherens / M. Bolus / A. Boroneanţ / Y. Boyadzhiev / A. Budnik / J. Burmaz / S. Chohadzhiev / N. J. Conard / R. Cottiaux / M. Čuka / C. Cupillard / D. G. Drucker / N. Elenski / M. Francken / B. Galabova / G. Ganetsovski / B. Gély / T. Hajdu / V. Handzhyiska / K. Harvati / T. Higham / S. Iliev / I. Janković / I. Karavanić / D. J. Kennett / D. Komšo / A. Kozak / D. Labuda / M. Lari / C. Lazar / M. Leppek / K. Leshtakov / D. Lo Vetro / Dž. Los / I. Lozanov / M. Malina / F. Martini / K. McSweeney / H. Meller / M. Menđušić / P. Mirea / V. Moiseyev / V. Petrova / T. D. Price / A. Simalcsik / L. Sineo / M. Šlaus / V. Slavchev / P. Stanev / A. Starović / T. Szeniczey / S. Talamo / M. Teschler-Nicola / C. Thevenet / I. Valchev / F. Valentin / S. Vasilyev / F. Veljanovska / S. Venelinova / E. Veselovskaya / B. Viola / C. Virag / J. Zaninović / S. Zäuner / P. W. Stockhammer / G. Catalano / R. Krauß / D. Caramelli / G. Zariņa / B. Gaydarska / M. Lillie / A. G. Nikitin / I. Potekhina / A. Papathanasiou / D. Borić / C. Bonsall / J. Krause / R. Pinhasi / D. Reich, The genomic history of southeastern Europe. Nature 555, 2018, 197–203. DOI: 10.1038/nature25778.

Mauro et al. 2003: A. Mauro / M. Arculeo / N. Parrinello, Morphological and molecular tools in identifying the Mediterranean limpets *Patella caerulea*, *Patella aspera* and *Patella rustica*. Journal Experimental Marine Biol. and Ecol. 295(2), 2003, 131–143. DOI: 10.1016/S0022-0981(03)00291-0.

Mayewski et al. 2004: P. A. Mayewski / E. J. Rohling / J. C. Stager / W. Karlén / K. Maasch / L. D. Meeker / E. Meyerson / F. Gasse / S. Van Kreveld / K. Holmgren / J. Lee-Thorp / G. Rosqvist / F. Rack / M. Staubwasser / R. R. Schneider / E. J. Steig, Holocene climate variability. Quaternary Research 62(3), 2004, 243–255. DOI: 10.1016/j.yqres.2004.07.001.

McBurney 1960: C. B. M. McBurney, The Stone Age of Northern Africa (Harmondsworth 1960).

1967: C. B. M. McBurney, The Haua Fteah (Cyrenaica) and the Stone Age of the Southeast Mediterranean (Cambridge 1967).

van der Meer et al. 2007: M. T. J. van der Meer / M. Baas / W. I. C. Rijpstra / G. Marino / E. J. Rohling / J. S. Sinninghe Damsté / S. Schouten, Hydrogen isotopic compositions of long-chain alkenones record freshwater flooding of the Eastern Mediterranean at the onset of sapropel deposition. Earth and Planetary Scien. Letters 262(3–4), 2007, 594–600. DOI: 10.1016/j.epsl.2007.08.014.

deMenocal 2001: P. B. deMenocal, Cultural Responses to Climate Change during the Late Holocene. Science 292(5517), 2001, 667–673. www.jstor.org/stable/3083336.

Menut et al. 2003: D. Menut / P. Fischet / J.-L. Lacour / A. Rivoallan / P. Mauchien, Micro-laser-induced breakdown spectroscopy technique: A powerful method for performing quantitative surface mapping on conductive and nonconductive samples. Applied Optics 42(30), 2003, 6063–6071. DOI: 10.1364/AO.42.006063.

Mercone et al. 2000: D. Mercone / J. Thomson / I. W. Croudace / G. Siani / M. Paterne / S. Troelstra, Duration of S1, the most recent sapropel in the eastern Mediterranean Sea, as indicated by accelerated mass spectrometry radiocarbon and geochemical evidence. Paleoceanography 15(3), 2000, 336–347. DOI: 10.1029/1999PA000397.

Mercuri/Sadori 2014: A. M. Mercuri / L. Sadori, Mediterranean Culture and Climatic Change: Past Patterns and Future Trends. In: S. Goffredo / Z. Dubinsky (eds), The Mediterranean Sea: Its History and Present Challenges (Dordrecht 2014) 507–527. DOI: 10.1007/978-94-007-6704-1_30.

Milano et al. 2017: S. Milano / B. R. Schöne / R. Witbaard, Changes of shell microstructural characteristics of *Cerastoderma edule* (Bivalvia) – A novel proxy for water temperature. Palaeogeogr., Palaeoclimatol., Palaeoecol. 465, 2017, 395–406. DOI: 10.1016/j.palaeo.2015.09.051.

Moundrea-Agrafioti 2003: A. Moundrea-Agrafioti, Mesolithic fish hooks from the Cave of Cyclope Youra. In: Galanidou/Perlès 2003, 131–141. www.jstor.org/stable/40960312.

2011: A. Moundrea-Agrafioti, The Mesolithic and Neolithic bone implements. In: A. Sampson (ed.), The Cave of the Cyclops. Mesolithic and Neolithic Networks in the Northern Aegean Basin. 2: Bone Tool Industries, Dietary Resources and the Paleoenvironment, and Archaeometrical Studies. Prehist. Monogr. 31 (Philadelphia 2011) 4–49.

Muñoz-Colmenero et al. 2012: M. Muñoz-Colmenero / P. Turrero / J. L. Horreo / E. Garcia-Vazquez, Evolution of limpet assemblages driven by environmental changes and harvesting in North Iberia. Marine Ecol. Progress Ser. 466, 2012, 121–131. DOI: 10.3354/meps09906.

Munro/Stiner 2015: N. D. Munro / M. C. Stiner, Zooarchaeological evidence for early Neolithic colonization at Franchthi Cave (Peloponnese, Greece). Current Anthr. 56(4), 2015, 596–603. www.jstor.org/stable/10.1086/682326.

2020: N. D. Munro / M. C. Stiner, A zooarchaeological history of the Neolithic occupations at Franchthi Cave and paralia in southern Greece. Journal Anthr. Arch. 58, 2020, 101162. DOI: 10.1016/j.jaa.2020.101162.

Mylona 2003: D. Mylona, The exploitation of fish resources in the Mesolithic Sporades: fish remains from the Cave of Cyclope, Youra. In: Galanidou/Perlès 2003, 181–188. www.jstor.org/stable/40960315.

2011: D. Mylona, Fish vertebrae. In: A. Sampson (ed.), The Cave of the Cyclops. Mesolithic and Neolithic Networks in the Northern Aegean Basin. 2: Bone Tool Industries, Dietary Resources and the Paleoenvironment, and Archaeometrical Studies. Prehist. Monogr. 31 (Philadelphia 2011) 237–266.

2021: D. Mylona, Catching tuna in the Aegean: biological background of tuna fisheries and the archaeological implications. Anthropozoologica 56(2), 2021, 23–37. DOI: 10.5252/anthropozoologica2021v56a2.

Mylonas 1929: G. E. Mylonas, Excavations at Olynthos. I: The Neolithic Settlement. Johns Hopkins Univ. Stud. Arch. 6 (Baltimore 1929).

Nanoglou 2001: S. Nanoglou, Social and monumental space in Neolithic Thessaly, Greece. European Journal Arch. 4, 2001, 303–322. DOI: 10.1179/eja.2001.4.3.303.

Nevin et al. 2012: A. Nevin / G. Spoto / D. Anglos, Laser spectroscopies for elemental and molecular analysis in art and archaeology. Applied Physics A 106, 2012, 339–361. DOI: 10.1007/s00339-011-6699-z.

Nigra et al. 2015: B. T. Nigra / K. F. Faull / H. Barnard, Analytical Chemistry in Archaeological Research. Analytical Chemistry 87(1), 2015, 3–18. DOI: 10.1021/ac5029616.

Ntinou 2014: M. Ntinou, Η φυσική βλάστηση και οι προϊστορικές κοινότητες της Μακεδονίας. Μια σύνθεση των πληροφοριών της ανθρακολογικής έρευνας. In: E. Stefani / N. Merousis / A. Dimoula (eds), 1912–2012. Εκατό Χρόνια Έρευνας στην Προϊστορική Μακεδονία [1912–2012. One hundred years of research in prehistoric Macedonia] (Thessaloniki 2014) 409–418.

Oksanen et al. 2022: J. Oksanen / G. L. Simpson / F. G. Blanchet / R. Kindt / P. Legendre / P. R. Minchin / R. B. O'Hara / P. Solymos / M. H. H. Stevens / E. Szoecs / H. Wagner / M. Barbour / M. Bedward / B. Bolker / D. Borcard / G. Carvalho / M. Chirico / M. De Caceres / S. Durand / H. B. Antoniazi / V. Evangelista / R. FitzJohn / M. Friendly / B. Furneaux / G. Hannigan / M. O. Hill / L. Lahti / D. McGlinn / M.-H. Ouellette / E. Ribeiro Cunha / T. Smith / A. Stier / C. J. F. Ter Braak / J. Weedon, vegan: Community Ecology Package. R package version 2.6-4 (2022). https://CRAN.R-project.org/package=vegan (02/05/2025).

Olalde et al. 2018: I. Olalde / S. Brace / M. E. Allentoft et al., The Beaker phenomenon and the genomic transformation of Northwest Europe. Nature 555, 2018, 190–196. DOI: 10.1038/nature25738.

Ortiz et al. 2015: J. E. Ortiz / I. Gutiérrez-Zugasti / T. Torres / M. Gonzàlez-Morales / Y. Sànchez-Palencia, Protein diagenesis in *Patella* shells: Implications for amino acid racemisation dating. Quaternary Geochronol. 27, 2015, 105–118. DOI: 10.1016/j.quageo.2015.02.008.

Oschmann 2009: W. Oschmann, Sclerochronology: editorial. Internat. Journal Earth Scien. 98, 2009, 1–2. DOI: 10.1007/s00531-008-0403-3.

Pannella/MacClintock 1968: G. Pannella / C. MacClintock, Biological and Environmental Rhythms Reflected in Molluscan Shell Growth. Mem. Paleont. Soc. 2, 1968, 64–80. www.jstor.org/stable/1315519.

Papathanasopoulos 1996: G. Papathanasopoulos, Νεολιθικός πολιτισμός στην Ελλάδα [Neolithic culture in Greece] (Athens 1996).

Pappa et al. 2004: M. Pappa / P. Halstead / K. Kotsakis / D. Urem-Kotsou, Evidence for large-scale feasting at Late Neolithic Makriyalos, Northern Greece. In: P. Halstead / J. C. Barrett (eds), Food, Cuisine and Society in Prehistoric Greece. Sheffield Stud. Aegean Arch. 5 (Oxford 2004) 14–44. DOI: 10.2307/j.ctvh1drns.6.

Parfitt et al. 2005: S. A. Parfitt / R. W. Barendregt / M. Breda / I. Candy / M. J. Collins / G. R. Coope / P. Durbidge / M. H. Field / J. R. Lee / A. M. Lister / R. Mutch / K. E. H. Penkman / R. C. Preece / J. Rose / C. B. Stringer / R. Symmons / J. E. Whittaker / J. J. Wymer / A. J. Stuart, The earliest record of human activity in northern Europe. Nature 438(7070), 2005, 1008–1012. DOI: 10.1038/nature04227.

Parker et al. 2020: W. Parker / Y. Yanes / E. Mesa Hernández / J. C. Hernández Marrero / J. Pais / N. Soto Contreras / D. Surge, Shellfish Exploitation in the Western Canary Islands Over the Last Two Millennia. Environmental Arch. 25(1), 2020, 14–36. DOI: 10.1080/14614103.2018.1497821.

Payne 1973: S. Payne, Animal bones. In: Jacobsen 1973a, 59–66. DOI: 10.2307/147470.

1975: S. Payne, Faunal change at Franchthi Cave from 20,000 B.C.–3000 B.C. In: A. Clason (ed.), Archaeozoological studies. Papers of the Archaeozoological Conference 1974, held at the Biologisch-Archaeologisch Instutuut of the State University of Groningen (Amsterdam 1975) 120–131.

1982: S. Payne, Faunal evidence for environmental/climatic change at Franchthi Cave (southern Argolid, Greece), 25,000 B.P.–5000 B.P. Preliminary results. In: J. L. Bintliff / W. van Zeist (eds), Palaeoclimates, Palaeoenvironments and Human Communities in the Eastern Mediterranean Region in Later Prehistory. BAR Internat. Ser. 133 (Oxford 1982) 133–137.

Pederzani/Britton 2017: S. Pederzani / K. Britton, Oxygen isotopes in bioarchaeology: Principles and applications, challenges and opportunities. Earth-Scien. Reviews 188, 2017, 77–107. DOI: 10.1016/j.earscirev.2018.11.005.

Peharda et al. 2021: M. Peharda / B. R. Schöne / B. A. Black / T. Corrège, Advances of sclerochronology research in the last decade. Palaeogeogr., Palaeoclimatol., Palaeoecol. 570, 2021, 110371. DOI: 10.1016/j.palaeo.2021.110371.

Penkman 2005: K. E. H. Penkman, Amino acid geochronology: a closed system approach to test and refine the UK model [PhD thesis, Univ. Newcastle 2005].

Penkman/Kaufman 2013: K. E. H. Penkman / D. S. Kaufman, Amino acid geochronology: Recent perspectives. Quaternary Geochronol. 16, 2013, 1–2. DOI: 10.1016/j.quageo.2012.12.007.

Penkman et al. 2008: K. E. H. Penkman / D. S. Kaufman / D. Maddy / M. J. Collins, Closed-system behaviour of the intra-crystalline fraction of amino acids in mollusc shells. Quaternary Geochronol. 3(1–2), 2008, 2–25. DOI: 10.1016/j.quageo.2007.07.001.

2011: K. E. H. Penkman / R. C. Preece / D. R. Bridgland / D. H. Keen / T. Meijer / S. A. Parfitt / T. S. White / M. J. Collins, A chronological framework for the British Quaternary based on *Bithynia opercula*. Nature 476(7361), 2011, 446–449. DOI: 10.1038/nature10305.

Perlès 1987: C. Perlès, Les industries lithiques taillées de Franchthi (Argolide, Grèce). I: Présentation générale et industries paléolithiques. Excav. Franchthi Cave, Greece 3 (Bloomington IN 1987).

1990: C. Perlès, Les industries lithiques taillées de Franchthi (Argolide, Grèce). II: Les industries du Mésolithique et du Néolithique Initial. Excav. Franchthi Cave, Greece 5 (Bloomington IN 1990).

1999: C. Perlès, Long-term perspectives on the occupation of the Franchthi Cave: continuity and discontinuity. In: Bailey et al. 1999, 311–318.

2001: C. Perlès, The Early Neolithic in Greece. The First Farming Communities in Greece (Cambridge 2001).

2004: C. Perlès, Les industries lithiques taillées de Franchthi (Argolide, Grèce). III: Du Néolithique ancien au Néolithique Final. Excav. Franchthi Cave, Greece 13 (Bloomington IN 2004).

2010: C. Perlès, Is the Dryas the culprit? Socio-economic changes during the Final Pleistocene and Early Holocene at Franchthi Cave (Greece). Journal Israel Prehist. Soc. 40, 2010, 113–129. www.jstor.org/stable/23385150.

Perlès/Vanhaeren 2010: C. Perlès / M. Vanhaeren, Black *Cyclope neritea* Marine Shell Ornaments in the Upper Palaeolithic and Mesolithic of Franchthi Cave, Greece: Arguments for Intentional Heat Treatment. Journal Field Arch. 35(3), 2010, 298–309. www.jstor.org/stable/24406710.

Perlès et al. 2013: C. Perlès / A. Quiles / H. Valladas, Early seventh-millennium AMS dates from domestic seeds in the Initial Neolithic at Franchthi Cave (Argolid, Greece). Antiquity 87(338), 2013, 1001–1015. DOI: 10.1017/S0003598X00049826.

Petraccioli et al. 2010: A. Petraccioli / F. M. Guarino / N. Maio / G. Odierna, Molecular cytogenetic study of three common Mediterranean limpets, *Patella caerulea*, *P. rustica* and *P. ulyssiponensis* (Archaeogastropoda, Mollusca). Genetica 138, 2010, 219–225. DOI: 10.1007/s10709-009-9412-9.

Peyron et al. 2011: O. Peyron / S. Goring / I. Dormoy / U. Kotthoff / J. Pross / J.-L. de Beaulieu / R. Drescher-Schneider / B. Vannière / M. Magny, Holocene seasonality changes in the central Mediterranean region reconstructed from the pollen sequences of Lake Accesa (Italy) and Tenaghi Philippon (Greece). Holocene 21(1), 2011, 131–146. DOI: 10.1177/0959683610384162.

Pierre 1999: C. Pierre, The oxygen and carbon isotope distribution in the Mediterranean water masses. Marine Geol. 153(1–4), 1999, 41–55. DOI: 10.1016/S0025-3227(98)00090-5.

Piperno/Tusa 1976: M. Piperno / S. Tusa, Relazione preliminare sulla seconda campagna di scavi alla Grotta dell'Uzzo. Sicilia Arch. 31, 1976, 39–42.

Piperno et al. 1980a: M. Piperno / S. Scali / A. Tagliacozzo, Mesolitico e Neolitico alla Grotta dell'Uzzo (Trapani). Primi dati per un'interpretazione paleoeconomica. Quaternaria 22, 1980, 275–300.

1980b: M. Piperno / S. Tusa / I. Valente, Campagne di scavo 1977 e 1978 alla Grotta dell'Uzzo, Trapani. Relazione preliminare e datazione dei livelli mesolitici e neolitici. Sicilia Arch. 42, 1980, 49–64.

Poulos et al. 1997: S. E. Poulos / P. G. Drakopoulos / M. B. Collins, Seasonal variability in sea surface oceanographic conditions in the Aegean Sea (Eastern Mediterranean): an overview. Journal Marine Systems 13(1–4), 1997, 225–244. DOI: 10.1016/S0924-7963(96)00113-3.

Prendergast/Schöne 2017: A. L. Prendergast / B. R. Schöne, Oxygen isotopes from limpet shells: implications for palaeothermometry and seasonal shellfish foraging studies in the Mediterranean. Palaeogeogr., Palaeoclimatol., Palaeoecol. 484, 2017, 33–47. DOI: 10.1016/j.palaeo.2017.03.007.

Prendergast et al. 2016: A. L. Prendergast / R. E. Stevens / T. C. O'Connell / A. Fadlalak / M. Touati / A. al-Mzeine / B. R. Schöne / C. O. Hunt / G. Barker, Changing patterns of eastern Mediterranean shellfish exploitation in the Late Glacial and Early Holocene: Oxygen isotope evidence from gastropod in Epipaleolithic to Neolithic human occupation layers at the Haua Fteah cave, Libya. Quaternary Internat. 407 B, 2016, 80–93. DOI: 10.1016/j.quaint.2015.09.035.

Preucel/Hodder 1996: R. W. Preucel / I. Hodder, Prologue. Communicating Present Pasts. In: R. W. Preucel / I. Hodder (eds), Contemporary Archaeology in Theory: A Reader (Cambridge 1996) 3–20.

Price 2000: T. D. Price (ed.), Europe's First Farmers (Cambridge 2000).

Pross et al. 2009: J. Pross / U. Kotthoff / U. C. Muller / O. Peyron / I. Dormoy / G. Schmiedl / S. Kalaitzidis / A. M. Smith, Massive perturbation in terrestrial ecosystems of the Eastern Mediterranean region associated with the 8.2 kyr BP climatic event. Geology 37(10), 2009, 887–890. DOI: 10.1130/G25739A.1.

R Team 2023: R Team, vegan: Community Ecology Package. R package version 2.6-4 (2023). https://CRAN.R-project.org/package=vegan (02/05/2025).

Rainsford et al. 2014: C. Rainsford / T. O'Connor / P. Miracle, Fishing in the Adriatic at the Mesolithic-Neolithic transition: Evidence from Vela Spila, Croatia. Environmental Arch. 19(3), 2014, 311–320. DOI: 10.1179/1749631414Y.0000000018.

Ravelo/Hillaire-Marcel 2007: A. C. Ravelo / C. Hillaire-Marcel, Chapter Eighteen: The Use of Oxygen and Carbon Isotopes of Foraminifera in Paleoceanography. In: C. Hillaire-Marcel / A. De Vernal (eds), Proxies in Late Cenozoic Paleoceanography. Developments Marine Geol. 1 (Amsterdam 2007) 735–764. DOI: 10.1016/S1572-5480(07)01023-8.

Refsnider et al. 2013: K. A. Refsnider / G. H. Miller / B. Frechette / D. H. Rood, A chronological framework for the Clyde Foreland Formation, Eastern Canadian Arctic, derived from amino acid racemization and cosmogonic radionuclides. Quaternary Geochronol. 16, 2013, 21–34. DOI: 10.1016/j.quageo.2012.05.009.

Reingruber 2018a: A. Reingruber, The Transition from the Mesolithic to the Neolithic in a Circum-Aegean Perspective: Concepts and Narratives. In: A. Sarris / E. Kalogiropoulou / T. Kalayci / E. Karimali (eds), Communities, Landscapes, and Interaction in Neolithic Greece. Internat. Monogr. Prehist. Arch. Ser. 20 (New York NY 2018) 8–26. DOI: 10.1515/9781789201468-006.

2018b: A. Reingruber, Geographical mobility and social motility in the Aegean before and after 6600 BC. Praehist. Zeitschr. 93(1), 2018, 1–24. DOI: 10.1515/pz-2018-0005.

Reisch 1976: L. Reisch, Beobachtungen an Vogelknochen aus dem Spätpleistozän der Höhle von Kephalari. Arch. Korrbl. 6, 1976, 261–265.

Renfrew 1973: C. Renfrew, Before Civilization: The Radiocarbon Revolution and Prehistoric Europe (New York 1973).

Renfrew/Bahn 2011: C. Renfrew / P. Bahn, Αρχαιολογία: Θεωρίες, Μεθοδολογία και Πρακτικές Εφαρμογές [Archaeology. Theories, Methods and Practice], trans. I. Karali Yannakopoulou (Athens [2]2011).

Rick et al. 2006: T. C. Rick / J. A. Robbins / K. M. Ferguson, Stable Isotopes from Marine Shells, Ancient Environments, and Human Subsistence on Middle

Holocene Santa Rosa Island, California, USA. Journal Island and Coastal Arch. 1(2), 2006, 233–254. DOI: 10.1080/15564890600935480.

de Rijk et al. 1999: S. de Rijk / A. Hayes / E. J. Rohling, Eastern Mediterranean sapropel S1 interruption: An expression of the onset of climatic deterioration around 7 ka BP. Marine Geol. 153(1–4), 1999, 337–343. DOI: 10.1016/S0025-3227(98)00075-9.

Roberts 1998: N. Roberts, The Holocene. An Environmental History (Oxford, New York ²1998).

Roberts et al. 2011: N. Roberts / W. J. Eastwood / C. Kuzucuoğlu / G. Fiorentino / V. Caracuta, Climatic, vegetation and cultural change in the eastern Mediterranean during the mid-Holocene environmental transition. Holocene 21(1), 2011, 147–162. DOI: 10.1177/0959683610386819.

Robinson et al. 2006: S. A. Robinson / S. Black / B. W. Sellwood / P. J. Valdes, A review of palaeoclimates and palaeoenvironments in the Levant and Eastern Mediterranean from 25,000 to 5000 years BP: setting the environmental background for the evolution of human civilisation. Quaternary Scien. Reviews 25(13–14), 2006, 1517–1541. DOI: 10.1016/j.quascirev.2006.02.006.

Rohling 2013: E. J. Rohling, Oxygen Isotope Composition of Seawater. In: S. A. Elias (ed.), The Encyclopedia of Quaternary Science. 2: Gla–Pal (Amsterdam ²2013) 915–922.

Rohling/Pälike 2005: E. J. Rohling / H. Pälike, Centennial-scale climate cooling with a sudden cold event around 8,200 years ago. Nature 434, 2005, 975–979. DOI: 10.1038/nature03421.

Rohling/de Rijk 1999: E. J. Rohling / S. de Rijk, Holocene Climate Optimum and Last Glacial Maximum in the Mediterranean: the marine oxygen isotope record. Marine Geol. 153(1–4), 1999, 57–75. DOI: 10.1016/S0025-3227(98)00020-6.

Rohling et al. 2002: E. J. Rohling / P. A. Mayewski / A. Hayes / R. H. Abu-Zied / J. S. L. Casford, Holocene atmosphere–ocean interactions: records from Greenland and the Aegean Sea. Climate Dynamics 18, 2002, 587–593. DOI: 10.1007/s00382-001-0194-8.

2019: E. J. Rohling / G. Marino / K. M. Grant / P. A. Mayewski / B. Weninger, A model for archaeologically relevant Holocene climate impacts in the Aegean–Levantine region (easternmost Mediterranean). Quaternary Scien. Reviews 208, 2019, 38–53. DOI: 10.1016/j.quascirev.2019.02.009.

Romanowska et al. 2021: I. Romanowska / C. D. Wren/ S. A. Crabtree, Agent-Based Modelling for Archaeology. Simulating the Complexity of Societies (Santa Fe NM 2021). DOI: 10.37911/9781947864382.

Rose 1995: M. Rose, Fishing at Franchthi Cave, Greece: changing environments and patterns of exploitation. Old World Arch. Newsletter 18(3), 1995, 21–26.

in prep.: M. Rose, The Fish Remains and Marine Engagement at Franchthi Cave, Greece. Excavations at Franchthi Cave, Greece (Bloomington IN, in prep.).

Runnels 1995: C. Runnels, Review of Aegean Prehistory IV: The Stone Age of Greece from the Palaeolithic to the Advent of the Neolithic. Am. Journal Arch. 99(4), 1995, 699–728. DOI: 10.2307/506190.

1996: C. Runnels, The Palaeolithic and Mesolithic remains from the Berbati–Limnes survey. In: B. Wells / C. Runnels (eds), The Berbati–Limnes Archaeological Survey 1988–1990. Skr. Utg. Svenska Inst. Athen 44 (Stockholm 1996) 23–35.

Sá-Pinto et al. 2005: A. Sá-Pinto / M. Branco / D. J. Harris / P. Alexandrino, Phylogeny and phylogeography of the genus *Patella* based on mitochondrial DNA sequence data. Journal Experimental Marine Biol. and Ecol. 325(1), 2005, 95–110. DOI: 10.1016/j.jembe.2005.04.025.

Sakellariou et al. 2015: D. Sakellariou / J. Beck / G. Rousakis / P. Georgiou / I. Panagiotopoulos / I. Morfis / K. Tsampouraki-Kraounaki / A. Zavitsanou, Submerged prehistoric landscapes off Franchthi Cave, East Argolic Gulf: preliminary results. In: 11th Panhellenic Symposium on Oceanography and Fisheries, Mytilene, Lesvos Island, Greece (Mytilene 2015) 993–996.

Sampson 2014: A. A. Sampson, Ikaria: proïstoria kai ethnoarchaiologia (Athēna 2014).

Sampson et al. 2012: A. Sampson / M. Kaczanowska / J. K. Kozłowski, Mesolithic occupations and environments on the island of Ikaria, Aegean, Greece. Folia Quaternaria 80 (Kraków 2012).

San et al. 2023: X. San / J. Hu / M. Chen / H. Niu / P. J. Smeets / J. Deng / K. Koo / R. D. Reis / V. P. Dravid / X. Hu, Unlocking the Structural Mystery of Vaterite $CaCO_3$. arXiv:2306.08228 (2023). DOI: 10.48550/arXiv.2306.08228.

Sanna et al. 2012: D. Sanna / G. L. Dedola / T. Lai / M. Curini-Galletti / M. Casu, PCR-RFLP: A practical method for the identification of specimens of *Patella ulyssiponensis s. l.* (Gastropoda: Patellidae). Italian Journal Zool. 79(1), 2012, 50–59. DOI: 10.1080/11250003.2011.620988.

Schöne 2008: B. Schöne, The curse of physiology – challenges and opportunities in the interpretation of geochemical data from mollusk shells. Geo-Marine Letters 28, 2008, 269–285. DOI: 10.1007/s00367-008-0114-6.

Schöne/Gillikin 2013: B. R. Schöne / D. P. Gillikin, Unravelling environmental histories from skeletal diaries – Advances in sclerochronology. Palaeogeogr., Palaeoclimatol., Palaeoecol. 373, 2013, 1–5. DOI: 10.1016/j.palaeo.2012.11.026.

Segre/Piperno 1975: A. Segre / M. Piperno, Scavi alla Grotta dell'Uzzo. Relazione preliminare. Sicilia Arch. 27, 1975, 11–16.

Shackleton 1973: N. J. Shackleton, Oxygen isotope analysis as a means of determining season of occupation of prehistoric midden sites. Archaeometry 15(1), 1973, 133–141. DOI: 10.1111/j.1475-4754.1973.tb00082.x.

Shackleton 1988: J. Shackleton, Marine molluscan remains from Franchthi Cave. Excav. Franchthi Cave, Greece 4 (Bloomington IN 1988).

Shanks/Tilley 1987: M. Shanks / C. Tilley, Re-constructing Archaeology: Theory and Practice (London 1987).

Sherratt 1983: A. Sherratt, The Neolithic period in Bulgaria in its European context. In: A. Pollster (ed.), Ancient Bulgaria. Papers presented to the International Symposium on the Ancient History and Archaeology of Bulgaria, University of Nottingham, 1981. Univ. Nottingham. Arch. Section: Monogr. Ser. 1 (Nottingham 1983) 188–198.

Skliris et al. 2011: N. Skliris / S. S. Sofianos / A. Gkanasos / P. Axaopoulos / A. Mantziafou / V. Vervatis, Long-term sea surface temperature variability in the Aegean Sea. Advances in Oceanography and Limnol. 2(2), 2011, 125–139. DOI: 10.1080/19475721.2011.601325.

Skourtanioti et al. 2023: E. Skourtanioti / H. Ringbauer / G. A. Gnecchi Ruscone / R. A. Bianco / M. Burri / C. Freund / A. Furtwängler / N. F. G. Martins / F. Knolle / G. U. Neumann / A. Tiliakou / A. Agelarakis / M. Andreadaki-Vlazaki / P. Betancourt / B. P. Hallager / O. A. Jones / O. Kakavogianni / A. Kanta / P. Karkanas / E. Kataki / K. Kissas / R. Koehl / L. Kvapil / J. Maran / P. J. P. McGeorge / A. Papadimitriou / A. Papathanasiou / L. Papazoglou-Manioudaki / K. Paschalidis / N. Polychronakou-Sgouritsa / S. Preve / E.-A. Prevedorou / G. Price / E. Protopapadaki / T. Schmidt-Schultz / M. Schultz / K. Shelton / M. H. Wiener / J. Krause / C. Jeong / P. W. Stockhammer, Ancient DNA reveals admixture history and endogamy in the prehistoric Aegean. Nature Ecol. & Evol. 7, 2023, 290–303. DOI: 10.1038/s41559-022-01952-3.

Smith/Dent 2005: E. Smith / G. Dent, Modern Raman Spectroscopy – A Practical Approach (Chichester 2005).

Sordinas 1969: A. Sordinas, Investigations of the prehistory of Corfu, dating 1964–1966. Balkan Stud. 10, 1969, 393–424.

1970: A. Sordinas, Stone Implements from Northwestern Corfu, Greece. Memphis State Univ. Anthr. Research Center, Occasional Papers 4 (Memphis TN 1970).

2003: A. Sordinas, The Sidarian: maritime Mesolithic non-geometric microliths in Western Greece. In: Galanidou/Perlès 2003, 89–97. www.jstor.org/stable/40960308.

Souvatzi 2008: S. G. Souvatzi, A Social Archaeology of Households in Neolithic Greece. An Anthropological Approach (Cambridge 2008).

Spann et al. 2010: N. Spann / E. M. Harper / D. C. Aldridge, The unusual mineral vaterite in shells of the freshwater bivalve *Corbicula fluminea* from the UK. Naturwissenschaften 97(8), 2010, 743–751. DOI: 10.1007/s00114-010-0692-9.

Starkovich 2017: B. Starkovich, Paleolithic subsistence strategies and changes in site use at Klissoura Cave 1 (Peloponnese, Greece). Journal Human Evol. 111, 2017, 63–84. DOI: 10.1016/j.jhevol.2017.04.005.

Stecher et al. 1996: H. A. Stecher / D. E. Krantz / C. J. Lord / G. W. Luther / K. W. Bock, Profiles of strontium and barium in *Mercenaria mercenaria* and *Spisula solidissima* shells. Geochimica et Cosmochimica Acta 60(18), 1996, 3445–3456. DOI: 10.1016/0016-7037(96)00179-2.

Stein et al. 2003: J. K. Stein / J. N. Deo / L. S. Phillips, Big Sites – Short Time: Accumulation Rates in Archaeological Sites. Journal Arch. Scien. 30(3), 2003, 297–316. DOI: 10.1006/jasc.2002.0838.

Stiner/Munro 2011: M. C. Stiner / N. D. Munro, On the evolution of diet and landscape during the Upper Paleolithic through Mesolithic at Franchthi Cave (Peloponnese, Greece). Journal Human Evol. 60(5), 2011, 618–636. DOI: 10.1016/j.jhevol.2010.12.005.

Surdez et al. 2018: M. Surdez / J. Beck / D. Sakellariou / H. Vogel / P. Birchler Emery / D. Koutsoumba / F. S. Anselmetti, Flooding a landscape: impact of Holocene transgression on coastal sedimentology and underwater archaeology in Kiladha Bay (Greece). Swiss Journal Geoscien. 111(3), 2018, 341–355. DOI: 10.1007/s00015-018-0309-4.

Surge/Barrett 2012: D. Surge / J. H. Barrett, Marine climatic seasonality during medieval times (10th to 12th centuries) based on isotopic records in Viking Age shells from Orkney, Scotland. Palaeogeogr., Palaeoclimatol., Palaeoecol. 350–352, 2012, 236–246. DOI: 10.1016/j.palaeo.2012.07.003.

Sykes et al. 1995: G. A. Sykes / M. J. Collins / D. Walton, The significance of a geochemically isolated intracrystalline organic fraction within biominerals. Organic Geochemistry 23(11–12), 1995, 1059–1065. DOI: 10.1016/0146-6380(95)00086-0.

Sykes et al. 2019: N. Sykes / M. Spriggs / A. Evin, Beyond curse or blessing: the opportunities and challenges of aDNA analysis. World Arch. 51(4), 2019, 503–516. DOI: 10.1080/00438243.2019.1741970.

Theocharis 1959: D. Theocharis, Εκ της προϊστορίας Ευβοίας και Σκύρου. Archeio Evoïkon Meleton 6, 1959, 279–328.

1967: D. Theocharis, Η αυγή της Θεσσαλικής προϊστορίας [The dawn of Thessalian prehistory] (Volos 1967).

1973: D. Theocharis, Νεολιθική Ελλάς [Neolithic Greece] (Athens 1973).

1981: D. Theocharis, Νεολιθικός πολιτισμός. Σύντομη επισκόπηση της Νεολιθικής στον ελλαδικό χώρο [Neolithic Culture: A brief overview of the Neolithic period in Greece] (Athens 1981).

Theodoraki 2018: D. Theodoraki, Marine resources in the Neolithic settlements of Northern Greece: Ethnoarchaeological research in the region of the Thermaic Gulf and the archaeomalacological assemblage of Paliambela [in Greek] [Master's thesis, Aristotle Univ. Thessaloniki 2018]. http://ikee.lib.auth.gr/record/301567/?ln=el (23/02/2026).

Theodoropoulou 2007: A. Theodoropoulou, L'exploitation des ressources aquatiques en Égée septentrionale aux périodes pré- et protohistoriques [PhD thesis, Univ. Paris I-Panthéon-Sorbonne 2007]. https://theses.fr/2007PA010629 (23/02/2026).

Thomas 1988: J. Thomas, Neolithic Explanations Revisited: The Mesolithic–Neolithic Transition in Britain and South Scandinavia. Proc. Prehist. Soc. 54(1), 1988, 59–66. DOI: 10.1017/S0079497X00005752.

Thomas 2015a: K. D. Thomas, Molluscs emergent, Part I: themes and trends in the scientific investigation of mollusc shells as resources for archaeological research. Journal Arch. Scien. 56, 2015, 133–140. DOI: 10.1016/j.jas.2015.01.024.

2015b: K. D. Thomas, Molluscs emergent, Part II: themes and trends in the scientific investigation of molluscs and their shells as past human resources. Journal Arch. Scien. 56, 2015, 159–167. DOI: 10.1016/j.jas.2015.01.015.

Thompson/Andrus 2011: V. D. Thompson / C. F. T. Andrus, Evaluating Mobility, Monumentality, and Feasting at the Sapelo Island Shell Ring Complex. Am. Ant. 76(2), 2011, 315–344. DOI: 10.7183/0002-7316.76.2.315.

Toniello et al. 2019: G. Toniello / D. Lepofsky / G. Lertzman-Lepofsky / A. K. Salomon / K. Rowell, 11,500 years of human–clam relationships provide long-term context for intertidal management in the Salish Sea, British Columbia. PNAS 116(44), 2019, 22106–22114. DOI: 10.1073/pnas.1905921116.

Tourloukis 2021: V. Tourloukis, Palaeolithic archaeology: a review of recent research. Arch. Reports 67, 2021, 61–79. DOI: 10.1017/S0570608421000041.

Tourloukis/Harvati 2018: V. Tourloukis / K. Harvati, The Palaeolithic record of Greece: A synthesis of the evidence and a research agenda for the future. Quaternary Internat. 466 A, 2018, 48–65. DOI: 10.1016/j.quaint.2017.04.020.

Towe 1980: K. M. Towe, Preserved organic ultrastructure: An unreliable indicator for Paleozoic amino acid biogeochemistry. In: P. E. Hare / T. C. Hoering / K. Jr. King (eds), The Biogeochemistry of Amino Acids; papers presented at a conference at Airlie House, Warrenton, Va., Oct. 29 to Nov. 1, 1978 (New York 1980) 65–74.

Trigger 1978: B. Trigger, Time and Tradition: Essays in Archaeological Interpretation (Edinburgh 1978).

2006: B. Trigger, A History of Archaeological Thought (Cambridge [2]2006).

Tsountas 1908: C. Tsountas, Αι προϊστορικαί ακροπόλεις Διμηνίου και Σέσκλου [Prehistoric acropoleis of Dimini and Sesklo] (Athens 2000 [1908]).

Urey 1947: H. C. Urey, The Thermodynamic Properties of Isotopic Substances. Journal Chemical Soc. 1947, 562–581. DOI: 10.1039/JR9470000562.

Vafidis et al. 2020: D. Vafidis / I. Drosou / K. Dimitriou / D. Klaoudatos, Population characteristics of the limpet *Patella caerulea* (Linnaeus, 1758) in Eastern Mediterranean (Central Greece). Water 12(1186), 2020, 1–19. DOI: 10.3390/w12041186.

Vander Putten et al. 2000: E. Vander Putten / F. Dehairs / E. Keppens / W. Baeyens, High resolution distribution of trace elements in the calcite shell layer of modern *mytilus edulis*: environmental and biological controls. Geochimica et Cosmochimica Acta 64(6), 2000, 997–1011. DOI: 10.1016/S0016-7037(99)00380-4.

Veropoulidou 2011: R. Veropoulidou, Shells from the settlements of the Thermaic Gulf: Reconstructing mollusc consumption during the Neolithic and the Bronze Age [in Greek] [PhD thesis, Aristotle Univ. Thessaloniki 2011]. https://ikee.lib.auth.gr/record/128457/?ln=en (23/02/2026).

Vignon 2015: M. Vignon, Extracting environmental histories from sclerochronological structures – Recursive partitioning as a mean to explore multi-elemental composition of fish otolith. Ecol. Informatics 30, 2015, 159–169. DOI: 10.1016/j.ecoinf.2015.10.002.

Vita-Finzi et al. 1970: C. Vita-Finzi / E. S. Higgs / D. Sturdy / J. Harriss / A. J. Legge / H. Tippett, Prehistoric economy in the Mount Carmel area of Palestine: site catchment analysis. Proc. Prehist. Soc. 36, 1970, 1–36. DOI: 10.1017/S0079497X00013074.

Vitelli 1989: K. D. Vitelli, Were pots first made for foods? Doubts from Franchthi. World Arch. 21(1), 1989, 17–29. www.jstor.org/stable/124482.

1993: K. D. Vitelli, Franchthi Neolithic Pottery. 1: Classification and ceramic phases 1 and 2. Excav. Franchthi Cave, Greece 8 (Bloomington IN 1993).

1999: K. D. Vitelli, Franchthi Neolithic Pottery. 2: The later Neolithic ceramic phases 3 to 5. Excav. Franchthi Cave, Greece 10 (Bloomington IN 1999).

Wace/Thompson 1912: A. J. B. Wace / M. S. Thompson, Prehistoric Thessaly. Being some account of recent excavations and explorations in North-Eastern Greece from Lake Kopais to the borders of Macedonia (Cambridge 1912).

Wang et al. 2012: T. Wang / D. Surge / S. Mithen, Seasonal temperature variability of the Neoglacial (3300–2500 BP) and Roman Warm Period (2500–1600 BP) reconstructed from oxygen isotope ratios of limpet shells (*Patella vulgata*), Northwest Scotland. Palaeogeogr., Palaeoclimatol., Palaeoecol. 317–318, 2012, 104–113. DOI: 10.1016/j.palaeo.2011.12.016.

Warburton et al. 2017: M. L. Warburton / M. R. Reid / C. H. Stirling / G. Closs, Validation of depth-profiling LA-ICP-MS in otolith applications. Canadian Journal Fisheries Aquatic Scien. 74(4), 2017, 572–581. DOI: 10.1139/cjfas-2016-0063.

Waselkov 1987: G. Waselkov, Shellfish gathering and shell midden archaeology. In: M. B. Schiffer (ed.), Advances in Archaeological Method and Theory 10 (San Diego 1987) 93–210. DOI: 10.1016/B978-0-12-003110-8.50006-2.

Watanabe et al. 2019: T. Watanabe / T. K. Watanabe / A. Yamazaki / S. Yoneta / K. Sowa / F. Sinniger / G. Eyal / Y. Loya / S. Harii, Coral Sclerochronology: Similarities and Differences in the Coral Isotopic Signatures Between Mesophotic and Shallow-Water Reefs. In: Y. Loya / K. Puglise / T. C. L. Bridge (eds), Mesophotic Coral Ecosystems. Coral Reefs of the World 12 (Cham 2019) 667–681. DOI: 10.1007/978-3-319-92735-0_36.

Wefer/Berger 1991: G. Wefer / W. H. Berger, Isotope paleontology: growth and composition of extant calcareous species. Marine Geol. 100(1–4), 1991, 207–248. DOI: 10.1016/0025-3227(91)90234-U.

Wehmiller 1993: J. F. Wehmiller, Applications of Organic Geochemistry for Quaternary Research. Aminostratigraphy and Aminochronology. In: M. H. Engel / S. A. Macko (eds), Organic Geochemistry: Principles and Applications (New York 1993) 755–783. DOI: 10.1007/978-1-4615-2890-6_36.

Wehmiller/Miller 2000: J. F. Wehmiller / G. H. Miller, Aminostratigraphic Dating Methods in Quaternary Geology. In: J. Stratton Noller / J. M. Sowers / W. R. Lettis (eds), Quaternary Geochronology: Methods and Applications. AGU Reference Shelf 4 (Washington, D.C. 2000) 187–222. DOI: 10.1029/RF004p0187.

Wehmiller et al. 1976: J. F. Wehmiller / P. E. Hare / G. A. Kujala, Amino acids in fossil corals: racemization (epimerization) reactions and their implications for diagenetic models and geochronological studies. Geochimica et Cosmochimica Acta 40(7), 1976, 763–776. DOI: 10.1016/0016-7037(76)90029-6.

Weiberg et al. 2016: E. Weiberg / I. Unkel / K. Kouli / K. Holmgren / P. Avramidis / A. Bonnier / F. Dibble / M. Finné / A. Izdebski / C. Katrantsiotis / S. R. Stocker / M. Andwinge / K. Baika / M. Boyd / C. Heymann, The socio-environmental history of the Peloponnese during the Holocene: Towards an integrated understanding of the past. Quaternary Scien. Reviews 136, 2016, 40–65. DOI: 10.1016/j.quascirev.2015.10.042.

Weiss et al. 1993: H. Weiss / M.-A. Courty / W. Wetterstrom / F. Guichard / L. Senior / R. Meadow / A. Curnow, The Genesis and Collapse of Third Millennium North Mesopotamian Civilization. Science 261(5124), 1993, 995–1004. DOI: 10.1126/science.261.5124.995.

Wells 1963: J. W. Wells, Coral Growth and Geochronometry. Nature 197, 1963, 948–950. DOI: 10.1038/197948a0.

Wells et al. 1993: B. Wells / C. N. Runnels / E. Zanger, In the shadow of Mycenae. Archaeology 46(1), 1993, 54–63. www.jstor.org/stable/41766249.

Weninger et al. 2006: B. Weninger / E. Alram-Stern / E. Bauer / L. Clare / U. Danzeglocke / O. Jöris / C. K. E. Claudia / G. R. F. Gary / H. Todorova / T. van Andel, Climate forcing due to the 8200 cal yr BP event observed at Early Neolithic sites in the eastern Mediterranean. Quaternary Research 66(3), 2006, 401–420. DOI: 10.1016/j.yqres.2006.06.009.

2014: B. Weninger / L. Clare / F. Gerritsen / B. Horejs / R. Krauß / J. Linstädter / R. Özbal / E. J. Rohling, Neolithisation of the Aegean and Southeast Europe during the 6600–6000 calBC period of rapid climate change. Doc. Praehist. 41, 2014, 1–31. DOI: 10.4312/dp.41.1.

West et al. 2018: C. West / M. Burchell / C. F. T. Andrus, Molluscs and paleoenvironmental reconstruction in island and coastal settings: Variability, seasonality, and sampling. In: C. M. Giovas / M. J. LeFebvre (eds), Zooarchaeology in Practice: Case Studies in Methodology and Interpretation in Archaeofaunal Analysis (Cham 2018) 191–208. DOI: 10.1007/978-3-319-64763-0_10.

West et al. 2019: G. West / D. S. Kaufman / F. Muschitiello / M. Forwick / J. Matthiansen / J. Wollenburg / M. O'Regan, Amino acid racemization in Quaternary foraminifera from the Yermak Plateau, Arctic Ocean. Geochronology 1(1), 2019, 53–67. DOI: 10.5194/gchron-1-53-2019.

Wheeler 2022: L. Wheeler, Towards an aminostratigraphy of foraminifera for Pleistocene sea-level records: Testing the intra-crystalline approach to amino acid racemisation dating [PhD thesis, Univ. York 2022]. https://etheses.whiterose.ac.uk/id/eprint/31704/ (23/02/2026).

Wijmstra et al. 1990: T. A. Wijmstra / R. Young / H. J. L. Witte, An evaluation of the climatic conditions during the Late Quaternary in northern Greece by means of multivariate analysis of palynological data and comparison with recent phytosociological and climatic data. Geol. en Mijnbouw 69, 1990, 243–251.

Willey/Phillips 1958: G. R. Willey / P. Phillips, Method and Theory in American Archaeology (Chicago, London 1958).

Wirtz/Lemmen 2003: K. A. Wirtz / C. Lemmen, A global dynamic model for the Neolithic transition. Climatic Change 59, 2003, 333–367. DOI: 10.1023/A:1024858532005.

Zanotti et al. 2015: A. Zanotti / R. Moussa / J. Dubouloz / J.-P. Bocquet-Appel, Exploring Scenarios for the First Farming Expansion in the Balkans Via an Agent-based Model. In: S. Campana / R. Scopigno / G. Carpentero / M. Cirillo (eds), CAA2015. Keep the Revolution Going: Proceedings of the 43rd Annual Conference on Computer Applications and Quantitative Methods in Archaeology (Oxford 2015) 773–780. DOI: 10.2307/jj.15135955.86.

van Zeist/Bottema 1982: W. van Zeist / S. Bottema, Vegetation history of the Eastern Mediterranean and the Near East during the last 20,000 years. In: J. L. Bintliff / W. van Zeist (eds), Palaeoclimates, Palaeoenvironments and Human Communities in the Eastern Mediterranean Region in Later Prehistory 2. BAR Internat. Ser. 133, 2 (Oxford 1982) 277–321.